Principles of Radar and Sonar Signal Processing

For a listing of recent titles in the *Artech House Radar Library,*
turn to the back of this book.

Principles of Radar and Sonar Signal Processing

François Le Chevalier

Artech House
Boston • London
www.artechhouse.com

Library of Congress Cataloging-in-Publication Data
Le Chevalier, François.
Principles of radar and sonar signal processing/François Le Chevalier.
p. cm. — (Artech House radar library)
Includes bibliographical references and index.
ISBN 1-58053-338-8 (alk. paper)
1. Radar. 2. Sonar. 3. Signal processing. I. Title. II. Series.
TK6575 .L44 2002
621.3848—dc21 2002019675

British Library Cataloguing in Publication Data
Le Chevalier, François
Principles of radar and sonar signal processing. — (Artech House radar library)
1. Radar 2. Sonar 3. Signal processing
I. Title
621.3'848

ISBN 1-58053-338-8

Cover design by Yekaterina Ratner

685 Canton Street
Norwood, MA 02062

International Standard Book Number: 1-58053-338-8
Library of Congress Catalog Card Number: 2002019675

10 9 8 7 6 5 4 3 2 1

Contents

Preface

As in most technical disciplines, the recent evolution of radar and sonar has been marked by the rapid increase in information processing capabilities. By enabling the implementation of signal processing procedures that are closer to theoretically optimum ones, this evolution has also revealed the arbitrary character of some of the underlying assumptions: single-scatterer target, additive white Gaussian noise. Moreover, in order to meet the ever increasing needs of the users, this evolution has led to the development of multifunction radars and sonars: detection in free space or in the presence of spurious echoes caused by the environment, tracking of detected targets, even identification—or at least rough classification—of targets. Finally, this evolution has enabled various types of processing developed in sonar to be adapted for radar—this relative prominence of sonar in the signal-processing field being explained both by the severe environmental conditions (propagation and noise) under which it must operate, and by the lower rates it has to deal with.

These three observations—unrealistic assumptions, multiple functions going as far as target identification, and transfer of sonar processing techniques to radar—are a sufficient indication of the necessity of taking into account the precise characteristics of target and background "signatures" in radar design. They also demonstrate that modern radar design often involves at least a partial knowledge of the processing techniques used in sonar.

In the specialized literature this recent evolution in radar and sonar is indicated by the publication of works dealing with adaptive antenna processing, target signatures, or high-resolution spectral analysis methods, which supplement the basic works on radar (Skolnik [1], Barton [2], Rihaczek [3], J. Darricau [4], and M. Carpentier [5]). However, to ensure the consistency of these new approaches with the fundamental principles of radar and sonar, and to introduce students to these modern techniques in a natural way, it appeared logical to incorporate these special aspects into this work devoted to radar and sonar processing principles—theoretical principles, but also operating principles.

By clarifying the relationships between these diverse approaches in the context of radar and sonar—whereas some are derived from seismic research, others come from speech recognition work and yet others from electromagnetics or acoustics—we can expect not only to gain a better understanding of their value in this field but also to determine their limits more precisely, since these limits are often determined by the validity of the basic assumptions. Furthermore, we will include in the same work the fundamental principles of signal processing and the essential physical characteristics of the targets from which these signals come. This will clearly show that it is necessary now, at the design stage of the detection system, to take into account the physics of the phenomena as well as signal theory or the inherent limitations of the transmitter or receiver.

These considerations point to the essential characteristics of our presentation: constant application of maximum likelihood criteria (to ensure the consistency of the presentation), detailed analysis of target signatures, justification of the principle and detailed description of the physical limitations of tracking systems, and parallel discussion of radar and sonar, both active and passive. They also explain the omission of various aspects not strictly specific to radar and sonar, for example track plotting of detected echoes, or passive signal recognition.

Finally, the concern for parallel description of radar and sonar has led to the choice of as general a formalism as possible so that the reasoning can apply equally well to antenna arrays—frequent in sonar—and to the Doppler processing generally used in radar. It will be seen that this generalization does not complicate the presentation, which is a fact already rare enough to be worth noting. Moreover, it enables the procedures for processing in colored noise to be described in a form that opens up their application to widely differing situations (e.g., tracking in the presence of external interference).

This book is divided into three parts. Chapters 1, 2, and 3 deal with processing in the presence of white noise—detection and location, ambiguity function, and tracking systems. Chapters 4 and 5 discuss reception in colored noise—various structures of the optimum receiver, adaptive processing, application to jammer suppression and to tracking, and location in passive listening mode. Chapters 6 and 7 deal with the physical characteristics of targets and backgrounds, their modeling, and the signal processing's taking them into account.

As far as possible, the various types of processing are illustrated by examples applied to real or simulated signals, thus giving an immediate feeling of the exact nature of the limitations encountered. However, in some cases—especially for passive listening—various industrial or military confidentiality requirements have restricted the extent of the available examples, and even of the methods described.

Focused on principles, this book is obviously aimed at students attending specialist courses; actually, it is derived from such a course, given at the Ecole Nationale Supérieure des Télécommunications (ENST) and the Ecole Nationale de l'Aéronautique et de l'Espace (ENSAE) in France for years. However, the sections on adaptive processing and on target and background signatures should also be of interest to engineers working on the design and evaluation of radar or sonar systems. This diversity in the potential readership has resulted in the provision of a detailed, equation-free summary at the beginning of each chapter. Each summary describes in the simplest possible terms the key points of the approach and the main results described in the chapter. This format should help the reader to understand the general approach of each chapter; he will then turn directly to the next, if he prefers, without getting lost in the details of the calculations of the previous ones. It is assumed that the reader is already familiar with the concepts of complex signal, Gaussian noise, correlation, and filtering (which can be found in [6], for instance). The other required concepts are explained in the appendixes or in the body of the text. In addition to some original analyses, concerning tracking, identification, or wideband radars in particular, this work obviously includes many standard or recent results; some essential references have been given in each case, but the number of references has been limited as much as possible, in order to avoid overwhelming the reader with a pile of documents for specialist use. As far as possible, the references are to recent articles or books so that readers can work their way up the citation tree of the field that interests them.

References

[1] Skolnik, M. I., *Radar Handbook*, New York: McGraw-Hill, 1990.

[2] Barton, D. K., *Modern Radar System Analysis*, Norwood, MA: Artech House, 1988.

[3] Rihaczek, A. W., *Principles of High-Resolution Radar*, New York: McGraw-Hill, 1969; Norwood, MA: Artech House, 1996.

[4] Darricau, J., *Physique et Théorie du Radar*, Sodipe, 1993.

[5] Carpentier, M. H., *Principles of Modern Radar Systems*, Norwood, MA: Artech House, 1988.

[6] Papoulis, A., *Probability, Random Variables, and Stochastic Processes*, New York: McGraw-Hill, 1965.

Acknowledgments

I am pleased to take this opportunity to thank various people in a few organizations whose contributions have been essential to the production of this book.

First, ENSTélécom, ENSAE, and ENSTA, where successive generations of students helped me, over the years, to correct and improve the radar processing lectures that constitute the core of this book.

Next, ONERA, where, in the dynamic team established around the late Jacques Dorey and C. Pouit, I have had the opportunity to deal with a wide range of problems. In particular, Mr. Vaizan provided priceless help in improving the original draft; in addition, he supplied some of the demonstrations in this book, as did Mr. Delhote (now with Thales Air Defense).

Le Centre Thomson d'Applications Radars (Thomson-LCTAR), where several specialists advised me on some useful finishing touches: Mr. Soiron, Mr. Zolesio, and both Mr. Grorud and Mr. Jurand, who improved both some technical points and my poor phrasing in English.

My colleagues at Thales Airborne Systems, Messrs. Lacomme, Kemkemian, Chamouard, Savy, Desvignes, Aulagnier, Julié, Normant, Becker, Mme. Schaub, and many others who are essential actors in the evolution of modern radar.

Finally, Frederique Ywanne at Thales Underwater Systems made an invaluable contribution in correcting the final manuscript—reviewing with such in-depth analysis that it would be better described as a collaboration.

Without the presence and support of my whole family, from their personal motivation and lively enthusiasm to their help with the diagrams, this work would have only been dreamed of. Thank you, and let us go on!

1

Radar and Sonar

1.1 Summary

A radar or an active sonar is a system that transmits a wave of known shape and receives the echoes returned by different obstacles or targets in the landscape. The transmitted wave can be a pure frequency ("pure tone") or its amplitude, phase, or frequency can be modulated. On reception, this wave must be amplified and analyzed in one way or another. This amplification has the effect of adding spurious noise ("thermal" noise) to the received wave; this noise can generally be considered as white and Gaussian in the frequency band used. A simple model of the situation is obtained by considering that the received signal consists of the transmitted signal, modified by propagation and by reflection from the target (which is assumed to be a single-point scatterer), to which white Gaussian noise is added (in the frequency band used) to take into account the noise added by the receiving amplifiers (and the ambient noise). If the target is stationary, the received signal is therefore a noisy replica of the transmitted signal, with a delay corresponding to the radar target two-way path and modified by a complex coefficient (amplitude and phase) resulting from the attenuation due to propagation and reflection from the target, and from the corresponding phase shift. This complex coefficient is unknown, and indeed is usually of no interest to the operator—but its existence must obviously be taken into account by signal processing.

If the target has a constant range-rate motion, the Doppler effect generates further distortion of the transmitted signal by a frequency shift—simply caused by the change in range and thus in delay in time—proportional to the radial velocity of the target and to the transmitted frequency. When the transmitted frequency band is sufficiently narrow—and this is generally the case with radar, though less often in sonar—the received signal is then a noisy, time-shifted and frequency-shifted replica of the transmitted signal, multiplied by an unknown complex coefficient.

For typical pulsed radars, the frequency shift can be interpreted as a phase decrease (or increase) from pulse to pulse due to the shortening (or lengthening) of the two-way path length, from pulse to pulse. During the repetition period T, the target motion amounts to a fraction of a wavelength.

When the narrowband assumption is not valid, the frequency distortion produced by the Doppler effect is more complicated (time compression, resulting in a homothetic frequency transformation, translation of each frequency component by a quantity proportional to its frequency) but remains perfectly known for a given radial velocity.

In general, the received signal in active radar or sonar can be modeled, to a first approximation, by a replica of the transmitted (and therefore known) signal, which is distorted in a known manner when the set of parameters (for example, speed and position) is known and multiplied by an unknown complex coefficient and to which Gaussian white noise has been added. The problem, when the set of parameters is unknown, is to determine its value from the received signal.

In the case of passive systems (passive listening), it is not possible to set up a general model of the received signals, as the situations encountered are too varied. The angular location aspect will be examined in Chapters 3 and 5, and an overview of the periodic signal identification methods will be given in Chapter 7.

1.2 Introduction

As their names indicate, Radio Detection And Ranging (RADAR) and Sound Navigation And Ranging (SONAR) are means for:

- Detecting targets or obstacles;
- Estimating some parameters of these targets including range, velocity, and bearing.

Radar and sonar have the same operating principle: transmission of an electromagnetic (radar) or acoustic (sonar) wave that is reflected by the target and captured by the reception system. The wavelengths generally used range from 1 cm to 1m in radar (frequency f: 300 MHz $< f <$ 30 GHz, wavelength $\lambda = c/f$, $c = 3 \times 10^8 \text{m} \times \text{s}^{-1}$) and from 3 mm to 3m in sonar (500 Hz $< f$ $<$ 500 kHz, c = 1,500m $\times$ s^{-1} in water).

These two types of systems, described as active, are complemented by passive systems, which have no transmitter but detect the inherent emission of the targets (e.g., radar transmitters, transmissions, or astronomical sources in the electromagnetic spectrum, or inherent noise radiated by ships in the acoustic spectrum). These passive acoustic systems often use longer wavelengths than the active systems, although this is not an absolute rule.

Whether active or passive, these systems are limited by the power of the signals used and by space constraints (antenna dimensions), which degrade the accuracy of measurements made in the presence of the following spurious signals or noise of different origins:

- The thermal noise of the receivers (most of which is generally caused by the mixer and the first amplifier in the reception chain), which can be considered as Gaussian and white within the frequency band concerned, is a fundamental limiting factor in radar. In sonar, ambient noise due to sea surface roughness (such as surface movements due to wind or swell) or ship traffic (especially under 100 Hz) are also taken into account by this additive noise.
- The various spurious signals caused by the environment. Jammers (intentional or otherwise) and spurious reflections on landscape elements (*clutter* in radar or *reverberation* in active sonar) can often be considered as additive noises, generally not correlated with the transmitted signal; in sonar and in low-frequency radars, these external, spurious signals are usually the predominant interfering factor.
- The various distortions due to the propagation (curvatures of the propagation rays) have a nonadditive—and often not immediately random—character that makes modeling difficult. This is the case with evaporation ducts over the sea in radar and of sound channels in sonar.

These spurious signals and distortions limit the estimation accuracy and the detection probability of the system in different ways, depending on the type of signal transmitted and on the structures of the transmission and reception systems. The design or improvement of a radar or sonar system therefore involves attempting to optimize these factors:

- The transmitted signals (waveform, carrier frequency, transducer positioning, or, in passive systems, frequency bands used);
- The structure of the receiver (sensor layout, processing of the received signals), in order to maximize the accuracy of target parameter estimation or the detection probability, given the characteristics of the noise and spurious signals.

In the field of signal processing—the subject of this book—the designer generally applies principles and arguments based on a random model of spurious signals (white noise or colored noise) and a point echo model of the target (single scattering center). This model will therefore be described in Section 1.3, and then discussed in detail in later chapters (external spurious signals are taken into account in Chapters 4 and 5, then precise target signatures in Chapters 6 and 7).

1.3 The Active Radar or Sonar Signal

Let $s(t)$ be the signal transmitted by the transducer of the radar or the sonar. This signal is generally narrowband (which will be discussed below) and can therefore be written in the form:

$$s(t) = u(t)e^{j\omega_0 t} \tag{1.1}$$

where

$u(t)$ is the complex envelope of the transmitted signal;
ω_0 is the angular frequency of the carrier.

Let us assume a point target with constant radial speed[1] located at a range $R(t) = R_0 - vt$ (v: radial velocity). The signal received at time t was reflected by the target at time $t - \tau(t)$ if $\tau(t)$ is the two-way delay time of the signal.

Consequently

1. This approximation is usually valid because of the short duration of the signal used in a period of coherent observation [see discussion of (1.6)].

$$c\tau(t) = 2R\left[t - \frac{\tau(t)}{2}\right]$$

$$R\left[t - \frac{\tau(t)}{2}\right] = R_0 - v\left(t - \frac{\tau(t)}{2}\right)$$

$$\Rightarrow \tau(t) = \frac{2R_0/c}{1 - v/c} - \frac{2tv/c}{1 - v/c}$$

The signal received at time t is therefore:

$$\alpha u(t - \tau(t))e^{j\omega_0(t - \tau(t))}$$

where α is an attenuation coefficient due to the two-way propagation. Let us examine separately the complex envelope and the carrier. The complex envelope is given by:

$$u[t - \tau(t)] = u\left[t - \frac{2R_0/c}{1 - v/c} + \frac{2v/c}{1 - v/c}t\right]$$

that is

$$u\left[t\left(1 + \frac{2v/c}{1 - v/c}\right) - \frac{2R_0/c}{1 - v/c}\right]$$

The target motion is thus represented by:

A delay $\frac{2R_0/c}{1 - v/c} \sim \frac{2R_0}{c}$ for the usual target velocities;

A change in the time scale, which is multiplied by $1 + \frac{2v/c}{1 - v/c}$.

This change in time scale is negligible in the vast majority of cases. In fact, if the band of the transmitted signal is B, the value of $u(t)$ does not change appreciably during a period $1/B$.

The error will therefore be negligible if:

$$T\frac{2v/c}{1 - v/c} << \frac{1}{B}$$

(T: duration of the transmitted signal.)

As $v/c << 1$, this condition can be rewritten as:

$$BT << \frac{c}{2v} \tag{1.2}$$

a condition generally valid in radar, but not always in sonar, since typical values of the ratio v/c are about 10^{-6} in radar, compared with 10^{-2} in sonar. We will see in Chapter 3 that this condition means that the target displacement vT during T is much smaller than the range resolution of the radar, $c/(2B)$, and wideband radars that violate this condition will be analyzed in Chapter 7.

Consequently, it can generally be considered that the motion of the target has no effect on the complex envelope apart from the delay due to the two-way path delay $2R_0/c$.

The received carrier is given by: $e^{j\omega_0(t-\tau(t))}$ that is,

$$e^{j\omega_0\left[1+\frac{2v/c}{1-v/c}\right]t}e^{-j\varphi}$$

where

$$\varphi = \frac{2\omega_0 R_0/c}{1-v/c} \sim 2\omega_0\frac{R_0}{c}$$

The carrier frequency is therefore shifted by:

$$\Delta f = \frac{2v}{\lambda_0}\left(\text{for } \frac{v}{c} << 1\right) \tag{1.3}$$

where λ_0 is the carrier wavelength.

The radial velocity of the target can thus be deduced from the measurement of this Doppler shift.

To summarize, if the transmitted signal is $u(t)e^{j\omega_0 t}$, the signal received from a target located at a range R_0, whose radial velocity is v, is:

$$Au(t-2R_0/c)e^{-j\varphi}e^{j\omega_0(1+2v/c)t}$$

if: $v/c << 1$

$$BT << \frac{c}{2v}$$

B: signal bandwidth
T: signal duration
A: received signal amplitude.

This signal can also be written:

$$\alpha u(t - 2R_0/c)e^{j\omega_0 2vt/c}e^{-j\varphi}e^{j\omega_0 t}$$

or again

$$\alpha a(t, \theta)e^{-j\varphi}e^{j\omega_0 t} \tag{1.4}$$

where θ is the vector of the parameters to be measured, equal to:

$$\begin{pmatrix} 2R_0/c \\ 2\omega_0 v/c \end{pmatrix}$$

and

$$a(t, \theta) = u(t - 2R_0/c)e^{+j\omega_0 2vt/c} \tag{1.5}$$

It is clear that the basic problem of radar or sonar is to measure the time of arrival and the frequency of a signal.

Another important characteristic of the received signal in radar or sonar is that this signal has an unknown phase shift (which means that the receiver must perform quadrature detection, and thus nonlinear processing).

In the case of (1.2), where limiting the product BT is not complied with, the received signal retains the general form (1.4), but the expression of $a(t, \theta)$ becomes more complicated. To keep the following chapters as general as possible, the arguments will only specify the form of $a(t, \theta)$ when this detail becomes necessary (essentially in Chapter 3, for the definition of the Doppler-range ambiguity function).

Moreover, the expression of $a(t, \theta)$ also becomes complicated when the radial velocity can no longer validly be considered constant. This situation occurs when the change in velocity during the period T of measurement,

γT, is no longer negligible compared with the velocity resolution, which, as we shall see in Chapter 3, is on the order of $\frac{1}{T}\frac{\lambda}{2}$. The condition is thus written:

$$\gamma << \frac{\lambda}{2T^2} = \frac{c}{2f_0T^2} \tag{1.6}$$

When this condition is not satisfied, the parameter set must include γ and the expression of $a(t, \theta)$ must be modified in consequence (phase modulation in t^2). However, this type of situation is rarely encountered in practice.

Finally, the preceding description of the received signal applies in the case of a single transmission-reception transducer; the case of a set of receivers will be examined in Chapters 3 and 5 and will enable a third quantity, the angular position (elevation and bearing), to be added to the two fundamental quantities of range and velocity.

1.4 Physical Interpretation

To fully appreciate the physical reality of the Doppler effect beyond the preceding equations, and to subsequently better understand the processing required to extract the speed information, it is instructive to examine the nature of the phenomenon in a typical case: that of a radar transmitting on a 10-GHz carrier (corresponding to a wavelength of 3 cm) periodic pulses of duration 1 μs every 100 μs (that is, at a repetition frequency F_r of 10 kHz). Figure 1.1 shows the signal generated by the reference oscillator and the transmitted and received signals.

In this figure all the pulses transmitted have a zero phase in relation to the reference oscillator at the instant of transmission:

$$\varphi_0 = 0, \text{ and } \varphi_0 + 2\pi f_0 T_r = \varphi_0 + 4\pi f_0 T_r = 0$$

which means that in this example, the transmission frequency f_0 is a multiple of the pulse repetition frequency F_r (which is often the case in practice, to ensure complete synchronization of the radar clocks). Note also the liberty taken with respect to the scale, for both the number of oscillations per pulse and the repetition period.

Within the signal received, shifted by τ during the two-way path, the first pulse has a phase φ_1 equal to π in the figure, in relation to the reference

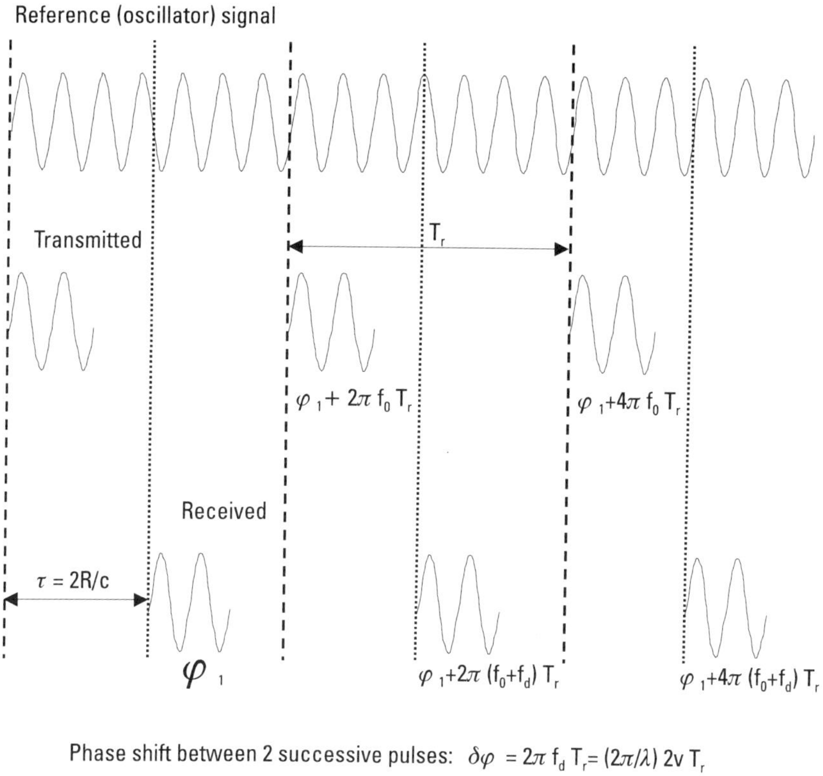

Figure 1.1 Doppler effect on a pulse radar.

oscillator. This shift is due, as already seen, to the reference oscillator rotation during the two-way path, and is therefore unknown and not used by the radar.

The second pulse has a phase φ_2, equal in the figure to $\varphi_1 - \pi/2$, that is, $\pi - \pi/2$; whereas, the transmitted pulses were all in phase, the second pulse received is observed to have a phase advance of $\pi/2$ with respect to the first. This means that during the period T, the radar–target–radar distance decreased by one-quarter of a wavelength; that is, the target got closer by one-eighth of a wavelength (which results in a change in the two-way path by one-quarter of a wavelength).

Similarly (as the target radial velocity is assumed to be constant), the third pulse is phase shifted by $-\pi/2$ in relation to the second, that is, by $-\pi$ in relation to the first, or by $\pi - \pi = 0$ in relation to the transmitted pulse.

It is then easy to deduce the target closing speed:

$$vT_r = \lambda_0/8, \text{ for example, } v = (\lambda_0/8)F_r = 37.5 \text{ ms}^{-1}$$

This phase shift from pulse to pulse is none other than the Doppler effect referred to previously. This Doppler effect results in a frequency shift f_d equal to: $f_d = 2v/\lambda$ and generates, from one pulse to the next, a phase shift $\delta\varphi$:

$$\delta\varphi = 2\pi f_d T_r = \frac{2\pi}{\lambda} 2vT_r$$

For $v = (\lambda/8)\ F_r$, the Doppler shift in frequency is equal to $F_r/4$ (that is, 2.5 kHz. The Doppler effect is negligible in relation to the 10-GHz carrier frequency), which corresponds to a phase shift of $-\pi/2$ from one pulse to the next. Thus, this analysis shows the following:

- The Doppler effect is due to the closing in, or moving away, of the target within the duration of the transmitted signal. When the target draws closer, the end of the transmitted signal covers a two-way distance shorter than the beginning of the signal so that the received signal seems to be compressed in relation to the transmitted signal, or to the reference oscillator.
- On a short-pulse radar, this effect takes the form of a phase shift from one pulse to the next; it is imperceptible through the duration of a pulse (in the chosen example, it would give a phase rotation 100 times lower than from one pulse to the next, as the repetition period is 100 times the duration of the pulse, that is, a phase shift of $-\pi/200 = 0, 90°$).

To summarize, the two-way path of the transmitted wave generally has the following effects on the signal received by the radar:

- A delay $2R_0/c$ (due to the out-and-return propagation);
- A phase shift (also due to this two-way propagation);
- A shift in the frequency received (Doppler effect $f_d = 2v/\lambda$, due to target closing), which results, for a pulse radar, in a phase shift $\delta\varphi = (4\pi/\lambda)vT_r$, from one received pulse to the next.

1.5 Passive Listening

The model discussed above does not apply to the case of passive listening (passive radar or sonar), because in this case the transmitted signals are, a priori, unknown—or poorly known—and of varied forms according to the type of source to be detected. For this reason, the passive listening receiver is usually broken down into two parts: angular location, which generally makes few assumptions about the temporal form of the signals searched for and which will be examined in Sections 3.5 and 5.7, and analysis of the shape of the received signal (identification), one aspect of which will be examined in Section 7.7.

An essential procedure for signal analysis consists of reducing the false-alarm rate by use of classification methods. In passive underwater acoustics, the signal is random but partially structured by spectral lines or transients, and the noise is random, colored, Gaussian—as with spurious signals.

The detection procedure uses several hierarchical levels (signals of interest, signals requiring further analysis, ship identification), so that the required performance can be achieved at low cost. In principle, this approach lies between optimal techniques, which introduce classification information at the detection step (high complexity), and techniques that combine a highly tolerant detection with a systematic classification of every false alarm. This compromise allows a near optimal performance if high gain processing (in terms of signal-to-noise) is performed on the raw signals, and if successive classification criteria progressively sharpen the identification.

This book will not cover all aspects of passive listening—which is nevertheless an essential operating mode of sonar—but will give at least a glimpse of the available techniques and their links with the processing used by the active systems. A more detailed analysis of passive systems would require both a very general description of the signal processing and pattern recognition procedures and a detailed discussion of operational signals, which cannot be done in the context of this book.

2

Optimum Reception in White Noise

2.1 Summary

An active radar or sonar can be used to answer two questions that are, a priori, very different:

- Is there a target?
- What are the range, the velocity, and other parameters of the target?

The first question corresponds to the problem of detection. There are two hypotheses to choose from—target present or absent. The second question raises a problem of estimation, or measurement, of the parameters of a target assumed to be present. As the questions are different, it might be expected that the structures of the appropriate radars or sonars would also be different and that, in particular, the processing of the received signals would be different. We shall see that this is not the case and that the structure of the optimum receiver is the same—except for the final decision criterion—whether the objective is detection or parameter estimation.

To determine the structure of the optimum receiver, an optimality criterion must first be chosen. For the estimation problem (measurement of, for example, range and velocity) many criteria can be considered, such as the least maximum error (minimax criterion) and the least mean square error. A natural choice consists of taking as the measurement result the value $\hat{\theta}$ of the parameter that would, with the highest probability, result in the signal actually being received. In other words, given the received signal r, the probability of receiving a signal such as r is calculated for each possible value θ of the parameters, and the value for which the probability of receiving r is the highest, is taken as the estimated value. This is a commonsense approach, proceeded by a query: if θ was the true value of the parameter, what would be the chances of the signal r actually being received? The best choice for θ, that is, $\hat{\theta}$, is then obviously that for which these chances are the greatest.

This method for determining θ is called the maximum likelihood method (the likelihood being defined as the probability of receiving r for a value θ of the parameters). Apart from its intuitive character, this method has an important advantage, related to the existence of the Cramer-Rao bounds. For a given estimation problem, that is, for a law describing the received signal as a function of the parameters θ to be measured, these bounds define an *absolute limit to the accuracy of any estimator,* whatever its principle, the only condition being that it gives on average the correct value of θ, which seems the least one could ask. These bounds are calculated from the likelihood mentioned previously. We then show that if a receiver achieves this limiting accuracy, then the receiver built according to the maximum likelihood criterion also achieves it. This very general result thus confirms the value of this maximum likelihood criterion. In addition, we show that in the case in which the noise is Gaussian, the structure of the receiver is the same for a wide variety of criteria, including the maximum likelihood and the least mean square error. This set of arguments—intuitive approach, Cramer-Rao bounds, equivalence with other criteria—is a strong incentive to use the maximum likelihood as a basic tool throughout the analysis of radar processing, even though in some cases the least squares criterion gives the same result using simpler calculations.

For the detection problem, a natural criterion appears rapidly if we put ourselves in the situation of the radar operator, who from time to time sees the equipment give an alert: detection.

This operator hopes:

- On the one hand, that the alert has not been triggered by mistake, that is, that the level of false alarms is low;
- On the other hand, that the targets are detected as reliably as possible, that is, that the detection probability is high.

These two hopes are obviously antithetical, since at the limit a zero false alarm rate can be achieved by a receiver that has failed, at the price of a detection probability that is also zero; whereas, a detection probability of

1 is obtained when the alert is triggered systematically at any time, at the price of a false alarm probability of 1. The operator seeks a compromise between these two extreme situations, and the natural approach consists of specifying that the false alarm probability P_{fa} does not exceed a given value—dependent on the extent of the area to be covered but also on the cost of the basic alert, which differs widely according to whether it is a radar triggering the automatic opening of a door or a ballistic missile detection radar. Under this condition of fixed P_{fa}, the operator wants to obtain the highest detection probability. The receiver optimization criterion is thus to maximize the detection probability at a fixed false alarm probability (Neyman-Pearson criterion).

This idea of a trade-off between the false alarm probability and the detection probability is fundamental in radar and sonar, and it is therefore important to know their orders of magnitude. For a radar to operate efficiently, the false alarm probability must generally be on the order of 10^{-5} to 10^{-8} (10^{-3} to 10^{-6} in sonar), and the detection probability (for a single scan of the target) must be on the order of 0.5 to 0.9 at maximum range. Although these figures concern only average situations, they clearly reveal the dissymmetry between the two requirements and the high stability necessary to obtain such false alarm probabilities.

Under these conditions (maximum likelihood for estimation and Neyman-Pearson criterion for detection), we show that the structure of the optimum receiver is the same for both problems, apart from the final decision logic. This receiver comprises consecutively: correlation with a replica of the expected signal (for each possible parameter θ), quadrature detection, normalization with respect to the replica power, and determination of the maximum on all the obtained values corresponding to all the possible parameters θ (in the case of estimation) (Figure 2.1), or comparison of these values with a threshold, defined according to the required false alarm probability (in the case of detection) (Figure 2.2).

Although not obvious, since the questions raised and the quality criteria chosen were not the same, this structural identity between the two receivers—detector or estimator—can be explained intuitively; in both cases, the receiver attempts to measure the similarity between the expected situation (target of parameter θ) and the observed situation (signal r received). It measures this similarity by correlating the received signal with the expected signal, performs square-law envelope detection to eliminate the phase—unknown—of the received signal, then normalizes so as not to distort this comparison by any multiplication factor of the replica (to avoid being "attracted" by the high-power replicas). Finally, it compares the results obtained for the different values of the parameters—estimation—or compares these results with a threshold above which it is considered that the similarity to the transmitted signal is not fortuitous but results from the presence of a target.

With respect to this intuitive argument, theory can define:

- The way to measure the similarity, by correlation;
- The way to eliminate the phase, by square-law envelope detection.

These two definitions are direct consequences of the Gaussian additive noise assumption. For other noise assumptions, these operations must be carried out in another way (correlation with the sign of the replica for Laplacian noise, for example). It thus appears that the structure of the receiver—square-law envelope detection, correlation, normalization, and decision—does not depend on the form of the modulation introduced by the parameters θ. Since it is assumed that the form is known, and it is thus possible to build replicas of the expected signal, the structure of the receiver is the same. This result was not obvious, since it might be expected that, depending on the type of modulation introduced by the parameters (e.g., phase, frequency, linear, quadratic), the structure of the receiver would also be different.

This argument thus gives the structure of the optimum receiver and demonstrates a characteristic approach of radar or sonar, consisting of scanning all possible hypotheses to make the decision (the approach will be different for tracking radars or sonars, which evaluate the value of the required parameter θ directly, in an explicit form, and no longer by scanning). This approach thus generates an image of the observed scene in the range-velocity plane, or in the elevation-bearing-range-velocity space if these are the required parameters. This image is built up from elementary points whose brightness represents the probability, in the widest sense, that the corresponding parameters are those of the target.

Like any imaging equipment, the radar or sonar receiver has an equipment function, which can be defined as the response to a point target of unit amplitude in the absence of noise. Ideally, this response should be equal to unity for the set of parameters corresponding to the position of the target, and zero for all other sets of parameters. In practice, the inherent limitations of the equipment—such as duration and form of the transmitted signal and layouts of the transmitters and receivers—result in an equipment function that, like the equipment function of an optical system, has two types of faults:

- The peak located at the position of the target is not infinitely narrow, but has the shape of a more-or-less pronounced maximum.
- Apart from this maximum, more-or-less bright secondary increases may exist for some values of the parameters.

In radar and sonar, this equipment function is called the ambiguity function (a strong secondary increase induces ambiguity in the position of the target) and it provides an indicator of the quality of the radar or sonar (Figure 2.3).

In particular, it is easy to comprehend that the accuracy of the parameter measurements is directly related to the more-or-less pronounced character of the main maximum. The theoretical analysis confirms this intuition, since it demonstrates (once again by means of the Cramer-Rao bounds) that when the additive noise is relatively low, the accuracy of the parameter measurements is proportional to:

- The inverse of the curvature of the ambiguity function at its maximum, that is, to the "sharpness" of this maximum;
- The ratio of the *energy* of the transmitted signal to the *spectral density* of the noise.

And this accuracy depends on the transmitted signal only by these two quantities. The ambiguity function thus completely characterizes the estimation problem, and this provides an initial explanation for the prime role that ambiguity plays in radar. Other explanations will appear in the next chapter, when the measured parameters (range and velocity) will be defined.

2.2 Principle

Let us consider the general problem of the reception of a signal $s(t, \theta)$, a known (possibly vectorial) function of time and of an unknown parameter (or parameter vector) θ, with additive Gaussian noise $n(t)$, whose characteristics will be defined later. The "reception" of this signal may involve two distinct problems [1, 2].

The first problem is detection, that is, the decision between two hypotheses: one, H_0, corresponding to the absence of a signal [the signal received then consists solely of noise $n(t)$], the other, H_1, for which the received signal is the sum of the desired signal $s(t, \theta)$ for a given value of θ and the noise $n(t)$:

$$\begin{cases} H_0: r(t) = n(t) \\ H_1: r(t) = s(t, \theta) + n(t) \end{cases} \tag{2.1}$$

The second problem is estimation, that is, the determination of the value θ of the parameter from the received signal $r(t)$:

$$\theta\ ?\ r(t) = s(t, \theta) + n(t) \tag{2.2}$$

The optimum receiver is obviously defined with respect to an "optimality criterion." The purpose of this chapter is to build the receiver for each of these two problems in terms of the maximum likelihood criterion, the use of which will be justified in Section 2.2.1, and to define the performance evaluation criteria of these receivers.

2.2.1 Estimation of a Parameter

Consider a signal $s(t, \theta)$, a known function of time t and of an unknown parameter θ, with additive noise $n(t)$.

The signal is described by

$$r(t) = s(t, \theta) + n(t)$$

The estimator $\hat{\theta}$ of the value θ of the observation parameter of the received signal $r(t)$ will be a function of the received signal and of the known assumed dependence law $s(t, \theta)$ of the signal as a function of the parameter θ. A natural optimality criterion consists of choosing the value $\hat{\theta}$ of the parameter that results with the highest probability in the signal actually received $r(t)$.

This involves calculating the probability $p_{r/\theta}$ of receiving the signal $r(t)$ for each possible value of the parameter θ, and determining $\hat{\theta}$ as the value of θ that maximizes $p_{r/\theta}$ for the signal $r(t)$ actually received:

$$\hat{\theta}: p_{r/\theta} \text{ maximum}/\theta$$

If this probability $p_{r/\theta}$ can be calculated, the value $\hat{\theta}$ will be a function of the received signal $r(t)$ and of the form of $s(t, \theta)$ which will contribute to $p_{r/\theta}$.

The efficiency of an estimator can be measured by its variance $\text{Var}(\hat{\theta} - \theta)$ (i.e., variance of the deviation from the actual value θ). In the case in which the mean value of the estimator is correct on average—unbiased—that is, where:

$$\text{E}(\hat{\theta}) = \theta, \qquad \text{E: mathematical expectation}$$

the lower limit of this variance is set by the Cramer-Rao bounds (demonstrated in Appendix 2A):

$$\text{Var}(\hat{\theta} - \theta) \geq 1/\text{E}\left[\left(\frac{\delta \log p_{r/\theta}}{\delta\theta}\right)^2\right] \tag{2.3}$$

or

$$\text{Var}(\hat{\theta} - \theta) \geq 1/\text{E}\left[\frac{\delta^2 \log p_{r/\theta}}{\delta\theta^2}\right] \tag{2.4}$$

These bounds thus constitute a lower limit to the variance of any unbiased estimator. When they are reached, the estimator $\hat{\theta}$ is said to be efficient.

It is then shown [(2A.1), described in Appendix 2A] that if an efficient estimator exists, then the maximum likelihood estimator is efficient. This very strong property is an additional reason for adopting this estimator hereafter.

2.2.2 Simultaneous Estimation of Several Parameters

The previous properties can easily be extended to the case of several parameters by defining θ as a vector of components θ_i [see (2.1)]. The Fisher matrix $\mathbf{J}$, of elements J_{ij}, is then defined as

$$\mathbf{J} = [J_{ij}] \text{ with } J_{ij} = \text{E}\left[\frac{\delta \log p_{r/\theta}}{\delta\theta_i}\frac{\delta \log p_{r/\theta}}{\delta\theta_j}\right] \tag{2.5}$$

or

$$\mathbf{J}_{ij} = -\text{E}\left[\frac{\delta^2 \log p_{r/\theta}}{\delta\theta_i\,\delta\theta_j}\right] \tag{2.6}$$

If J_{ij} is the current element of the inverse matrix $\mathbf{J}^{-1}$:

$$\mathbf{J}^{-1} = [J^{ij}] \tag{2.7}$$

then

$$\mathrm{Var}(\hat{\theta}_i - \theta_i) \geq J^{ii} \tag{2.8}$$

2.2.3 Optimum Detection

The problem of detecting a signal $s(t)$ in noise $n(t)$ consists of choosing between two hypotheses, depending on the received signal $r(t)$:

$$\mathrm{H}_1: \quad r(t) = s(t) + n(t)$$
$$\mathrm{H}_0: \quad r(t) = n(t)$$

Various criteria can be used; the most natural, corresponding to the approach actually used in practice, consists of considering that two different types of error can be made:

- Deciding that there is a target (H_1) when only noise is present (H_0): this is a false alarm, with probability P_{fa} for a given decision rule;
- Deciding that there is no target (H_0) when hypothesis H_1 is in fact true: this is a nondetection, of probability P_{nd}.

It is clear that in practice, the two errors are antithetical. Minimizing the probability of one of these errors will automatically increase the probability that the other will occur. The practical criterion used, that of Neyman-Pearson, consists of setting the false alarm probability P_{fa} to a specified value α and minimizing the nondetection probability:

$$\begin{cases} P_{fa} = \alpha \\ \mathrm{Min}\ P_{nd} \end{cases} \tag{2.9}$$

Let us consider the case in which the received signal is a vector $\mathbf{r}$ of measurement samples. Defining a decision rule means defining an area Z_0 in the space of the vectors $\mathbf{r}$ such that if $\mathbf{r}$ lies within this area, hypothesis H_0 will be chosen, and if not, hypothesis H_1 will be chosen. The false alarm and nondetection probabilities are then given by:

$$P_{fa} = 1 - \int_{Z_0} p_{r/H_0}(r)dr \tag{2.10}$$

$$P_{nd} = 1 - \int_{Z_0} p_{r/H_1}(r)dr \tag{2.11}$$

where p_{r/H_0}, and p_{r/H_1}, are the probability densities for r in the absence and presence of a target, respectively.

The minimization problem (2.9) that the radar receiver must solve can be expressed as follows, using the Lagrange multiplier:

$$\begin{cases} \text{Min } F = P_{nd} + \lambda(P_{fa} - \alpha) \\ \text{with } P_{fa} = \alpha \end{cases}$$

by adjustment of the boundary of the area Z_0. This is written:

$$\text{Min } F = \lambda(1 - \alpha) + \int_{Z_0} [p_{r/H_1}(r) - \lambda p_{r/H_0}(r)]dr$$

F will be minimal if Z_0 is chosen so that the term into brackets is negative at all points r of Z_0. The decision rule is therefore to decide H_0 if:

$$\frac{p_{r/H_1}(r)}{p_{r/H_0}(r)} < \lambda \tag{2.12}$$

the threshold value being set by the condition: $P_{fa} = \alpha$.

The receiver formally compares p_{r/H_1} and p_{r/H_0} and decides that a target is present if the former is more than λ times greater than the latter. The structure of the receiver for detection is thus identical to that of the receiver for estimation (calculation of p_{r/H_1}, and calculation of $p_{r/\theta}$), the determination of the maximum being replaced here by a comparison with a threshold. This identity of form between the optimum receiver for detection (according to the Neyman-Pearson criterion) and the optimum receiver for estimation according to the maximum likelihood is another reason for adopting this maximum likelihood criterion.

2.3 Optimum Receiver

2.3.1 Optimum Estimator

The optimum receiver is thus the one that maximizes the probability $p_{r/\theta}(r)$ with respect to θ for the received signal r; the definition of the optimum receiver therefore requires the calculation of this probability $p_{r/\theta}(R)$.

The received signal is given by:

$$r(t) = s(t, \theta) + n(t) \tag{2.13}$$

Let us express the signals in the form obtained in Chapter 1:

$$s(t, \theta) = \alpha a(t, \theta)e^{j\omega_0 t} \tag{2.14}$$

$$r(t) = z(t)e^{j\omega_0 t} \tag{2.15}$$

$$n(t) = b(t)e^{j\omega_0 t} \tag{2.16}$$

where ω_0 is the angular frequency of the carrier (the choice of this carrier in the band occupied by the signal will be explained later), α is an *unknown* complex coefficient accounting for the attenuation and the phase-shift during the two-way path, and $a(t, \theta)$ is the complex envelope of the received signal including any Doppler effect.

Let us assume that $b(t)$ is "band-limited complex white noise," that is, limited to a frequency band $[-F_{max}, +F_{max}]$ (bandwidth of the receivers), with a correlation function $B(t)$:

$$B(t) = N_0 \operatorname{sinc}(2F_{max} t)$$

where

$$\begin{cases} \operatorname{sinc} t = \dfrac{\sin \pi t}{\pi t} \\ N_0 = E|b(t)|^2 \end{cases} \tag{2.17}$$

where N_0 is the power spectral density of the noise.

Let us sample this signal at the rate given by Shannon's criterion $\Delta t = 1/(2F_{max})$. The noise samples $b(t_i)$ obtained are Gaussian and uncorrelated since $B(k/(2F_{max})) = 0$ for all nonzero integer values of k; they are therefore independent, and the probability density of the set of samples is therefore the product of the probability densities of the individual samples:

$$\Lambda = \prod_i \frac{1}{\pi N_0} \exp\left[-(z_i^* - \alpha^* a_i^*(\theta)) \frac{1}{N_0} (z_i - \alpha a_i(\theta)) \right] \qquad (2.18)$$

Strictly speaking, as the noise spectrum is limited in bandwidth, the duration of $z(t)$ is infinite. However, we shall assume that this duration is finite ($2K$ samples), and then make the number K tend to infinity. It can be shown that, under these conditions of band-limited white noise, the required convergence conditions are satisfied, because the expansion of $z(t)$ into a set of samples $z(t_i)$ is a Karhunen-Loève expansion (Appendix 4A).

The maximum of Λ will be obtained by taking the logarithm and retaining only terms that are functions of θ, that is, maximizing the following expression:

$$\text{Max}/_\theta \sum_{-K}^{+K} (z_i^* \alpha a_i(\theta) + z_i \alpha^* a_i^*(\theta)) - \sum_{-K}^{+K} |\alpha a_i(\theta)|^2 \qquad (2.19)$$

This contains scalar product expressions, which can be rewritten by letting K tend to infinity and returning to continuous signals. The optimum receiver must therefore determine the value of θ and α, which maximize this expression for a received signal $r(t)$ (or $z(t)$ after demodulation):

$$\text{Max}/_{\theta,\alpha} 2\int \text{Re}[\alpha z(t) a^*(t, \theta)]dt - |\alpha|^2 \int |a(t, \theta)|^2 dt \qquad (2.20)$$

$$\text{Max}_{/\theta,\alpha} - \int |a(t,\theta)|^2 dt \left[\left| \alpha - \frac{\int z(t) a^*(t,\theta) dt}{\int |a(t,\theta)|^2 dt} \right|^2 - \left| \frac{\int z(t) a^*(t,\theta) dt}{\int |a(t,\theta)|^2 dt} \right|^2 \right]$$

Consequently

$$\hat{\theta}: \frac{\left|\int z(t)a^*(t,\ \theta)dt\right|^2}{\int |a(t,\ \theta)|^2 dt} \text{ Maximum}/\theta \tag{2.21}$$

$$\hat{\alpha} = \frac{\int z(t)a^*(t,\ \hat{\theta})dt}{\int |a(t,\ \hat{\theta})|^2 dt} \tag{2.22}$$

In general, the value of α is not determined by the radar, which just determines θ, satisfying condition (2.21).

The expressions (2.21) and (2.22) thus constitute the solution to the very general problem described by (2.2) of estimating the parameters of a signal of known form and unknown amplitude and phase in the presence of white Gaussian noise (narrowband). The most direct approach to this estimation procedure is to give a set of discrete neighboring values and determine the maximum with respect to θ by a systematic exploration of all these values. The procedure, illustrated by the diagram in Figure 2.1, consists of correlating the received signal $z(t)$ after demodulation of the carrier $e^{j\omega_0 t}$, with a set of replicas of the expected signal $a(t,\ \theta_i)$, detecting and normalizing each channel, and choosing $\hat{\theta}$ corresponding to the highest result.

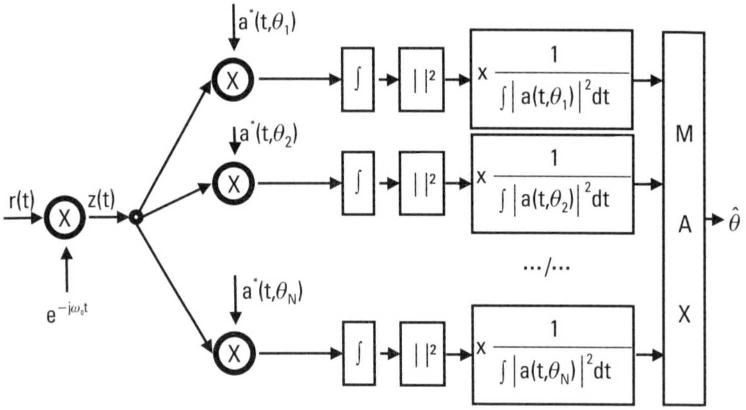

Figure 2.1 Optimum estimator in white noise.

2.3.2 Optimum Detector

It is then easy to derive the optimum detection procedure from the estimation procedure; given the analysis performed in Section 2.2.3, the structure of the receiver is in fact identical. In other words, if one wants to detect a target characterized by its given parameters, θ, p_{r/H_1} (for this value θ of the parameters) and p_{r/H_0} must be calculated. In fact, p_{r/H_1} is the quantity called $p_{r/\theta}$ in the context of the estimation problem (probability of r if a signal of parameter θ is present).

After demodulation, the signals can be expressed as follows:

$$s(t, \theta) = \alpha a(t, \theta) e^{j\omega_0 t}, \ \alpha \text{ unknown}$$

$$r(t) = z(t) e^{j\omega_0 t}$$

$$n(t) = b(t) e^{j\omega_0 t}$$

In the sampled situation described in the previous paragraph:

$$p_{r/\mathrm{H}_1} = \prod_i \frac{1}{\pi N_0} \exp\left[-(z_i^* - \alpha^* a_i^*(\theta)) \frac{1}{N_0} (z_i - \alpha a_i(\theta)) \right]$$

In this same situation, p_{r/H_0} is given by:

$$p_{r/\mathrm{H}_0} = \prod_i \frac{1}{\pi N_0} \exp\left[-\frac{1}{N_0} |z_i|^2 \right]$$

and the decision rule consists, in accordance with (2.12), of comparing the quantity:

$$\frac{p_{r/\mathrm{H}_1}}{p_{r/\mathrm{H}_0}}$$

or its logarithm:

$$\log \frac{p_{r/\mathrm{H}_1}}{p_{r/\mathrm{H}_0}} = \sum_{-K}^{+K} (z_i^* \alpha a_i(\theta) + z_i \alpha^* a_i^*(\theta)) - \sum_{-K}^{+K} |\alpha a_i(\theta)|^2 \qquad (2.23)$$

with a threshold (function of the desired false alarm rate).

Equation (2.23) can be recognized as the quantity whose maximum had to be determined in the estimation problem (2.19).

Letting K tend to infinity, as previously, and returning to continuous signals, we obtain as decision rule the comparison of the following quantity with a threshold:

$$2\int \mathrm{Re}[\alpha z(t)a^*(t,\ \theta)]dt - |\alpha|^2 \int |a(t,\ \theta)|^2 dt \tag{2.24}$$

This quantity cannot be used as such because it includes the unknown quantity α (it is assumed that θ is known; we are trying to determine, by choosing between hypotheses H_0 and H_1, whether a target of parameter θ is present). The problem is solved by using the maximum likelihood criterion to estimate this parameter α. If a target is present (hypothesis H_1), then its parameter α is that given by the maximum likelihood criterion, that is, maximizing, in accordance with the argument of the preceding paragraph, the quantity (2.24) expressed as:

$$\Rightarrow \int |a(t,\ \theta)|^2 dt \left[\left| \alpha - \frac{\int z(t)a^*(t,\ \theta)dt}{\int |a(t,\ \theta)|^2 dt} \right|^2 - \left| \frac{\int z(t)a^*(t,\ \theta)dt}{\int |a(t,\ \theta)|^2 dt} \right|^2 \right] \tag{2.25}$$

$$\Rightarrow \hat{\alpha} = \frac{\int z(t)a^*(t,\ \theta)dt}{\int |a(t,\ \theta)|^2 dt} \tag{2.26}$$

The optimum detection rule then compares (2.25) with a threshold, with $\alpha = \hat{\alpha}$

$$\frac{\left| \int z(t)a^*(t,\ \theta)dt \right|^2}{\int |a(t,\ \theta)|^2 dt} \underset{H_0}{\overset{H_1}{\gtrless}} T \tag{2.27}$$

and, if the hypothesis H_1 is confirmed (target detected), the estimated coefficient $\hat{\alpha}$ of the received signal is:

$$\hat{\alpha} = \frac{\int z(t)a^*(t, \theta)dt}{\int |a(t, \theta)|^2 dt} \tag{2.28}$$

In practice, the target presence or absence question must be raised for a set of possible values of the parameters θ; consequently, the detection test will be carried out for all these possible values of the parameters. This results in a detector structure (shown in Figure 2.2) that is similar to that of the estimator: correlation of the received signal $z(t)$ with a set of replicas $a(t, \theta_i)$, square-law envelope detection and normalization of each channel, and comparison with a threshold (the normalization of each channel being equivalent to modifying the threshold value).

It is interesting to note that in the optimum receiver (for estimation or for detection) with the structure described above:

- Square-law envelope detection is a consequence of the need to estimate α (or at least to eliminate the unknown α). Without this need, the optimum receiver would simply calculate the real part of the correlation between received signal and replica, in accordance with (2.20). This same need for quadratic detection would result if

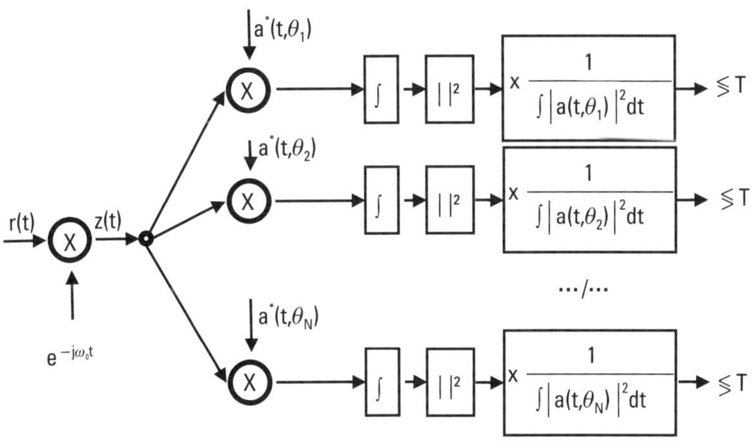

Figure 2.2 Optimum detector in white noise.

we assumed a random signal $s(t, \theta)$, which is another way of modeling the complete lack of knowledge of the phase of the received signal.

- The correlation, or *matched filtering,* is a consequence of the Gaussian character of the noise. It is the Gaussian character that generates the double product terms containing $z_i a_i^*(\theta)$ in the function to be maximized, (2.19).

Furthermore, the optimum estimator in Figure 2.1 determines the maximum of the function (2.21) by systematic exploration of the parameter θ. This procedure may appear unsophisticated—and we shall see in our analysis of tracking radars, that it is sometimes possible to calculate the estimated value $\hat{\theta}$ in the form of an analytical expression—but it is natural in radar and sonar, in which the question asked is "Where are the targets?" that is, in a situation where the following operations must be performed simultaneously:

- Detection of any target in any possible position;
- Estimation of its position (parameter θ).

2.3.3 Estimation Performance: The Ambiguity Function

Let us apply (2.21) to be maximized to obtain $\hat{\theta}$ in the case in which the received signal corresponds to a value θ_0 of the parameters:

$$z(t) = \alpha_0 a(t, \theta_0) + b(t) \text{ with } |\alpha_0|^2 = A_0 \tag{2.29}$$

The expression to be maximized with respect to θ becomes:

$$Q(\theta, \theta_0) = \frac{1}{\int |a(t, \theta)|^2 dt} \left| \alpha_0 \int a(t, \theta_0) a^*(t, \theta) dt + \int b(t) a^*(t, \theta) dt \right|^2$$

$$Q(\theta, \theta_0) = \frac{A_0}{\int |a(t, \theta)|^2 dt} \left| \int a(t, \theta_0) a^*(t, \theta) dt + \frac{1}{\alpha_0} \int b(t) a^*(t, \theta) dt \right|^2$$

Let

$$\chi(\theta,\ \theta_0) = \int a(t,\ \theta_0)a^*(t,\ \theta)dt \tag{2.30}$$

and

$$X(\theta) = \frac{1}{\alpha_0}\int b(t)a^*(t,\ \theta)dt \tag{2.31}$$

We shall now examine the case in which $\int |a(t,\ \theta)|^2 dt = 1$, for all values of θ (in this case, A_0 is the *energy* of the received signal). This situation is the one generally encountered in radar, and the argument can easily be generalized to the case in which the signal energy depends on the parameter θ.

$$Q(\theta,\ \theta_0) = A_0|\chi(\theta,\ \theta_0) + X(\theta)|^2 \tag{2.32}$$

$X(\theta)$ is a zero-mean Gaussian random variable resulting from the passage of the white Gaussian noise $b(t)$ through the filter whose impulse response is $\frac{1}{\alpha_0}a^*\ (-t,\ \theta)$. The power of $X(\theta)$ is N_0/A_0 since:

$$E[X(\theta)X^*(\theta)] = \frac{1}{A_0}\int a^*(t,\ \theta)E[b(t)b^*(u)]a(u,\ \theta)du\,dt = \frac{N_0}{A_0}$$

Consequently, the average value of receiver output $Q(\theta,\ \theta_0)$ is:

$$E[Q(\theta,\ \theta_0)] = A_0|\chi(\theta,\ \theta_0)|^2 + N_0 \tag{2.33}$$

It is useful to interpret these quantities: $Q(\theta,\ \theta_0)$ is the receiver output "at θ" when a target of parameters θ_0 is present; $|\chi(\theta,\ \theta_0)|^2$ is called the *ambiguity function.* It is the receiver output in the absence of noise. This function has a maximum for $\theta = \theta_0$.

$$|\chi(\theta,\ \theta_0)|^2 \le |\chi(\theta_0,\ \theta_0)|^2 = 1 \tag{2.34}$$

given the normalization to 1 of the energy of $a(t, \theta)$. Consequently, in the absence of noise, the estimated value $\hat{\theta}$ is unbiased: $\hat{\theta} = \theta_0$.

In general, this ambiguity function has multiple local maxima, which, in the presence of noise, may be taken for the absolute maximum. This confusion corresponds to a genuine ambiguity—confusion between two completely distinct values of the parameters θ. An example of an ambiguity function is shown in Figure 2.3, for a situation in which χ has two parameters t and f.

If the noise is not too high, it can be assumed that the value $\hat{\theta}$ is located "on the main peak" of this ambiguity function; it is clear that in this case the more pronounced the peak, that is, the higher the curvature near the maximum, the better the estimation.

This intuitive assessment of the quality of the estimation as a function of the shape of the ambiguity function can be expressed rigorously. In [2], it is shown that asymptotically (for high signal-to-noise ratios A_0/N_0):

- The estimator is unbiased;
- The estimator is efficient; that is, the Cramer-Rao bounds are reached;
- These bounds are expressed directly as a function of the second derivatives of the ambiguity function at $\theta = \theta_0$: If $\mathbf{J}$ is the Fisher matrix (2.5) with general element J_{ij}:

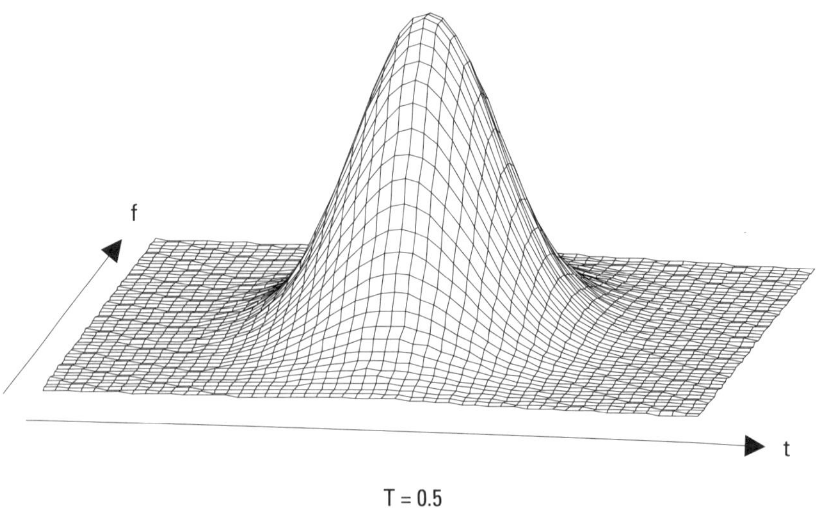

Figure 2.3 Ambiguity function.

$$\mathbf{J} = |J_{ij}|$$

$$J_{ij} = -\frac{A_0}{N_0}\frac{\delta^2}{\delta\theta_i\,\delta\theta_j}|\chi(\theta,\ \theta_0)|^2 \text{ for } \theta = \theta_0 \tag{2.35}$$

$$\mathrm{E}|(\hat{\theta}_k - \theta_{0k})(\hat{\theta}_1 - \theta_{01})| = J^{kl} \tag{2.36}$$

J^{kl} being the general element of matrix $\mathbf{J}^{-1}$, inverse of $\mathbf{J}$. The demonstration is performed neglecting $|X(\theta)|^2$ in the expansion of $Q(\theta,\ \theta_0)$—in (2.32)—and expanding Q in the neighborhood of θ_0.

This important general result—not specific to the case of radar or sonar signals—shows that the ambiguity function $|\chi(\theta,\ \theta_0)|^2$, which is the output of the optimum receiver (see Figure 2.1) in the absence of noise when a target is present at $\theta = \theta_0$, characterizes the entire estimation problem—true ambiguities (when the presence of noise causes a secondary maximum of the ambiguity function to be chosen for $\hat{\theta}$) and variance of the estimation about the actual value θ_0 [by (2.35) and (2.36)].

More intuitively, this ambiguity function can also be interpreted as an "equipment function," which transforms the target landscape into a "degraded" landscape. This intuitive representation gives an idea of the importance of this notion of ambiguity function, which, moreover, depends only on the form of the signals $s(t,\ \theta)$.

Finally, the form of this function depends on the number of channels in the optimum receiver (Figures 2.1 or 2.2). This number is determined by the sampling interval of θ (which was not specified until now); this interval must obviously be chosen so that the ambiguity function is correctly sampled, that is, so that there are at least a few reception channels on the width of a peak of this function. Any reduction in the number of channels will lead to a degradation in performance, and the performance calculated previously is, strictly speaking, obtained only for continuous exploration of θ.

In addition, it is important to note that the estimation accuracy given by (2.35) and (2.36) is a function of the ratio of the received signal energy, A, to the noise spectral density, N_0.

2.3.4 Detection Performance

To calculate the performance of the optimum detector defined by (2.27), consider the situation in which the energy of $a(t,\ \theta)$ is normalized:

$$\int |a(t, \theta)|^2 dt = 1$$

The optimum detector then performs the following operation for a given value θ of the parameter.

$$x = \left| \int z(t) a^*(t, \theta) dt \right|^2 \underset{H_1}{\overset{H_0}{\lesseqgtr}} T \tag{2.37}$$

Just as the natural detector optimization criterion consisted of maximizing the detection probability for a given false alarm probability, the detector performance will be measured by the change in the detection probability P_d as a function of the false alarm probability P_{fa}.

These two quantities are easily calculated from the two probability densities:

- p_{x/H_1}: Probability density of the quantity x defined by (2.37) under the hypothesis in which the target is present:

$$H_1: z(t) = \alpha a(t, \theta) + b(t), \text{ with } |\alpha|^2 = A: \text{energy of the received signal} \tag{2.38}$$

- p_{x/H_0}: Probability density of the same quantity x under the hypothesis in which no target is present:

$$H_0: z(t) = b(t) \tag{2.39}$$

The detection probability is then the probability of exceeding the threshold T, under hypothesis H_1:

$$P_d = \int_T^{+\infty} p_{x/H_1}(x) dx \tag{2.40}$$

Similarly, the false alarm probability is the probability of exceeding the threshold under hypothesis H_0:

$$P_{fa} = \int_{T}^{+\infty} p_{x/H_0}(x)dx \tag{2.41}$$

These two probabilities are represented in Figure 2.4 by the hatched areas. This figure clearly shows that a change in the threshold T can improve P_d only at the price of an increase in P_{fa}.

The false alarm probability can be calculated in the Gaussian case under hypothesis H_0. The variable x is given by:

$$x = \left| \int z(t)a^*(t, \theta)dt \right|^2 = |b_1|^2 \tag{2.42}$$

where

$$b_1 = \int b(t)a^*(t, \theta)dt \tag{2.43}$$

is a complex zero-mean Gaussian noise

$$E|b_1|^2 = E|b(t)|^2 = N_0 \tag{2.44}$$

Thus, x is the square of the modulus of a complex zero-mean Gaussian variable. Its probability density is therefore, according to a classic result, an exponential law [3]:

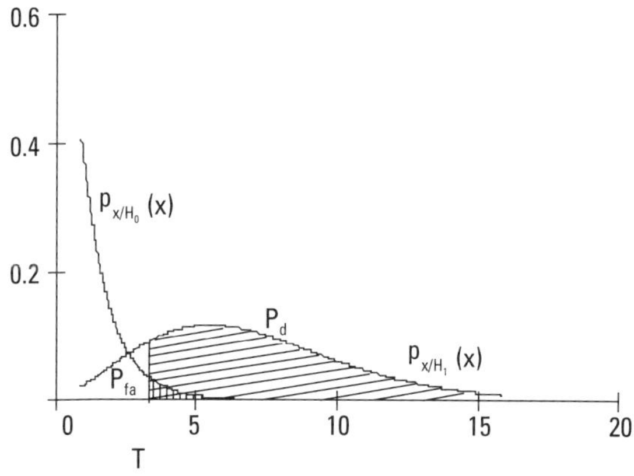

Figure 2.4 Detection probability P_d and false alarm probability P_{fa}.

$$p_{x/\mathrm{H}_0}(x) = \frac{1}{N_0} e^{-\frac{x}{N_0}} \tag{2.45}$$

when the false alarm probability:

$$P_{fa} = \frac{1}{N_0} \int_T^{+\infty} e^{-\frac{x}{N_0}} dx = e^{-\frac{T}{N_0}} \tag{2.46}$$

This simple result enables the value of the threshold T to be determined for a required false alarm probability P_{fa} when N_0 is known.

Similarly, under the target-present hypothesis, H_1, the variable x is given, from (2.37) and (2.38), by:

$$x = \left| \int z(t) a^*(t, \theta) dt \right|^2 = |\alpha + b_1|^2 \tag{2.47}$$

where b_1 is defined by (2.43).

Thus, x is the square of the modulus of the sum of a constant (nonfluctuating target) and a Gaussian complex variable. It is shown in [2] and [3] that the probability density of such a quantity is given by:

$$p_{x/\mathrm{H}_1}(x) = \frac{1}{N_0} e^{-\frac{x+A}{N_0}} I_0\left(\frac{2\sqrt{xA}}{N_0}\right) \tag{2.48}$$

where I_0 is the zero-order Bessel function.

The calculation of the detection probability P_d by integration of $p_{x/\mathrm{H}_1}(x)$ according to (2.40) must be done numerically (various analytical approximations of this probability are given in [4]).

The characteristics of the receiver can be summarized by the graphs plotted in Figure 2.5, known as the receiver operating curves (ROC). This figure illustrates the trade-off between detection probability and false alarm probability. The practical application to calculation of radar range is examined in Appendix 3B, since it requires prior analysis of radar resolution in angle, range, and Doppler (Chapter 3).

2.3.5 Conclusion

Summarizing, in the presence of white noise, the optimum estimation procedure consists of correlating the received signal with a set of replicas corre-

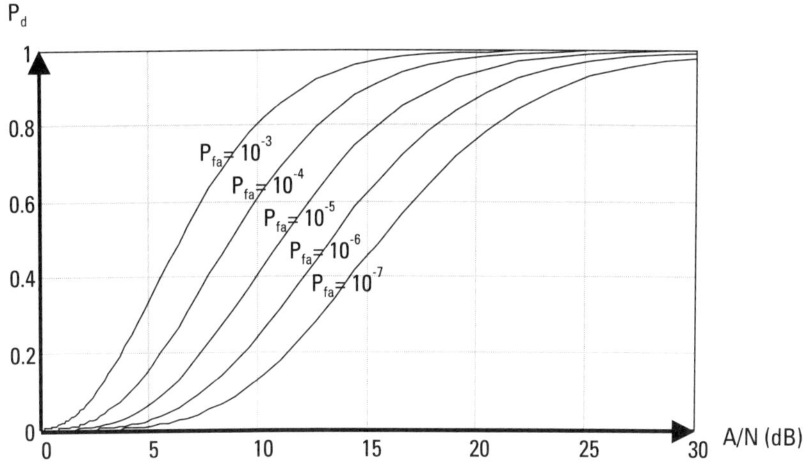

Figure 2.5 Detection performance, detection on a single pulse, and nonfluctuating target.

sponding to the different possible values of the parameters, detecting and normalizing these correlation results with respect to the energy of the replica, and choosing the parameter corresponding to the channel that gives the maximum result. The optimum detector has the same structure, the determination of the maximum being replaced by comparison with a threshold determined by the desired false alarm probability.

This optimum maximum likelihood estimator is also asymptotically efficient, which means that asymptotically, when the signal-to-noise ratio tends to infinity, the Cramer-Rao bounds are reached—and thus, at the limit, the probability estimator also minimizes the variance of the estimation error. This property is an a posteriori justification of our choice of the maximum likelihood criterion. Furthermore, examination of the performance of the estimator reveals the importance of the role of the ambiguity function $|\chi(\theta, \theta_0)|^2$:

- The curvature of this surface near the maximum $\theta = \theta_0$ limits the variance of the estimator.
- The "spurious" increases on this surface (local maxima) define the possible ambiguities in the presence of noise.

Finally, remember that the ambiguity function is the output of the optimum receiver shown in Figure 2.1 for all values of θ when the received signal is noise free (2.33). It is the equipment function of the receiver.

Throughout this chapter, the form of the received signal, which determines the form of the ambiguity function, has not been specified. The results obtained are thus applicable to any problem involving the estimation of the parameters of a signal of unknown amplitude and phase—or the detection of such a signal whatever its bandwidth and, above all, whatever the form of its dependence on θ. The only assumption is that this form is known. It thus appears that this single assumption, together with the additive Gaussian character of the noise, is sufficient to define the structure of the optimum estimator and detector by correlation and square-law envelope detection.

References

[1] Van Trees, H., *Detection, Modulation, and Estimation Theory,* New York: Wiley, 1971.

[2] Levine, M., *Fondements Théoriques de la Radiotechnique Statistique,* Moscow: Mir (Translation from Russian), 1973.

[3] Rice, S. O., "Mathematical Analysis of Random Noise," *Bell System Technical Journal,* Vol. 23, No. 3, July 1944, pp. 282–332, and Vol. 24, No. 1, Jan. 1945, pp. 46–156.

[4] Darricau, M., *Physique et Théorie du Radar,* Paris, France: Sodipe, 1994.

Appendix 2A: Cramer-Rao Equality

Consider the case in which the received signal consists of a measurement sample vector **r**. Let $p_{r/\theta}(r)$ be the probability density of r, conditional to the value θ of the parameters. Let $\hat{\theta}(r)$ be an estimator of θ based on the observation r. If $\hat{\theta}(r)$ is unbiased, then:

$$\mathrm{E}[\hat{\theta}(r) - \theta] = 0$$

$$\Leftrightarrow \int p_{r/\theta}(r)\,[\hat{\theta}(r) - \theta]\,dr = 0$$

Differentiating this equation with respect to 0:

$$\int \frac{\delta p_{r/\theta}(r)}{\delta\theta}[\hat{\theta}(r) - \theta]dr = 1$$

Moreover, the following identity can be written:

$$\Rightarrow \frac{\delta p_{r/\theta}(r)}{\delta\theta} = \frac{\delta \log p_{r/\theta}(r)}{\delta\theta} \cdot p_{r/\theta}(r)$$

$$\Rightarrow \int \left[\frac{\delta \log p_{r/\theta}(r)}{\delta\theta} \sqrt{p_{r/\theta}(r)} \right] \{\sqrt{p_{r/\theta}(r)}\,[\hat{\theta}(r) - \theta]\} dr = 1$$

Applying Schwarz's inequality, we obtain:

$$\mathrm{E}[\,\hat{\theta}(r) - \theta]^2 \geq 1/E\left[\frac{\delta \log p_{r/\theta}(r)}{\delta\theta} \right]^2$$

Another expression of this inequality can be obtained by differentiating the equality:

$$\int p_{r/\theta}(r) dr = 1$$

$$\Rightarrow \int \frac{\delta p_{r/\theta}(r)}{\delta\theta} dr = \int \frac{\delta \log p_{r/\theta}(r)}{\delta\theta} p_{r/\theta}(r) dr = 0$$

Differentiating a second time, we obtain:

$$\int \frac{\delta^2 \log p_{r/\theta}(r)}{\delta\theta^2} p_{r/\theta}(r) dr + \int \left(\frac{\delta \log p_{r/\theta}(r)}{\delta\theta} \right)^2 p_{r/\theta}(r) dr = 0$$

That is

$$\mathrm{E}\left[\frac{\delta^2 \log p_{r/\theta}(r)}{\delta\theta^2} \right] = -\mathrm{E}\left[\frac{\delta \log p_{r/\theta}(r)}{\delta\theta} \right]^2$$

Consequently

$$\mathrm{E}[\,\hat{\theta}(r) - \theta]^2 \geq -1/\mathrm{E}\left[\frac{\delta^2 \log p_{r/\theta}(r)}{\delta\theta^2} \right]$$

Furthermore, Schwarz's inequality becomes an equality when the two functions under the integral are proportional:

$$\frac{\delta \log p_{r/\theta}(r)}{\delta\theta} = [\hat{\theta}(r - \theta)]k(\theta) \tag{2A.1}$$

However, if $\hat{\theta}_{mv}(r)$ is the estimator in terms of maximum likelihood, then by definition:

$$\frac{\delta \log p_{r/\theta}(r)}{\delta\theta} = 0 \text{ for } \theta = \hat{\theta}_{mv}(r)$$

Substituting in the inequality (2A.1), we obtain:

$$\hat{\theta}(r) - \hat{\theta}_{mv}(r) = 0$$

Thus if $\hat{\theta}(r)$ is an efficient estimator, then it is also the maximum likelihood estimator.

3

Application to Radar

3.1 Summary

Let us consider the case of a radar or an active sonar composed of an omnidirectional transmitter-receiver. According to Chapter 1, the received signal is then a replica of the transmitted signal, shifted in time by the two-way time delay and in frequency by the Doppler effect (throughout this chapter, the transmitted signal is assumed to be of narrow relative bandwidth, so the Doppler effect reduces to a shift in frequency; wideband systems will be considered in Chapter 7), multiplied by an unknown complex coefficient due to propagation, and with additive Gaussian white noise. Hence, this signal contains information on the target range and radial velocity by its delay and frequency shift with respect to the known transmitted signal. The range and velocity measurements are thus reduced to time and frequency measurements. It can be expected that the accuracy of measurements will depend, among other factors, on the duration of the transmitted signal and on the width of the frequency band it uses.

Let us first assume that the observed target is not moving and that, consequently, only its range is to be determined. One way of operating consists of transmitting a brief pulse and in measuring the time elapsed between transmission and reception. In order to improve the accuracy of the measurement, the exact moment of arrival of the pulse's leading edge will be measured. The steeper the leading edge, that is, the wider the frequency bandwidth occupied by the signal, the more accurate the measurement. This simple example illustrates the fact that the range measurement accuracy is proportional to the bandwidth of the transmitted signal. In the case of a phase-, frequency-, or amplitude-modulated pulse, theory makes it possible to extend this result. The operation to carry out is not a leading edge detection but, as shown in Chapter 2, a correlation between the received signal and a set of time-shifted replicas of the transmitted signal, followed by square-law envelope detection and normalization. The target range is then given by the index of the replica for which the result obtained is the maximum. Here, the ambiguity function is the result of the correlation-detection-normalization with these successive replicas (producing as many points on the ambiguity function as the number of replicas used), and the measurement accuracy is inversely proportional to the curvature of the ambiguity function at its maximum (i.e., proportional to the main peak sharpness). This curvature is shown here to be the same as the signal bandwidth and to depend on the shape of the transmitted signal only because of this parameter—the bandwidth. (The accuracy is also, as always, proportional to the ratio of the energy of the received signal to the spectral density of the noise.) This very simple result was not obvious a priori in the case of a signal of any shape, and therefore generalizes the very special case of the brief pulse.

Then, and symmetrically, let us consider the case of the measurement of the radial velocity of a target assumed to be at a known range. A simple method of proceeding consists of transmitting a nonmodulated wave (pure frequency) during a time T, and in measuring the frequency of the received signal—for instance, by measuring the rate of zero-crossing of the received signal, or by any other means. It seems intuitive that the measurement accuracy of this frequency is inversely proportional to the duration T of this signal. The shorter the signal is, the more difficult it is to measure its frequency. Here again, theory allows extending this result to signals of any shape; the optimal receiver carries out a correlation-detection-normalization with a set of replicas shifted in frequency, and the measurement accuracy is proportional to the inverse of the signal duration and depends on the shape of the transmitted signal only on account of this parameter (accuracy being also always proportional to the ratio of the energy of the received signal to the spectral density of the noise).

Generally, the target range and velocity are both unknown; the measurement accuracy for each of them will consequently be better only if one of them is known. The optimum receiver is still composed of a set of correlations-detections-normalizations with time-shifted and frequency-shifted replicas. In the absence of noise, it delivers a range-velocity ambiguity function, every point of which—at range r and velocity v—is, in terms of intensity (or altitude in the case of a three-dimensional plot), the result of the correlation-detection-normalization with the corresponding replica.

The measurement accuracy is, here again, proportional to the curvature of the ambiguity function at its main peak. It is then demonstrated that this accuracy depends on the duration and bandwidth of the transmitted signal but also on a third parameter characterizing this signal: the modulation index, which is zero when the signal is modulated in amplitude only, but which is higher as the frequency (or phase) modulation of the signal is greater. This index measures the correlation between the time and the frequency in the transmitted signal, so one can then imagine that the higher this index is, the closer the coupling between the quantities to be measured will be, and consequently, the more the uncertainty about one will increase the uncertainty about the other. This is what more rigorous reasoning shows, because the accuracy of time and frequency (or range and velocity) measurement is—when this modulation index is zero (signals modulated in amplitude only)—equal to the measurement accuracy of each quantity considered separately. When this index is not zero, the measurement accuracy of each quantity worsens [see (3.26) and (3.27)].

Lastly, a fundamental result is obtained by calculating, for any signal, the total volume under the Doppler-range ambiguity function. This volume is constant whatever the shape of the transmitted signal. This theorem sets a limit to the dreams of the radar operator, who would like to select a signal with an ambiguity function of the "thumbtack" type: a very narrow peak at the target position, and zero everywhere else. This is not possible, and what can be gained in terms of range accuracy will be lost in Doppler accuracy or in secondary peaks (ambiguities). This result, which is a consequence of the special nature of the quantities measured (time and frequency, related by Fourier transform), is an essential constraint when defining a radar or a sonar.

This necessary compromise between range accuracy and Doppler accuracy becomes very clear when the transmitted signal is a nonphase-modulated pulse (a pure-frequency fragment, or pure tone) and when its amplitude is modulated by a Gaussian pattern. The bandwidth of such a pulse is inversely proportional to its duration because, should this duration be reduced, the steepness of the leading edge would be accentuated in proportion, thus widening the bandwidth. As range accuracy is proportional to the bandwidth, and velocity accuracy to duration, the product of these accuracies remains a constant; improving the one results in making the other worse. There are two ways to avoid this dilemma (partly because the volume invariance of the ambiguity function must be respected): (1) modulation of the transmitted pulse, in phase or frequency, which distorts the main peak of the ambiguity function, and (2) repetition—periodical or not—of the basic pulse in order to sharpen the main peak, but at the cost of the generation of sidelobes.

The modulation of the transmitted pulse eliminates another important drawback of the single unmodulated pulse—to obtain good range accuracy, the duration of the pulse must be reduced in order to increase the bandwidth. In order to keep the received energy constant when doing this, the peak power to transmit must be increased. This technique is then limited by the capacities of the transmitters. Consequently, instead of being increased by reducing the duration of transmission, the bandwidth is increased by the phase or frequency modulation of the transmitted pulse.

Linear frequency modulation is often used because of its ease of application. The bandwidth of the transmitted pulse is then given by the frequency deviation produced by this modulation. However, this type of modulation has the disadvantage of introducing a Doppler-range ambiguity, which is simple to interpret: Such a modulated pulse is much like the same pulse, slightly delayed like the same pulse slightly shifted in frequency, because its frequency (instantaneous) varies linearly with time. Consequently, distinguishing a short delay from a small Doppler shift will prove difficult; the Doppler and range measurements are thus coupled, and the range measurement will be correct if the Doppler is known, and vice versa.

Another type of modulation used often is phase coding, which, in its simplest version, consists of breaking down the basic pulse into consecutive "subpulses," multiplied by a sequence—or code—of +1 and −1, and selected in order to obtain the desired ambiguity function. This kind of modulation enables the generation of an ambiguity function with a rather narrow main peak, at the cost of sidelobes that are difficult to reduce in the Doppler mode.

However, the means used most often in radar to remedy the two limitations of the unmodulated pulse mentioned previously (peak power limitation when the required range accuracy is increased, and existence of a strict relationship—constant product—between the range accuracy and the velocity accuracy) consists of transmitting repeated pulses at a certain repetition frequency F_r. For this signal that is composed of a train of periodic pulses, the frequency bandwidth remains approximately that of the basic pulse, whereas the duration evidently increases with the number of pulses transmitted. The modulation index remains zero (since the signal is not phase-modulated), and the range and velocity accuracies therefore remain uncoupled. As a consequence, the range accuracy (inverse of the bandwidth) will remain that of the basic pulse, whereas the velocity accuracy (inverse of the duration) will improve in proportion to the number of pulses processed.

However, in accordance with the theorem of invariance of the volume under the ambiguity function, this radical improvement in the sharpness of the main peak of the ambiguity function is possible only at the cost of secondary sidelobes, which, in this case of a periodic pulse train, correspond to precise range and velocity values. It is clear that the similarity—measured by correlation—between the signal delayed by a certain value and the same signal delayed by $+1/F_r$, is great, since these signals are identical except for the end pulses. As well, the similarity with the signal delayed by $2/F_r$ remains great (identical but for the two end pulses), and so on. For the ambiguity function, this results in periodic peaks at time intervals of $1/F_r$ (or range intervals of $c/2F_r$). The same reasoning can be applied for frequency (or velocity), since the spectrum of a periodic pulse train is, in a first analysis, composed of lines located at regular intervals F_r; the ambiguity function will therefore also include periodic peaks at frequency intervals F_r (or velocity intervals $(\lambda/2)\,F_r$). The ambiguity function for a periodic pulse train therefore looks like a nail board, the width of the peaks being the inverse of the bandwidth of the basic pulse in time and the inverse of the total duration of the pulse train in frequency. If the search domain in range and velocity is less than the spacing between the peaks, this ambiguity function is thus close to the ideal thumbtack-shaped ambiguity function; this situation is quite frequent with short-range radars. On the contrary, when the required range increases, the pulses must be spaced out to avoid any ambiguity in range, at the cost of bringing velocity ambiguities closer and thus restricting the unambiguous field in velocity. In Chapter 7 we shall see that in some cases, the extent of the search domain, in range and velocity, is such that further processing of the signal must be carried out in order to remove these ambiguities.

Although range and velocity are generally data that both radar and sonar directly use, there are also situations in which the system uses this data indirectly. Such is the case of synthetic aperture radar, which measures, from an aircraft, the relative radial velocity of targets assumed to be motionless in order to deduce their position in azimuth. Indeed, for this type of radar, installed on the side of the aircraft (side-looking), an echo on the ground has a different Doppler effect according to whether it comes from a point located exactly abeam of the aircraft, hence with zero Doppler; in front of the aircraft, hence with positive Doppler; or behind it, hence with negative Doppler. The Doppler data can thus be transformed directly into target azimuth data. This operation provides images of the observed ground with a much higher resolution than that obtained using only the antenna directivity—provided that the received echoes do actually originate from motionless targets.

Besides the range and radial velocity data, the radar or sonar generally also deliver data relating to the direction of the target. Such data can be obtained only by using several sensors or an antenna of sufficiently large dimensions with respect to the wavelength. Physically, it is obtained by the measurement of the delay—or phase shift—between the signals received by the different sensors (or by the different components of the antenna); this delay, equal to zero when the echo originates from a direction perpendicular to the straight line between the sensors, increases with the inclination relative to this direction.

Adopting the approach described in Chapter 2 (Section 2. 1), the radar or sonar, here again, examine all possible assumptions to make a decision. From a practical point of view, this amounts to compensating the delays between sensors for a given direction, summing the signals thus obtained from the different sensors, and then performing the optimum Doppler-range processing by correlation-detection-normalization with a set of replicas shifted in time and frequency. This operation is then repeated consecutively (or in parallel) for every possible direction. Obviously, the process thus breaks down into spatial (or angular) processing of delays between sensors and summing, followed by the Doppler-range processing.

This spatial processing can be carried out in different ways, two of which are quite generally used. One, very general in radar nowadays, consists of using a parabolic reflector as an antenna, which reflects the signals coming from the axial direction "in phase" to its focal point (which is a geometrical property of the parabola). Such an antenna, with its paraboloid of revolution form, thus carries out the spatial processing directly if a reception horn is installed at its focal point, behind which the Doppler-range processing is carried out. Then all that is needed is to make this antenna rotate in azimuth (angular scan) to perform the optimal range-angle-Doppler processing. The second technique, often used with sonars, consists of using an array of independent sensors (in sonar, a hydrophones array, in radar, an array of radiating elements—dipoles—referred to as a *phased array*) and performing the delays and summings corresponding to the angular processing by processing the signals thus obtained: this is the sonar "beamformer," each beam corresponding to the delay and summing operation for a given observation direction.

In the case of tracking radars or sonars, the antenna must be slaved to the direction of the target. One method consists of looking in two very similar directions—in parallel or consecutively—and comparing the level of the signals received from these two directions. By trying to equalize these two levels, the steering in the direction of this target will be maintained. This is referred to as the amplitude monopulse technique (the monopulse term comes from the fact that this processing is based on the comparison of levels received in echo to one transmitted pulse). A similar technique consists of comparing the phases of the signals received on two antennas steered in the same direction; this phase difference is zero if the steering direction is that of the target. These two techniques are shown to be equivalent, as far as principles are concerned. They enable measurement of the target steering offset angle—angle error deviation—with respect to the antenna line of sight. Still using the same criterion of maximum likelihood, we obtain, for small target steering offset angles in relation to the inherent directivity of the antennas, the estimated steering offset angle θ as a function of the signals received on the two antennas (or of the sum or difference of these signals), in a simple form: θ is simply proportional to the ratio of the difference between the signals received on these two antennas to their sum—or more exactly, the real part of this ratio, since θ is a real quantity. This simple relationship is rather intuitive, since the difference between the two signals is actually zero when the target is on the axis—dividing by the sum cancels the influence of propagation attenuation and of target reflection, which affects both the sum and the difference.

The volume and speed required for radar or sonar signal processing, together with the real-time constraints involved, generally lead to multiprocessor implementations of the algorithms. The appropriate methodology for development of those complex architectures and softwares involves three successive steps: functional design, multiprocessor machine architecture design, and finally hardware and software implementation. Special attention must be devoted to the selection of an efficient and evolutive architecture for systematic signal processing, with a maximal use of processors and communications capacities, while maintaining clear programming structures, at each stage of the product life.

3.2 Doppler-Range Ambiguity

An essential characteristic of the radar problem is that the useful signals are parameterized by two values conjugated by Fourier transform: time (or range) and frequency (or velocity). This specificity induces certain properties of the ambiguity function that are decisive in the design of the radar systems. These properties will be examined in this chapter and then illustrated by various nonconventional examples: pulse signals, pulse trains, and synthetic aperture radar principles.

As the radar must also carry out target angular location, the principles of the optimum angular location of targets will be described in Section 3.5 and applied to the case of tracking radars (angle error detection).

3.2.1 Definitions: Time, Duration, Frequency, and Bandwidth Modulation

Let $a(t)$ be a complex signal such that $\int |a(t)|^2 dt = 1$ (signal normalized with respect to energy).

$|a(t)|^2$ can then be considered as the energy density of the signal over time. It is then natural to define the time $\bar{t}$ of the signal as a "mean occurrence instant," according to the formula:

$$\bar{t} = \int t|a(t)|^2 dt \tag{3.1}$$

Its equivalent duration $\overline{\Delta t}$ is then defined as a standard deviation about this mean instant (see Figure 3.1):

$$(\overline{\Delta t})^2 = \int (t^2 - \bar{t}^2)|a(t)|^2 dt \tag{3.2}$$

As the signal is normalized with respect to energy, its spectrum is also normalized (Parseval relation):

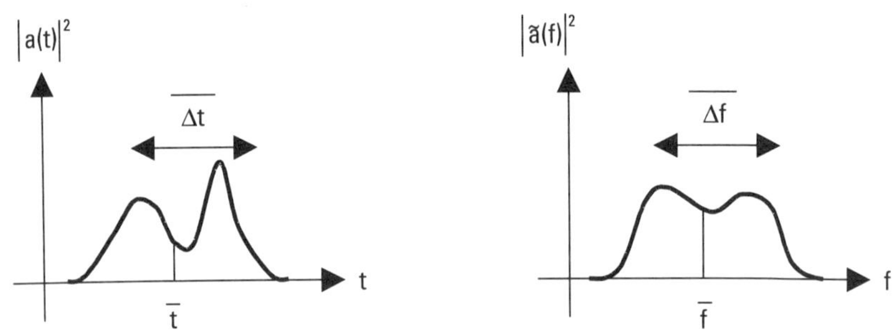

Figure 3.1 Temporal and spectral descriptions of a signal: (a) time and duration; (b) average frequency and bandwidth.

$$\int_{-\infty}^{+\infty} |\tilde{a}(f)|^2 df = 1, \text{ with } \tilde{a}(f) = \int_{-\infty}^{+\infty} a(t)e^{-2\pi jft}dt$$

The mean frequency and the equivalent bandwidth can then be defined by analogy with time and duration:

$$\bar{f} = \int f|\tilde{a}(f)|^2 df \tag{3.3}$$

$$(\overline{\Delta f})^2 = \int (f^2 - \bar{f}^2)|\tilde{a}(f)|^2 df \tag{3.4}$$

It is shown in Appendix 3A that equivalent duration and bandwidth are related by:

$$\overline{\Delta t}\ \overline{\Delta f} \geq \frac{1}{4\pi} \tag{3.5}$$

which means that the duration and bandwidth of a signal cannot be reduced simultaneously; the bandwidth of a very short signal is necessarily relatively wide, or conversely, a very narrow-band signal necessarily lasts a relatively long time.

Another quantity will prove useful in what follows, which is the modulation index m defined by:

$$m = \text{Im} \int ta(t)[a'(t)]^* \tag{3.6}$$

where $a'(t)$ is the time derivative of $a(t)$.

This quantity is simple to interpret if the signal is written in the form of an amplitude modulation and of a phase modulation:

$$a(t) = \rho(t)e^{j\varphi(t)},\ \rho(t) \geq 0 \tag{3.7}$$

$$m = \int_{-\infty}^{+\infty} t\frac{d\varphi(t)}{dt}\rho^2(t)dt \tag{3.8}$$

where $\rho^2(t)$ is the energy density defined previously, and $d\varphi(t)/dt$ is the instantaneous pulsation (angular rate of the vector representing the complex signal). The parameter m can thus be interpreted as the correlation between time and instantaneous pulsation: a signal will be more modulated the higher the correlation between time and instantaneous pulsation. For instance, for a signal modulated in amplitude only, $\varphi(t)$ is constant and therefore $m = 0$.

For a signal linearly modulated in frequency, and thus quadratically in phase, of duration T:

$$a(t) = \frac{1}{\sqrt{T}} e^{jbt^2}$$

$$\Rightarrow m = \frac{1}{T} \int_0^T t(2bt)dt = 2b\frac{T^2}{3}$$

m is consequently proportional to the modulation slope $2b$.

These five quantities $\bar{t}$, $\overline{\Delta t}$, $\bar{f}$, $\overline{\Delta f}$, and m will be used to calculate the performance limits of the Doppler-range radars and sonars.

3.2.2 Measurement of the Time Delay or of the Frequency of Signal

As seen in Chapter 1, the basic problem of the radar is to measure the range of a target according to the time required by the signal to propagate along the two-way path between the antenna and the target. Let us first assume that both radar and target are fixed so as to avoid any Doppler effect.

Let $u(t)e^{2\pi jf_0 t}$ be the transmitted signal. The received signal is written (see Chapter 1):

$$\begin{aligned} r(t) &= s(t, \tau) + n(t) = \alpha u(t - \tau)e^{2\pi jf_0 t} + n(t) \\ &= \alpha a(t, \tau)e^{2\pi jf_0 t} + n(t) \end{aligned} \tag{3.9}$$

where α is an unknown complex coefficient and τ the delay, which is to be estimated.

According to Chapter 2, the optimum processing consists of correlating the received signal with the time-shifted replicas of the transmitted signal, detecting each channel (square-law envelope detection) and selecting the channel delivering the maximum output.

The ambiguity function associated with this receiver is, according to the definition (2.30), the square of the modulus of $\chi(\tau, \tau_0)$:

$$\chi(\tau, \tau_0) = \int a(t, \tau_0)a^*(t, \tau)dt = \int u(t - \tau_0)u^*(t - \tau)dt \quad (3.10)$$

According to (2.36), the variance of the estimator of τ thus defined is limited by the Cramer-Rao bound:

$$\mathrm{Var}(\tau - \hat{\tau}) \geq -\frac{N_0}{A}\left[\frac{\delta^2}{\delta\tau^2}|\chi(\tau, \tau_0)|^2_{(\tau=\tau_0)}\right]^{-1} \quad (3.11)$$

Let us clarify this expression, taking (3.10) into account:

$$\mathrm{Var}(\tau - \hat{\tau}) \geq -\frac{N_0}{A}[2\,\mathrm{Re}(\chi^*\chi'' + |\chi'|^2)]^{-1}_{(\tau=\tau_0)}$$

with

$$\chi(\tau, \tau_0) = \int |\tilde{u}(f)|^2 e^{2\pi jf(\tau-\tau_0)}df$$

hence

$$j\chi'_{(\tau=\tau_0)} = -2\pi\int f|\tilde{u}(f)|^2 df, \text{ and } -\chi''_{(\tau=\tau_0)} = (2\pi)^2\int f^2|\tilde{u}(f)|^2 df$$

therefore

$$\mathrm{Var}(\tau - \hat{\tau}) \geq \frac{N_0}{2A}\frac{1}{(2\pi)^2\,\overline{\Delta f^2}} \quad (3.12)$$

As a result, the standard deviation of the time delay estimator of a signal is (asymptotically) inversely proportional to the bandwidth of this signal and hence does not directly depend on the duration of this signal—except for the fact that, according to (3.5), the product (bandwidth) × (duration) of a signal has a lower limit and, consequently, a short signal necessarily has a wide bandwidth.

Likewise, by a transposition of this reasoning into the spectral domain, one can show that the standard deviation of the optimum estimator of the frequency of a signal is inversely proportional to the duration of this signal:

$$\mathrm{Var}(f-\hat{f}) \geq \frac{N_0}{2A} \frac{1}{(2\pi)^2\,\overline{\Delta t^2}} \tag{3.13}$$

For example, for a signal of constant amplitude and duration T:

$$a(t) = 1/\sqrt{T},\ t \in \left[-\frac{T}{2}, +\frac{T}{2}\right]$$

straightforward calculations show that $\overline{\Delta t^2} = \frac{T^2}{12}$, so that:

$$\mathrm{Var}(f-\hat{f}) \geq \frac{N_0}{2A} \frac{1}{\pi^2 T^2}$$

Both these fundamental results already show the influence of the form of the transmitted signal on the accuracy of estimation. Therefore, when only the time delay (respectively only the frequency) of the received signal is to be determined, the form of the signal intervenes only through parameter $\overline{\Delta f}$ (respectively, $\overline{\Delta t}$ for frequency measurement).

3.2.3 Simultaneous Measurement of the Time Delay and the Doppler Shift

Let $u(t)e^{2\pi j f_0 t}$ be the signal transmitted by the radar. The signal backscattered by a moving target, characterized by its Doppler ν and delay τ due to the two-way path, is written (with $\nu = 2v/\lambda$ and $\tau = 2R/c$, according to Chapter 1):

$$s(t, \tau, \nu) = \alpha u(t-\tau)e^{2\pi j\nu(t-\tau)}e^{2\pi j f_0(t-\tau)} \tag{3.14}$$

with α being an unknown complex scalar, the unknown phase term $e^{-2\pi j f_0 \tau}$ can be incorporated into it. The received signal, in absence of noise, can then be written:

$$s(t,\ \tau,\ \nu) = \alpha u(t-\tau)e^{2\pi j\nu(t-\tau)}e^{2\pi jf_0t} \tag{3.15}$$

Defining, as previously:

$$s(t,\ \tau,\ \nu) = \alpha a(t,\ \tau,\ \nu)e^{2\pi jf_0t} \tag{3.16}$$

with

$$a(t,\ \tau,\ \nu) = u(t-\tau)e^{2\pi j\nu(t-\tau)} \tag{3.17}$$

the ambiguity function, defined by (2.30), can be expressed as:

$$\chi(\tau-\tau_0,\ \nu-\nu_0) = \int a(t,\ \tau_0,\ \nu_0)a^*(t,\ \tau,\ \nu)dt \tag{3.18}$$

$$\chi(\tau-\tau_0,\ \nu-\nu_0) = e^{2\pi j\nu(\tau-\tau_0)}\int u^*(t-\tau+\tau_0)u(t)e^{-2\pi j(\nu-\nu_0)t}dt \tag{3.19}$$

It is also written, in the spectral domain:

$$\chi(\tau-\tau_0,\ \nu-\nu_0) = e^{2\pi j\nu(\tau-\tau_0)}\int \hat{u}^*(f-\nu+\nu_0)\tilde{u}(f)e^{+2\pi jf(\tau-\tau_0)}df \tag{3.20}$$

Hence, the ambiguity function is written:

$$|\chi(\tau-\tau_0,\ \nu-\nu_0)|^2 = \left|\int u(t-\tau+\tau_0)u^*(t)e^{2\pi j(\nu-\nu_0)t}dt\right|^2 \tag{3.21}$$

$$|\chi(\tau-\tau_0,\ \nu-\nu_0)|^2 = \left|\int \tilde{u}(f-\nu+\nu_0)\tilde{u}^*(f)e^{-2\pi jf(\tau-\tau_0)}df\right|^2 \tag{3.22}$$

The ambiguity function, originally introduced by Ville [1], has a certain number of properties that will be examined in Section 3.2.4.

The accuracy of the simultaneous estimation of the delay and the Doppler effect is limited by the Cramer-Rao bounds, according to (2.35) and (2.36), based on the Fisher matrix:

$$\mathbf{J} = -\frac{A}{N_0}\begin{bmatrix} \dfrac{\delta^2|\chi|^2}{\delta\tau^2} & \dfrac{\delta^2|\chi|^2}{\delta\tau\delta\nu} \\ \dfrac{\delta^2|\chi|^2}{\delta\nu\delta\tau} & \dfrac{\delta^2|\chi|^2}{\delta\nu^2} \end{bmatrix} \tag{3.23}$$

Long but elementary calculations enable calculation of [2]:

$$\mathbf{J} = -\frac{2A}{N_0}(2\pi)^2\begin{bmatrix} \overline{\Delta f^2} & \dfrac{m}{2\pi} - \overline{f}\cdot\overline{t} \\ \dfrac{m}{2\pi} - \overline{f}\cdot\overline{t} & \overline{\Delta t^2} \end{bmatrix} \tag{3.24}$$

If the time origin is chosen so that $\overline{t} = 0$, and the carrier frequency so that $\overline{f} = 0$

$$\mathbf{J} = -\frac{2A}{N_0}(2\pi)^2\begin{bmatrix} \overline{\Delta f^2} & \dfrac{m}{2\pi} \\ \dfrac{m}{2\pi} & \overline{\Delta t^2} \end{bmatrix} \tag{3.25}$$

Hence, by inverting the matrix, the Cramer-Rao bounds on the time and frequency estimates:

$$\mathrm{Var}(\tau - \hat{\tau}) \geq \frac{\overline{\Delta t^2}}{\overline{\Delta t^2}\,\overline{\Delta f^2} - \left(\dfrac{m}{2\pi}\right)^2}\frac{N_0}{2A(2\pi)^2} \geq \frac{1}{\overline{\Delta f^2}}\frac{N_0}{2A(2\pi)^2} \tag{3.26}$$

$$\mathrm{Var}(\nu - \hat{\nu}) \geq \frac{\overline{\Delta f^2}}{\overline{\Delta t^2}\,\overline{\Delta f^2} - \left(\dfrac{m}{2\pi}\right)^2}\frac{N_0}{2A(2\pi)^2} \geq \frac{1}{\overline{\Delta t^2}}\frac{N_0}{2A(2\pi)^2} \tag{3.27}$$

It is thus clear that the accuracy of the joint estimation of Doppler and range is not as good as that of the estimation of only one of these quantities, the other being known, which is intuitively reasonable. In contrast, in the case where the modulation index is zero (signal not frequency-modulated), the performance of the joint estimation of both these values is identical to that of the estimation of each value, and the other is assumed to be known. The time and frequency estimations are consequently coupled via the phase or frequency modulation.

Finally, it appears that the estimation accuracy depends only on (for good signal-to-noise ratios, since these bounds are only asymptotically achieved):

- The ratio of signal energy A to the power spectral density N_0, of the noise;
- The three parameters $\overline{\Delta t}$, $\overline{\Delta f}$, and m (duration, bandwidth, and modulation) of the signal, which summarize all the properties of the transmitted signal with respect to this problem of time and frequency estimation.

We shall now describe the properties of the Doppler-range ambiguity function in more detail, before making these notions concrete by means of some examples.

3.2.4 Doppler-Range Ambiguity Function

The Doppler-range ambiguity function [output of the optimum receiver in (τ, ν)], when a target is located in (τ_0, ν_0), defined by (3.21) and (3.22), has a certain number of important properties. To examine them, let us assume $\tau_0 = \nu_0 = 0$ (the target is located at the origin, which does not restrict the generality of the discussion):

$$|\chi(\tau, \nu)|^2 = \left| \int u(t - \tau) u^*(t) e^{2\pi j \nu t} dt \right|^2 \tag{3.28}$$

The fundamental property of this function is that the volume contained under the ambiguity function is constant and independent of the form of the transmitted signal $u(t)$.

$$\int |\chi(\tau, \nu)|^2 d\tau d\nu = 1 \tag{3.29}$$

This property, easily derived from the definition (3.28), sets a limit to the quality of the Doppler-range estimation. Indeed, taking the previous analyses into account, the ideal radar would have an ambiguity function such that:

- The main peak, at $\tau = \nu = 0$, is infinitely narrow, so as to minimize the variance of the estimation;
- No spurious peak occurs at $(\tau, \nu) \neq (0, 0)$, in order to avoid any ambiguity.

This ideal ambiguity function would thus be of the "thumbtack" type; (3.29) shows that such a function cannot be an ambiguity function because the volume it would contain would be practically zero. The selection of the "best" signal, for a given application, will be the result of a compromise between range resolution, Doppler resolution, and level of ambiguities (secondary peaks of the function for τ and ν different from zero). This property—which is itself a consequence of the fact that the quantities to be estimated are conjugated by Fourier transform—is at the origin of the highly moral nature of radar, for which "what is gained on one side is paid for on another."

Various other properties result from definition (3.28); these are mainly symmetry properties due to the conjugation between time and frequency. For instance, if we define $\Theta(\tau, \nu) = |\chi(\tau, \nu)|^2$, we obtain:

$$\Theta(\tau, \nu) = \Theta(-\tau, -\nu) \tag{3.30}$$

or again, the ambiguity function is its own Fourier transform:

$$\int \Theta(\tau, \nu) e^{2\pi j(v\nu - u\tau)} d\tau d\nu = \Theta(v, u) \tag{3.31}$$

It would be interesting to be able to determine the signal $u(t)$ to be transmitted, in order to obtain a given ambiguity function $\Theta(\tau, \nu)$, but there is unfortunately no such procedure. This is the reason why several examples of signals and ambiguity functions are described in the following paragraph, in order to give intuition useful references to design a system.

3.3 Examples of Doppler-Range Ambiguity Functions

A detailed description of various types of ambiguity function can be found in [3]. The purpose of this section is simply, following the approach of [2], to reveal the main properties of the ambiguity functions of common signals: the unmodulated pulse, the linearly frequency-modulated pulse, the phase-encoded pulse, and the consecutive pulse train.

In the following, the ambiguity function will be defined perfectly symmetrically by:

$$\Theta(\tau, \nu) = |\phi(\tau, \nu)|^2 = \left| \int_{-\infty}^{+\infty} u\left(t - \frac{\tau}{2}\right) u^*\left(t + \frac{\tau}{2}\right) e^{2\pi j \nu t} dt \right|^2 \quad (3.32)$$

3.3.1 Gaussian Pulse

Let us consider the case in which the pulse transmitted by the radar $u(t)e^{2\pi j f_0 t}$ is a signal modulated in amplitude by a Gaussian pattern and unmodulated in frequency:

$$u(t) = \left(\frac{1}{\pi T^2}\right)^{1/4} \exp\left(-\frac{t^2}{2T^2}\right) \quad (3.33)$$

This waveform is a good approximation of a short pulse, and the Gaussian form simplifies the expression of the ambiguity function, which is written:

$$\phi(\tau, \nu) = \left(\frac{1}{\pi T^2}\right)^{1/2} \int \exp\left[-\frac{(t - \tau/2)^2}{2T^2} - \frac{(t + \tau/2)^2}{2T^2} + 2\pi j \nu t\right] dt \quad (3.34)$$

$$\phi(\tau, \nu) = \exp\left[-\frac{1}{4}\left(\frac{\tau^2}{T^2} + (2\pi)^2 T^2 \nu^2\right)\right]$$

$$\Rightarrow \Theta(\tau, \nu) = \exp\left[-\frac{1}{2}\left(\frac{\tau^2}{T^2} + (2\pi)^2 T^2 \nu^2\right)\right] \quad (3.35)$$

The equivalent bandwidth and duration of this signal are:

$$(\overline{\Delta f})^2 = \frac{1}{8\pi^2 T^2}, \text{ and } (\overline{\Delta t})^2 = \frac{T^2}{2}$$

The contour lines of this surface are ellipses with axes proportional to T and $1/2\pi T$ (see Figure 3.2). According to (3.26) and (3.27), the standard deviation of the estimation of τ has a lower limit $\sqrt{\frac{N_0}{A}}T$, and the standard deviation of the estimation of ν has a lower limit $\sqrt{\frac{N_0}{A}}\frac{1}{2\pi T}$: The time and frequency measurement accuracies evolve with T, one the inverse of the other.

Using this kind of signal, if good range resolution is required, it is necessary to use a very short pulse. As the measurement accuracy is inversely proportional to the energy of the transmitted signal, concentrating this energy over a short duration results in transmitting high powers. One then comes up against various technological limitations, which in practice limit the range of unmodulated pulse radars.

3.3.2 Pulse with Linear Frequency Modulation (Chirp)

A means to overcome these limitations consists in using a frequency-modulated pulse (parameter m different from zero). The simplest case—commonly used in practice—is linear frequency modulation (see Figure 3.3).

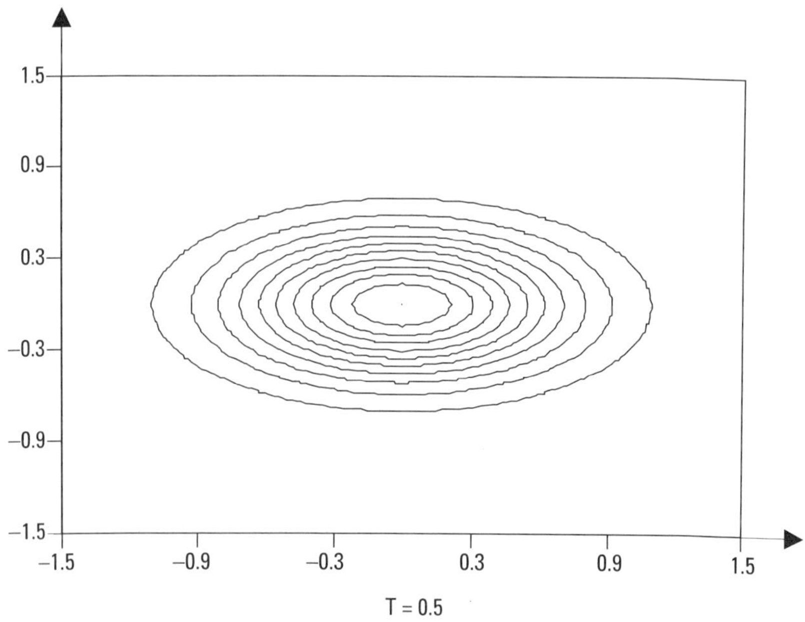

Figure 3.2 Contour lines of a Gaussian pulse ambiguity function (10 contours equally spaced between the maximum and minimum values).

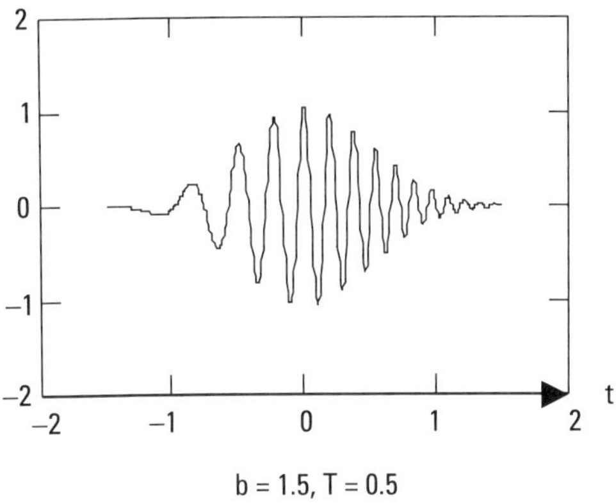

Figure 3.3 Gaussian pulse with linear frequency modulation.

It is easy to show that if $\phi_1(\tau, \nu)$ is associated with $u_1(t)$, then the signal $u_2(t) = u_1(t)e^{2\pi jbt^2}$ is associated with $\phi_2(\tau, \nu)$, with:

$$\phi_2(\tau, \nu) = \phi_1(\tau, \nu - 2b\tau) \tag{3.36}$$

The ambiguity function of the linearly frequency-modulated (and therefore quadratically in phase) Gaussian pulse:

$$u(t) = \left(\frac{1}{\pi T^2}\right)^{1/4} \exp\left(-\frac{t^2}{2T^2} + 2\pi jbt^2\right) \tag{3.37}$$

is then directly deduced, from that of the unmodulated pulse (3.35):

$$\Theta(\tau, \nu) = \exp\left[-\frac{1}{2}\left(\frac{\tau^2}{T^2} + (2\pi)^2 T^2 (\nu - 2b\tau)^2\right)\right] \tag{3.38}$$

The contour lines of this surface [see Figure 3.4(a)] are always ellipses whose equation is:

$$T^2\nu^2 - 4bT^2\nu\tau + \left(4b^2T^2 + \frac{1}{(2\pi)^2T^2}\right)\tau^2 = c^2 \tag{3.39}$$

Now, the parameters $\overline{\Delta t}$, $\overline{\Delta f}$, and m characteristic of the pulse are given by:

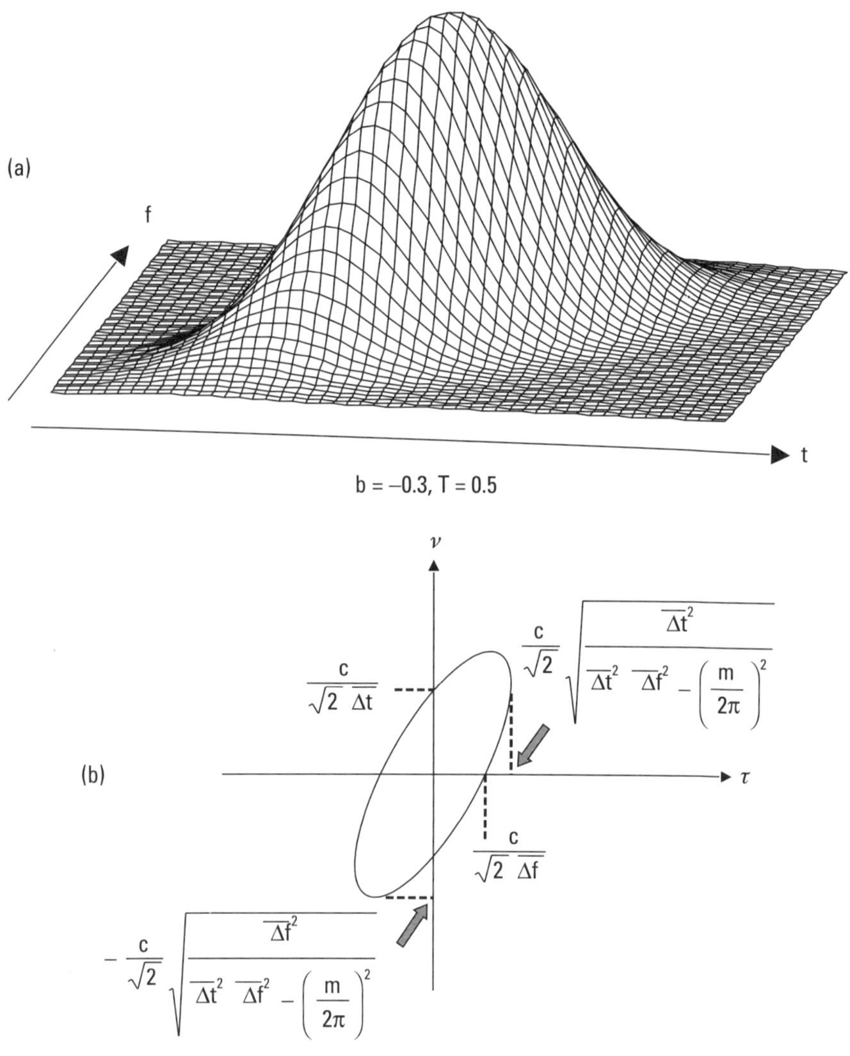

Figure 3.4 Ambiguity function of a linearly frequency-modulated Gaussian pulse: (a) 3D representation; (b) contour line.

$$\overline{\Delta t}^2 = \frac{T^2}{2} \tag{3.40}$$

$$\overline{\Delta f}^2 = \frac{1}{2(2\pi)^2 T^2} + 2b^2T^2 \tag{3.41}$$

$$m = 2\pi bT^2 \tag{3.42}$$

The ellipses formed by the contour lines [see Figure 3.4(b)] have the following equation:

$$\overline{\Delta t}^2\nu^2 - 2\frac{m}{2\pi}\nu\tau + \overline{\Delta f}^2\tau^2 = c^2/2 \tag{3.43}$$

If the target velocity is known ($\nu = 0$), the time measurement accuracy is $\sqrt{\frac{N_0}{2A}\frac{1}{(2\pi)\overline{\Delta f}}}$; similarly, if the range is known ($\tau = 0$), the Doppler measurement accuracy is $\sqrt{\frac{N_0}{2A}\frac{1}{(2\pi)\overline{\Delta t}}}$.

If both velocity and range are unknown, an area of ambiguity between Doppler and time remains along the ellipse major axis (this can be understood intuitively, since delaying the pulse or increasing its frequency produce very similar signals, except for edge effects).

3.3.3 Pulse Compression

The radars and sonars using a modulation (of phase or frequency) of the transmitted pulse so that the modulation index is different from zero are referred to as pulse compression systems. This term originates from the fact that the optimum receiver, correlating the received signal $u(t - \tau_0)e^{2\pi j\nu_0 t}$ with replicas $u(t - \tau)e^{2\pi j\nu t}$ produces, at the output of this correlation, a compressed pulse, that is, a pulse of shorter duration. This becomes obvious when the correlation is interpreted in the spectral domain:

$$\int \tilde{u}(f + \nu_0)\tilde{u}^*(f + \nu)e^{-2\pi jf(\tau_0 - \tau)}df$$

The effect of the correlation is then to "set in phase the lines of the transmitted spectrum" when $\tau = \tau_0$ and $\nu = \nu_0$, that is to say, for the replica corresponding to the received signal. More precisely, the correlator output, as a function of time, for $\nu = \nu_0$ (the replica corresponding to the target Doppler), is:

$$\int \tilde{u}(f + \nu_0)\tilde{u}^*(f + \nu_0)e^{-2\pi jf(\tau_0 - \tau)}df = \int |\tilde{u}(f + \nu_0)|^2 e^{-2\pi jf(\tau_0 - \tau)}df$$

If modulation is such that the spectrum is nearly flat in the bandwidth $\overline{\Delta f}$—which is the general case—this correlator output is a pulse at $\tau = \tau_0$, whose duration is the inverse of $\overline{\Delta f}$ (inverse Fourier transform of $|\tilde{u}(f + \nu_0)|^2$); as shown in Figure 3.5, the correlation converts the initial modulated pulse of duration $\overline{\Delta t}$ into a pulse of shorter duration, inversely proportional to $\overline{\Delta f}$. This property of the correlation with a replica to compress the duration of a modulated pulse is essential in radar and sonar. It intuitively explains the power gain provided by modulation.

3.3.4 Encoded Pulse Sequence

Another way of improving the resolution obtained, without increasing the transmitted power, consists of modulating the phase of the transmitted

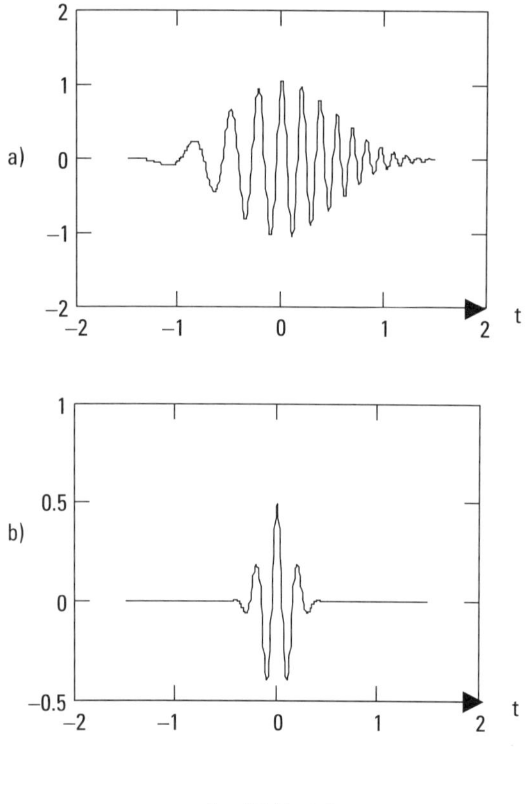

b = 1.5, T = 0.5

Figure 3.5 Pulse compression effect: (a) long pulse; (b) compressed pulse.

pulse. An often used technique because it is easy to implement, consists of transmitting a pulse composed of adjacent "subpulses" whose initial phase is different (phase encoding); the simplest encoding is performed by multiplication of the transmitted signal by a sequence of +1 and −1 (0–π encoding), as shown in Figure 3.6. The encoding sequence is chosen in order to obtain a required ambiguity function. The ideal would be to optimize this encoding sequence in order to optimize the whole ambiguity function $\Theta(\tau, \nu)$. In practice, for want of a global optimization technique, one generally tries to optimize the range response $\Theta(\tau, 0)$ (for a Doppler that is assumed to be known), and then one makes sure that the appearance of $\Theta(\tau, \nu)$ is not too bad for $\nu \neq 0$.

If T_s is the duration of the elementary subpulse, N is the number of subpulses composing the transmitted pulse, and $T = NT_s$ is the duration of this transmitted pulse, we would like the ambiguity function (see 3.32):

$$\Theta(\tau, 0) = \left| \int u\left(t - \frac{\tau}{2}\right) u^*\left(t + \frac{\tau}{2}\right) dt \right|^2$$

(which is just the square of the modulus of the autocorrelation of $u(t)$) to have a narrow main peak at $\tau = 0$, and sidelobes as low as possible.

Barker coding makes it possible, for instance, to produce coded sequences with 1 as the maximum autocorrelation value (because of the normalization of the signal) and sidelobes lower than I/N^2. These codes exist for $N = 3, 4, 5, 7, 11$, and 13. By way of illustration, Figure 3.7 shows the autocorrelation function obtained by Barker code over 13 subpulses (coding: 1, 1, 1, −1, −1, +1, −1) for which the sidelobes of the ambiguity function are therefore 7^2, that is 17 dB below the main peak. This figure

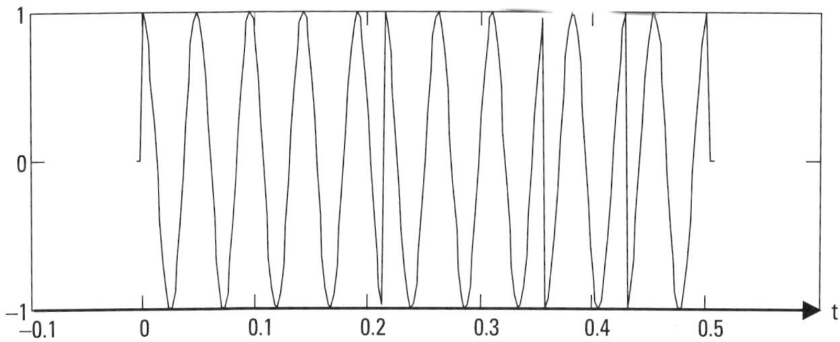

Figure 3.6 Phase coding (Barker code of length 7).

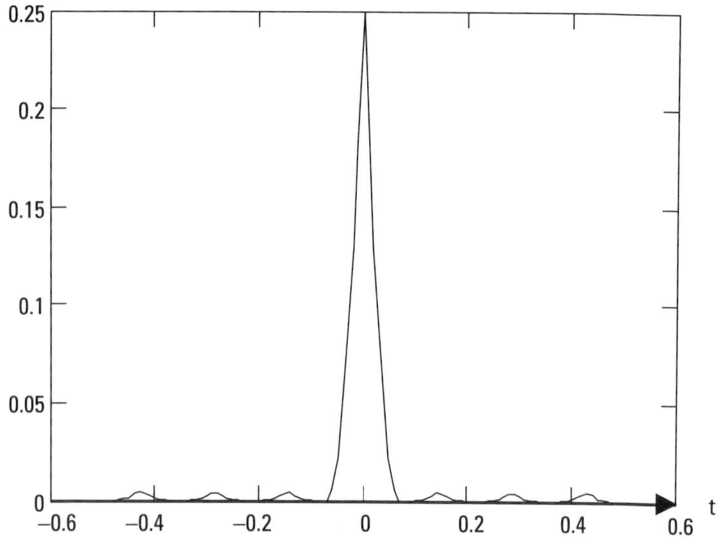

Figure 3.7 Autocorrelation function of a Barker code of length 7.

clearly shows the pulse compression effect described in the previous paragraph: the main peak of the autocorrelation is 7 times narrower when the pulse is modulated by this 7 "moments" binary code.

Figure 3.8 shows in contrast that spurious peaks of $\Theta(\tau, \nu)$ at much higher levels, occur for $\nu \neq 0$. It can be shown that these higher sidelobes occur for values of ν such that the Doppler is sufficiently high for significantly changing the phase coding, that is such that $\nu \geq 1/T$.

Owing to its relative simplicity of use, for transmission as well as reception, this phase encoding of the transmitted pulses has been often used and thoroughly studied. Various processes for reducing the sidelobes of the ambiguity function have been studied [4]. A summary of the different properties of these types of modulation can be found in [5].

3.3.5 Pulse Train

In practice, most radars transmit repeated pulses at a given repetition frequency F, generally fixed. For processing, either each pulse is processed separately[1]—and ambiguity functions of the previous types are obtained—or the pulse train is processed coherently as a whole. According to (3.26)

1. That is the general case in sonar, because of the low sound velocity, which does not allow enough time for successive pulses to be transmitted, propagated, and received.

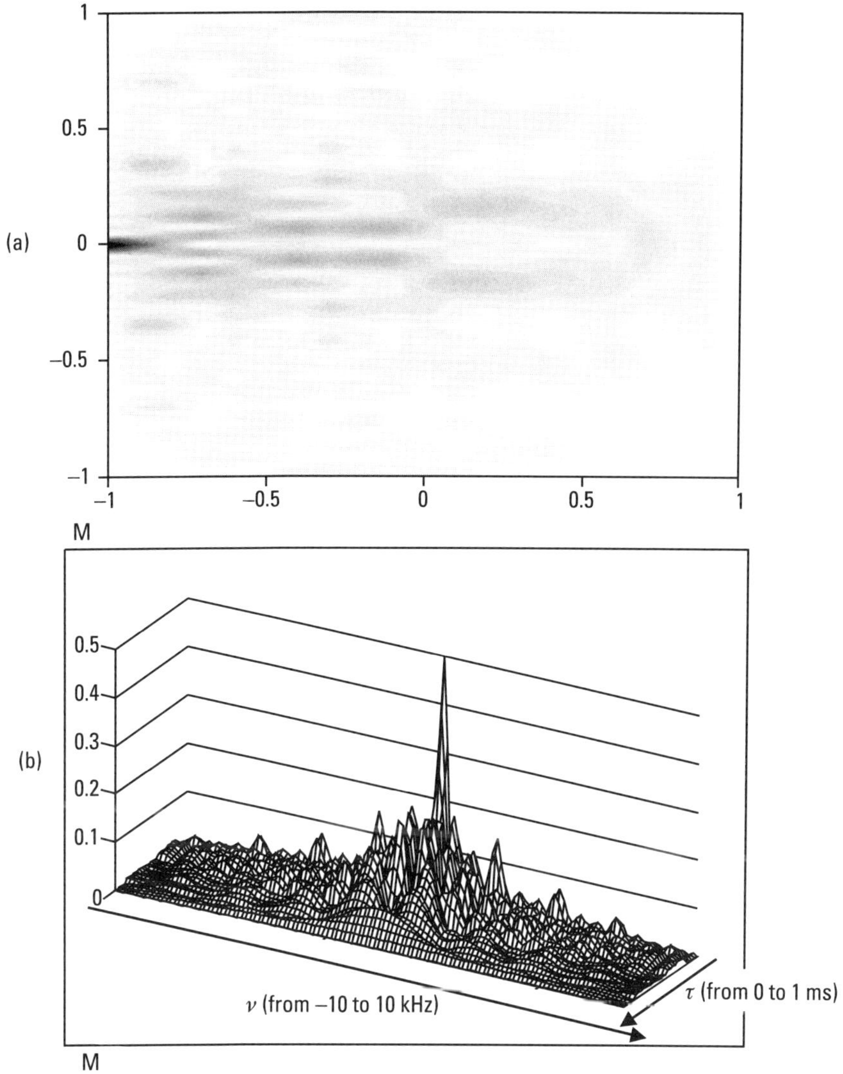

Figure 3.8 Ambiguity function of a 7-moments Barker code: (a) gray level presentation; (b) perspective view of the same ambiguity function.

and (3.27) giving the estimator variance, it is evidently much more advantageous to adopt the latter solution when enough illumination time is available, as it makes it possible to use a signal whose duration $\overline{\Delta t}$ is longer, whereas the bandwidth $\overline{\Delta f}$ and the modulation index m remain unchanged (or slightly changed).

Let us consider a train of $2n + 1$ pulses with complex envelope $u_e(t)$, transmitted at the repetition frequency $F_r = 1/T_r$. Let T be the pulse duration and $T << T_r$; the total duration of the pulse train is: $T_t = 2nT_r + T$.

The complex envelope of the pulse train is:

$$u(t) = \frac{1}{\sqrt{2n+1}} \sum_{k=-n}^{+n} u_e(t - kT_r) \tag{3.44}$$

where $u_e(t)$ is the complex envelope of the elementary transmitted pulse.

Now let us calculate the ambiguity function $\phi(\tau, \nu)$, and first of all for $|\tau| < T$. According to (3.32), we obtain:

$$\phi(\tau, \nu) = \frac{1}{2n+1} \sum_{k=-n}^{+n} \int_{kT_r - \frac{1}{2}(T-|\tau|)}^{kT_r + \frac{1}{2}(T-|\tau|)} u_e\left(t - kT_r - \frac{\tau}{2}\right)$$

$$u_e^*\left(t - kT_r + \frac{\tau}{2}\right) e^{2\pi j \nu t} dt, \text{ for } |\tau| < T$$

That is, defining $z = t - kT_r$

$$\phi(\tau, \nu) = \frac{1}{2n+1} \sum_{k=-n}^{+n} e^{2\pi j \nu k T_r} \int_{-\frac{T}{2}+\frac{|\tau|}{2}}^{\frac{T}{2}-\frac{|\tau|}{2}} u_e\left(z - \frac{\tau}{2}\right) u_e^*\left(z + \frac{\tau}{2}\right) e^{2\pi j \nu z} dz$$

that is

$$\phi(\tau, \nu) = \frac{1}{2n+1} \frac{\sin[2\pi\nu(n + 1/2)T_r]}{\sin(2\pi\nu T_r/2)} \phi_{u_e}(\tau, \nu) \tag{3.45}$$

with

$$\phi_{u_e}(\tau, \nu) = \int_{-\frac{T}{2}+\frac{|\tau|}{2}}^{\frac{T}{2}-\frac{|\tau|}{2}} u_e\left(z - \frac{\tau}{2}\right) u_e^*\left(z + \frac{\tau}{2}\right) e^{2\pi j \nu z} dz$$

It is clear that repeating the pulses results in multiplying, for $|\tau| < T$, the original ambiguity function of the pulse $u_e(\tau, \nu)$ by the function:

$$\frac{\sin[2\pi\nu(n + 1/2)T_r]}{\sin(2\pi\nu T_r/2)}$$

independent of τ, whose first zero is obtained for:

$$\nu = \frac{1}{(2n + 1)T_r} \sim \frac{1}{T_t}$$

and whose secondary peaks are obtained for $\nu = K/T_r$ (K being an integer). This function is shown in Figure 3.9, for $n = 6$ and $T_r = 1$. The Doppler width of the main peak of the ambiguity function is then on the order of $1/(2n + 1)T_r$, the inverse of the total duration of the pulse train.

Now let us examine what happens for $|\tau| > T$. If the secondary lobes of the ambiguity function of the basic pulse are not taken into account,

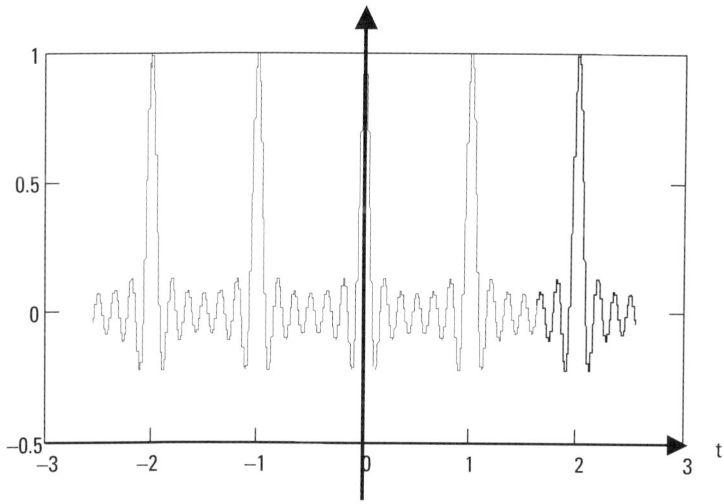

$$u(\nu) := \frac{1}{2 \cdot n + 1} \cdot \frac{\sin\left[2 \cdot \pi \cdot \left(n + \frac{1}{2}\right) \cdot \nu \cdot Tr\right]}{\sin\left(2 \cdot \pi \cdot \nu \cdot \frac{Tr}{2}\right)}$$

Figure 3.9 Ambiguity function Doppler gating ($n = 6$, $T_r = 1$).

nothing happens until $\tau = T_r - T$ (since $\phi(\tau, \nu) \cong 0$). From that point, the previous situation is restored ($|\tau| < T$), except that one pulse less is used (there are only $2n$ pulses overlapping in the product $u(t - \tau/2)$ $u^*(t + \tau/2)$ instead of $2n + 1$ previously).

Consequently

$$\phi(\tau, \nu) = \frac{1}{2n+1} \sum_{k=-n+1}^{+n} e^{2\pi j\nu kT_r} \phi_{u_e}(\tau - T_r, \nu)$$

$$\phi(\tau, \nu) = \frac{1}{2n+1} \frac{\sin 2\pi\nu nT_r}{\sin 2\pi\nu T_r/2} \phi_{u_e}(\tau - T_r, \nu) e^{2\pi j\nu T_r/2} \tag{3.46}$$

When comparing with (3.45), it is clear that $\phi(\tau, \nu)$ has the same appearance in the vicinity of $\tau = T_r$ as in the vicinity of $\tau = 0$, the level being reduced in the ratio $2n/(2n + 1)$.

The same phenomenon occurs when τ is close to $2T_r$, [with a level $(2n - 1)/(2n + 1)$], $3T_r$, and so on.

The repetition of the basic pulse u_e with a repetition period T_r thus has the following main effects.

- *Repeating* the ambiguity function of the basic pulse by translation, parallel to the axis τ, every T_r, with a level decreasing linearly with τ;
- *Gating* the function thus obtained by multiplication by a function of the form $\frac{\sin nx}{n \sin x}$, whose main lobe width in Doppler is the inverse of the total duration of the pulse train and whose secondary lobes are spaced at $1/T_r$ intervals in ν.

The second effect (Doppler gating) is the effect sought by the pulse repetition: an improvement in the Doppler estimation accuracy. The cost, in accordance with the considerations of Section 3.2.4 (conservation of the ambiguity volume), is the creation of additional range ambiguities.

3.3.6 Reception of a Pulse Train

The structure of the optimum receiver, described in Chapter 2 (Figures 2.1 and 2.2), is significantly simplified in the case of the reception of a train of periodic pulses $u_e(t)$ (pulse duration T, repetition period T_r). In the general case, the receiver included N parallel branches, corresponding to the values that can be taken by the parameter pairs (τ, ν) of the type illustrated in Figure 3.10: correlation of the received signal $z(t)$ (after demodulation of the carrier $e^{2\pi j f_0 t}$) with the replica $a(t, \theta_i) = u(t - \tau_i)e^{2\pi j \nu_i (t-\tau_i)}$, square-law envelope detection and normalization, with respect to the energy of the replica.

Hence the correlation leads to the calculation of C:

$$C = \int z(t)u^*(t - \tau_i)e^{2\pi j \nu_i (t-\tau_i)}\,dt \text{ with } \nu_i = \frac{2v_i f_0}{c} \tag{3.47}$$

If the transmitted signal is a train of basic pulses $u_e(t)$:

$$u(t) = \sum_{k=-n}^{+n} u_e(t - kT_r)$$

the correlation is written:

$$C = \sum_{k=-n}^{+n} \int z(t)u_e^*(t - \tau_i - kT_r)e^{-2\pi j \nu_i (t-\tau_i)}\,dt \tag{3.48}$$

Let $x = t - kT_r - \tau_i$

$$C = \sum_{k=-n}^{+n} e^{2\pi j \nu_i k T_r} \int_{\frac{T}{2}}^{\frac{T}{2}} z(x + kT_r + \tau_i)u_e^*(x)e^{-2\pi j \nu_i x}\,dx$$

In practice, $\nu_i T$ is generally very small in radar, which means that the Doppler frequency does not shift the phase significantly during the duration

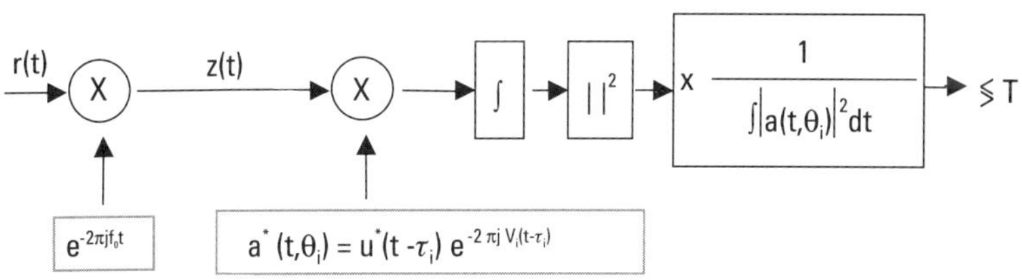

Figure 3.10 Optimum receiver branch.

of the basic pulse, but only from one pulse to the next (which is another way of saying that the duration of the basic pulse is too short to resolve the Doppler of the various targets: $\nu_i << \frac{1}{T}$). For example, typically:

$$T = 10^{-6}\text{ s};\ f_0 = 10^{10}\text{ Hz};\ \nu_i = 300\text{ m s}^{-1} \Rightarrow \nu_i = 2 \cdot 10^4\text{ Hz}$$

$$2\pi\nu_i T = 2 \cdot 10^{-2}\text{ rad} \cong 7.2°\text{, which is negligible.}$$

Therefore, the correlation with the replica can be approximated by:

$$C = \sum_{k=-n}^{+n} e^{2\pi j\nu_i kT_r} \int_{-\frac{T}{2}}^{\frac{T}{2}} z(x + kT_r + \tau_i)u_e^*(x)dx \tag{3.49}$$

It is composed of three consecutive basic operations:

1. Matched filtering, or correlation with the basic pulse;
2. Sampling at times $\tau + kT_r$;
3. Discrete Fourier transform of the sequence of samples obtained, for each value of the parameter ν_i.

The diagram of the receiver obtained is shown in Figure 3.11. Reception is seen to be appreciably simplified, since a single correlator (or matched filter) is enough, instead of the set of correlators required in the general case (this correlator performing the pulse compression described in Section 3.2.3). The range analysis is carried out simply by sampling the compressed signal at the different possible times ("range gates") and the Doppler analysis by Fourier transform of the series of $2n + 1$ values sampled in each range gate. In practice, the matched filtering can be done analogically (a simple filter in the case of phase- or frequency-unmodulated pulses, or "compression line" in the case of linearly frequency-modulated pulses) or digitally (encoded pulse sequence), the rest of the processing then generally being performed digitally [Fast Fourier Transform (FFT) algorithm].

This receiver structure is used in most present-day pulse Doppler radars. For each direction of observation of the antenna, the receiver, after demodulating the carrier, carries out the filtering matched to the pulse, the

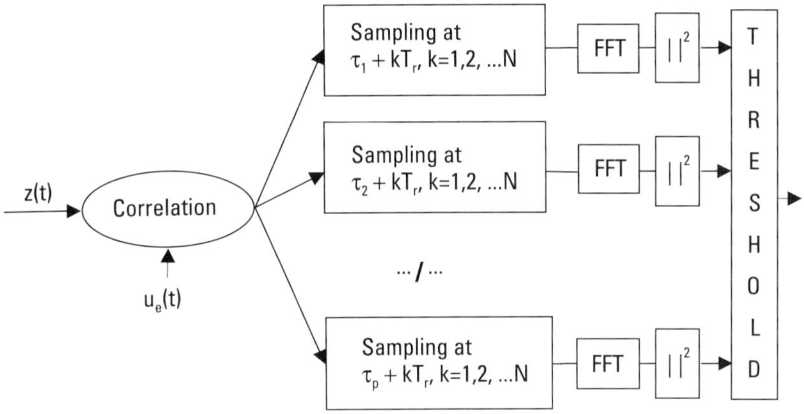

Figure 3.11 Optimum receiver in white noise (pulse radar).

sampling of the range gates, and Fourier transforms corresponding to the Doppler analysis in each range gate. Given the generally large number of directions of observation (100), of range gates (often more than 100), and of Doppler bins (between 10 and 100), the practical importance of the progress made in the FFT algorithms [6] is obvious.

3.3.7 Range and Velocity Ambiguities

The considerations developed in Section 3.3.5 show that a radar transmitting a train of periodic pulses (period T_r) is subject to two types of ambiguity:

- Range ambiguities, caused by the repetition of the basic ambiguity function every T_r;
- Doppler ambiguities, caused by the gating of the basic ambiguity function by the quasi-periodical function $\dfrac{\sin[2\pi\nu(n+1/2)T_r]}{\sin(2\pi\nu T_r/2)}$, every $F_r = \dfrac{1}{T_r}$.

A periodic pulse radar thus has an ambiguous range $D_a = \dfrac{c}{2}T_r$, and an ambiguous velocity $V_a = \dfrac{\lambda}{2T_r}$.

Consequently, the unambiguous domain is limited by these two quantities, the product of which is constant for a given wavelength:

$$D_a V_a = \frac{\lambda c}{4} = \frac{c^2}{4f_0}$$

For instance, for $f_0 = 10$ GHz, we obtain:

$$D_a V_a = 2.25 \cdot 10^6 \text{ m}^2 \text{ s}^{-1}$$

If an unambiguous range of 10 km is required, an ambiguous velocity of 225 ms^{-1} must be accepted. The modern requirement for fast-moving target location at long range often leads to the use of fundamentally ambiguous radars if the simple technology of the periodic pulse trains is to be retained. This is why many radars have several operating modes (varied repetition frequencies, various modulations and lengths of the basic pulses transmitted) in order to meet the various needs (long-range search, tracking, detection in ground clutter, for example). Moreover, various processes for removing ambiguities (transmission of consecutive trains at different repetition frequencies, hence with different ambiguities, for example) are used to interpret data delivered during consecutive radar observations. This aspect of the processing, which is essential, particularly for airborne radars, will be examined in Section 7.3.4, taking the environment signature (clutter) into consideration.

Finally, it should be pointed out that, although the use of this type of pulse train signal is quite frequent in radar, it is much less so in sonar, for which the relative slowness of the propagation velocity would lead to the use of pulse trains whose duration is too long with respect to the required response times and for which the Doppler is generally measurable on one pulse duration.

3.3.8 Constant False Alarm Rate Detection

In the previous analysis, it was assumed that the noise power was known. In actual practice, this power has to be estimated on the received signals, since different limitations due to radar components' imperfections (phase noise, see Appendix 3C, and nonlinearities) and to the presence of clutter (see Chapters 6 and 7) tend to forbid the use of a fixed threshold determined in advance. Instead, an estimate of the "ambient" noise and interference levels is formed around each range-Doppler cell, and this estimate is used to set the threshold. Depending on the type of interference and the characteristics of the radar, different detection schemes can be used. Basically, most of them amount to:

- Defining an area in range and Doppler (and possibly angle) surrounding the cell under test (and usually excluding the neighboring cells), over which the ambient noise level will be estimated;
- Calculating, on the basis of the signals received in the selected reference cells, the value of the threshold;
- Comparing the signal level in the cell under test to this threshold and declaring a target present if the threshold is exceeded.

For the first step (see Figure 3.12), the selected area usually consists of adjacent cells in range, at the same Doppler as the cell under test [range constant false alarm rate (CFAR), striped area on Figure 3.12]. For radars highly ambiguous in range [high pulse repetition frequency (high PRF) radars], the area consists of adjacent cells in Doppler, at the same range as the cell under test (Doppler CFAR, gray area on Figure 3.12). A combination of range and Doppler reference cells can be used, especially when situations with high target density can be encountered.[2]

The threshold is then calculated, usually by averaging the levels obtained in the ambient noise area (cell-averaging CFAR) and setting a contrast level giving the appropriate probability of false alarm. In some situations, the ambient noise area is divided into two parts, and the level of the threshold is set according to the noisiest part (greatest-of CFAR, or GO-CFAR). This

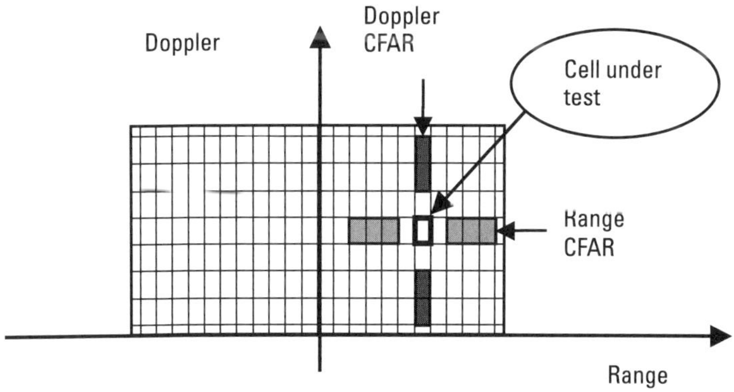

Figure 3.12 Range and Doppler CFAR.

2. For instance, selection of a diagonal area in the range-Doppler plane is especially appropriate for certain situations of raid or convoy observations, or against certain types of jammers.

mode of computation guarantees that isolated clutter returns, or borders between different clutter regions, do not produce excessive false alarms—at the cost of a slight loss in detection probability.

More sophisticated techniques, called censored CFAR [7, 8], can also be used for the calculation of the threshold. They basically consist of ranking the samples of the ambient noise area according to their level, and leaving out the highest samples before averaging the remaining ones (cell-averaging or greatest-of) for setting the contrast level. Such procedures can guarantee high detection probability in spiky clutter and multiple targets situations—at the cost of higher computation load.

Finally, it should be noted that, for many applications where residual clutter is present, it will be preferred to operate on the logarithms of the levels obtained in each cell. As will be seen in Section 7.4.1, this is due to the fact that for clutter echoes due to large number of scatterers in each cell (Rayleigh clutter) the standard deviation of the fluctuations are then independent from the RCS level.

3.4 Radar Mapping: Synthetic Aperture Radar

Let us consider an aircraft flying horizontally in a straight line at constant velocity V, observing to either side the ground flown over. If its mission is to generate an image of the land at long range, it can analyze the ground in range thanks to the range resolution of the radar, but its cross-range resolution is limited by the angular beamwidth of the antenna (see Figure 3.13). This is usually inadequate, since the angular resolution of an antenna is determined by its dimension according to: $\theta_a = \frac{\lambda}{L}$, where λ is the wavelength, L is the antenna dimension, and θ_a is the angular resolution in radians (a proof of this relation will be given in Section 3.4.2).

For a radar operating at a wavelength of 3 cm, and with an antenna length of 1m, we obtain a transverse resolution, at 50-km range, of 1.5 km, which is obviously insufficient for imaging the ground.

In this situation, the principle of *synthetic aperture* radar consists of exploiting the fact that, as all points of the land are fixed, their apparent radial velocity with respect to the aircraft—and consequently the Doppler effect of the signal they return—differs according to their azimuth relative to the aircraft course. It is thus possible to separate the points in azimuth, at a given range, by Doppler analysis of the received signal.

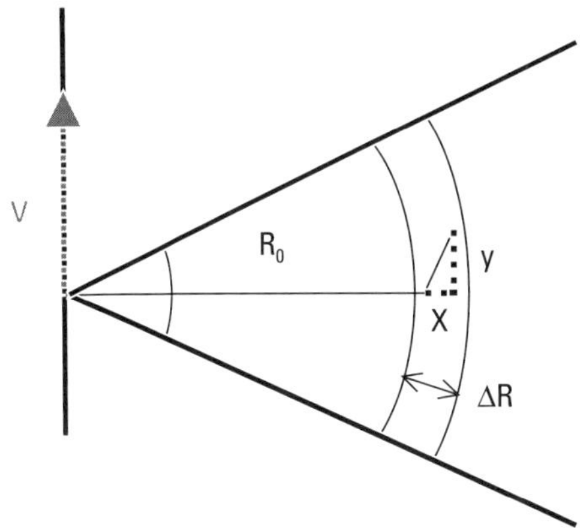

Figure 3.13 Side-looking radar.

Let $I(x, y)$ be the complex backscattering coefficient of point x, y of the ground (Figure 3.13). If R denotes the distance from the point of coordinates x, y to the radar at instant t, and $u(t)e^{2\pi j f_0 t}$ is the signal transmitted by the radar, the received signal is written:

$$u\left(t - \frac{2R}{c}\right)e^{2\pi j f_0 (t - 2R/c)} I(x, y) \tag{3.50}$$

As can be seen in Figure 3.13:

$$R = \sqrt{(R_0 + x)^2 + (y - vt)^2}$$

where $R_0 + x >> y - vt$ (observation at long range),

$$R \cong R_0 + x - \frac{vyt}{R_0} + \frac{y^2}{2R_0} \tag{3.51}$$

assuming that:

$$v^2 t^2 << 2vyt$$

$$vt << 2y \tag{3.52}$$

a condition that will be discussed later.

The received signal is then written:

$$I(x, y)\, u\left(t - \frac{2(R_0 + x)}{c}\right) e^{-4\pi j f_0 \frac{R_0 + x + y^2/2R_0}{c}} e^{4\pi j \frac{f_0}{c} \frac{vy}{R_0} t} e^{2\pi j f_0 t} \tag{3.53}$$

The term $\exp \dfrac{-4\pi j f_0 (R_0 + x + y^2/2R_0)}{c}$ is a phase term that can be included in the phase, not significant in radar, of $I(x, y)$.

The term $e^{4\pi j \frac{f_0}{c} \frac{vy}{R_0} t}$ represents the Doppler of point (x, y) used in synthetic aperture radar to measure y. Then the optimum receiver calculates, according to Section 3.1.3:

$$\left| I(x, y)\, \chi\left(\tau - \frac{2(R_0 + x)}{c},\ \nu - \frac{2f_0}{c} \frac{vy}{R_0}\right)\right|^2 \tag{3.54}$$

and the radar assigns the reflectivity $|I(x, y)^2|$ to the point whose x and y coordinates are respectively given by:

$$\text{delay: } \tau = \frac{2(R_0 + x)}{c} \tag{3.55}$$

$$\text{Doppler: } \nu = \frac{2f_0}{c} \frac{v}{R_0} y \tag{3.56}$$

Using this law for conversion of the delay and Doppler into x and y coordinates on the ground, the radar response as a function of τ and ν is, hence, an image of the ground reflectivity $|I(x, y)|^2$.

If the transmitted pulse $u(t)$ is modulated in neither phase nor frequency ($m = 0$), the location accuracy is limited by (3.26) and (3.27):

$$\mathrm{Var}(\tau - \hat{\tau}) \geq \frac{1}{\overline{\Delta f}^2} \frac{N_0}{2A(2\pi)^2}$$

$$\mathrm{Var}(\nu - \hat{\nu}) \geq \frac{1}{\overline{\Delta t}^2} \frac{N_0}{2A(2\pi)^2}$$

from which the asymptotic accuracies σ_x and σ_y are deduced, for low $N_0/2A$, in x and y (according to 3.55 and 3.56):

$$\sigma_x = \frac{1}{2\pi\overline{\Delta f}} \frac{c}{2} \sqrt{\frac{N_0}{2A}} \tag{3.57}$$

$$\sigma_y = \frac{1}{2\pi\overline{\Delta t}} \frac{c}{2f_0} \frac{R_0}{v} \sqrt{\frac{N_0}{2A}} \tag{3.58}$$

It is then possible to interpret the validity condition (3.52), which implies that the relation $vt << 2y$ remains valid for any t, hence, in order of magnitude:

$$v\overline{\Delta t} << 2\delta y \tag{3.59}$$

where δy is the resolution in y.

It therefore sets a minimum value to the resolution in y. This limit can be estimated from (3.58), assuming that the resolution is equal to the precision for a signal-to-noise ratio equal to unity:[3]

$$(\delta y)^2 >> \frac{c}{4f_0(2\pi)} R_0 \tag{3.60}$$

$$\delta y >> \frac{1}{2}\sqrt{\frac{R_0\lambda}{2\pi}} \tag{3.61}$$

Numerical application, for $\lambda = 3$ cm, $R_0 = 50$ km

$$\delta y >> 7.7\text{m for } \frac{N_0}{2A} = 1$$

Thus, neglecting the frequency modulation of the received pulse (terms in t^2, expressing the change in the Doppler of each scattering center as the aircraft advances) limits the resolution by (3.61).

Furthermore, (3.59) simply expresses the fact that during the observation time $\overline{\Delta t}$ the carrier platform should not have moved by a distance larger than the resolution required in y (parallel to the carrier course).

3. This is, of course, a very rough approximation; a more suitable approach could be to define resolution as the 3-dB width of the main lobe of the Doppler response, for instance.

When better resolution is required, the duration of the pulse train analyzed must be increased, taking into account the change in Doppler of the observed point during the analysis duration.

Figure 3.14 shows the evolution of the instantaneous Doppler frequency of a point scatterer on the ground with time, from $+2v/\lambda$, when this point is far away in front of the aircraft, to $-2v/\lambda$, when it disappears at the back. The unfocused processing is sufficient as long as the Doppler widening (due to the Doppler evolution during the integration time) is smaller than the required Doppler resolution. When this condition is not met, it becomes necessary to use a filter matched to the received signal—with its frequency modulation approximately linear due to the change in Doppler of the observed point—and no longer to the transmitted signal. The radar receiver using this principle is said to be "focused synthetic aperture" and enables the theoretical resolution given by (3.58) to be approached at long distances.

The synthetic aperture radar principle (unfocused) is currently used, for example, in maritime surveillance to detect the presence of oil slicks or in airborne ground surveillance. In the focused version, it is also used in satellites (Seasat or space shuttle, for example). To further illustrate this, Figure 3.15 shows a picture of a stationary ship, obtained by an ONERA 94-GHz unfocused synthetic aperture radar, and Figure 3.16 shows an image obtained by a THALES focused synthetic aperture radar operating in X band (10 GHz). On these figures, the gray levels represent the energy of the received signals (reflectivity of the target elements).

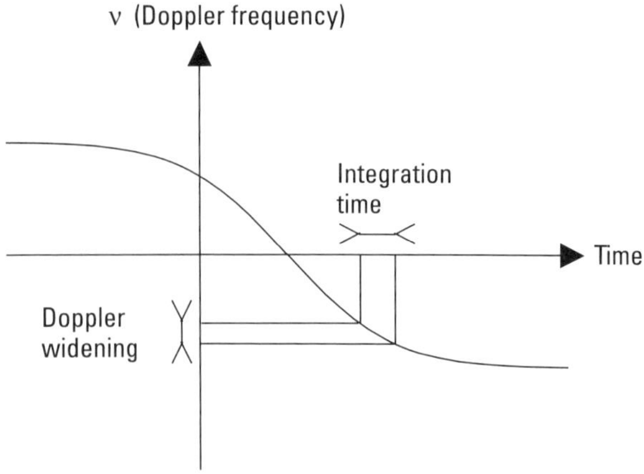

Figure 3.14 Evolution of the Doppler frequency of a point scatterer on the ground.

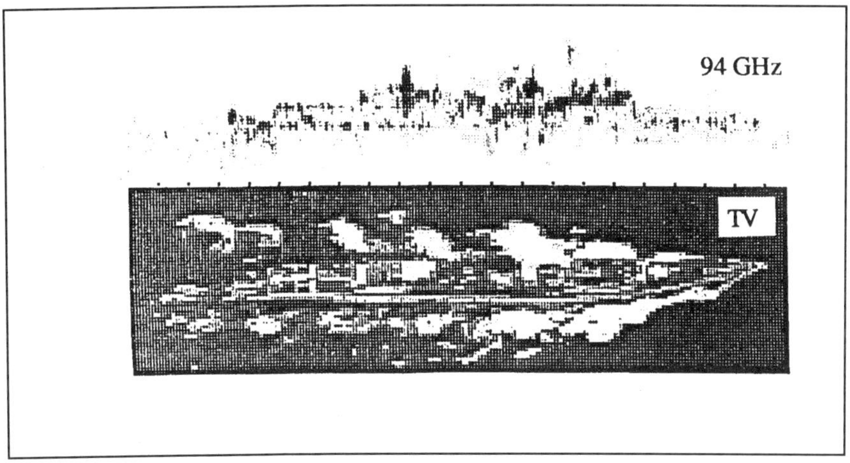

Figure 3.15 Radar image of a ship (ONERA 94-GHz radar).

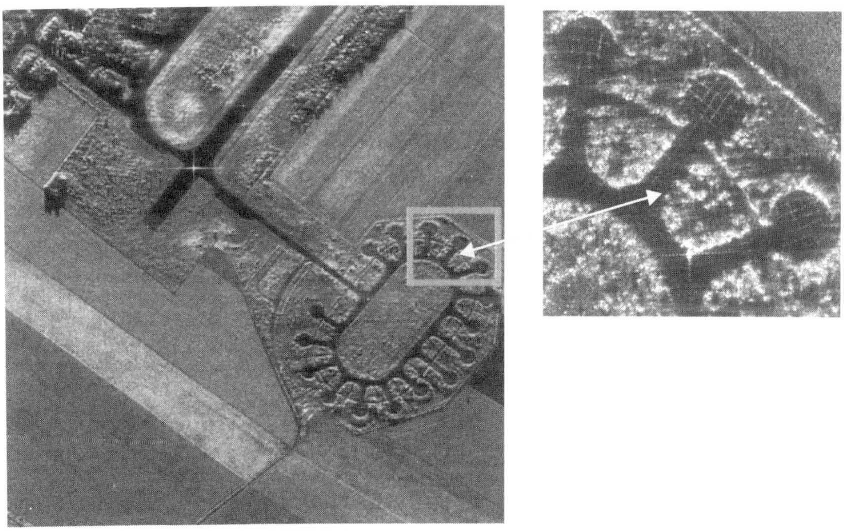

Figure 3.16 SAR images: Ground image obtained by a THALES synthetic aperture radar.

3.5 Angular Location

In the preceding discussion we have assumed that the angular position of the target with respect to the radar did not modify the received signal (in the case of an omnidirectional antenna, or when the target remains in the antenna "beam" or cone illuminated by the radar). In practice, the angular position of the targets is also unknown, and must therefore be measured by the radar. To do this, the radar must obviously be fitted with a directional antenna. Antenna theory shows—and a proof will be given in Section 3.5.2, and equation (3.74)—that the resolution of an antenna is determined by its dimension L, expressed in wavelengths, according to the relation: $\theta_a = \lambda / L$ (in radians). Consequently, to be able to measure an angular position, the radar must occupy a certain portion of space; the antenna is assumed to consist of elementary sensors distributed along the axis Ox (the generalization to a surface or volume distribution of sensors will be done in Section 3.5.4), so that the received signal is a function of two variables: the time and the abscissa x of the sensor.

The optimum receiver in angle, range, and Doppler will first be defined. Then the problem of the tracking radar, which must remain slaved to the target direction, will be examined.

3.5.1 Angle-Range-Doppler Optimum Receiver

Let us first assume that the transmitter is omnidirectional, located at O, and that the receivers are located along the axis Ox. A point target located at long range in direction θ returns a wave that can be considered planar on the antenna (see Figure 3.17). All sensors thus receive the same signal, delayed or phase-shifted proportionally to the sensor abscissa:

$$s(x, t) = \alpha u(t - \tau) e^{2\pi j \nu (t-\tau)} e^{2\pi j \frac{x}{\lambda} \sin\theta} e^{2\pi j f_0 t} \tag{3.62}$$

$$s(x, t) = \alpha a(t, \tau, \nu) e^{2\pi j \frac{x}{\lambda} \sin\theta} e^{2\pi j f_0 t}$$

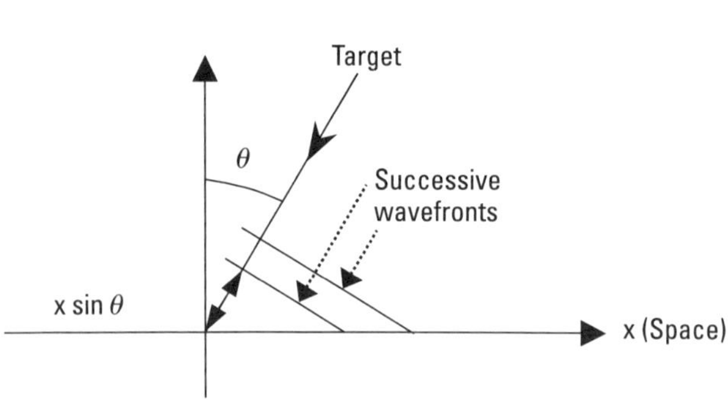

Figure 3.17 Angular reception: Geometry.

Strictly speaking, the equivalence between delay and phase shift is true only at the given frequency f. When the signal occupies a certain bandwidth Δf, the phase shift, from one frequency to the other, varies at the most by:

$$\Delta\varphi = 2\pi \frac{\Delta f}{c} x \sin\theta$$

This phase error will be negligible if $\Delta\varphi << 2\pi$, that is, for the extreme case ($x \sin\theta = L$, antenna dimension):

$$\frac{\Delta f}{f} << \frac{\lambda}{L} = \theta_a \tag{3.63}$$

antenna resolution (in radians): see (3.73).

This condition, generally satisfied in radar, is not always satisfied in sonar, owing to the wide relative bandwidths used; in wideband situations, two approaches can be used:

1. Preliminary spectral analysis of the signal received in each channel, followed by narrowband angular (spatial) analysis, and then coherent summation of the subbands to recover the time signal, for each direction;
2. Angular analysis by using time delays rather than phase shifts.

Under the narrowband assumption, it can be demonstrated that the optimum estimator derived in Section 2.3.1 for time-domain processing [(2.21) and (2.22)] can be generalized to functions of two variables, x and t, so that the optimum estimator of τ, ν, θ, and α performs:

$$\begin{cases} \hat{\tau},\ \hat{\nu},\ \hat{\theta} \text{ maximize } \dfrac{\left|\displaystyle\iint r(x,\ t)s^*(x,\ t)dxdt\right|^2}{\displaystyle\iint |s(x,\ t)|^2 dxdt} \\ \hat{\alpha} = \dfrac{\displaystyle\iint r(x,\ t)s^*(x,\ t)dxdt}{\displaystyle\iint |s(x,\ t)|^2 dxdt} \quad \begin{matrix}\text{for replica } s(x,\ t) \\ \text{corresponding to the triplet } \hat{\tau},\ \hat{\nu},\ \hat{\theta}\end{matrix} \end{cases} \tag{3.64}$$

That is, after carrier demodulation ($z(t) = r(t)e^{-2\pi j f_0 t}$) and in the case of a signal of unit energy $\int |a(t, \tau, \nu)|^2 dt = \int |u(t)|^2 dt = 1$:

$$\begin{cases} \hat{\tau}, \hat{\nu}, \hat{\theta} \text{ maximize} \left| \int \left[\int z(x, t) e^{-2\pi j \frac{x}{\lambda} \sin \hat{\theta}} dx \right] a^*(t, \tau, \nu) dt \right|^2 \\ \hat{\alpha} = \int \left[\int z(x, t) e^{-2\pi j \frac{x}{\lambda} \sin \hat{\theta}} dx \right] a^*(t, \hat{\tau}, \hat{\nu}) dt \end{cases} \tag{3.65}$$

Putting aside the estimation of α—which is not the essential need—it is clear that the optimum receiver calculates:

$$\left| \int y(t) a^*(t, \tau, \nu) dt \right|^2 \tag{3.66}$$

with

$$y(t) = \int z(x, t) e^{-2\pi j \frac{x}{\lambda} \sin \theta} dx \tag{3.67}$$

and determines which ($\hat{\tau}$, $\hat{\nu}$, $\hat{\theta}$) triplet maximizes the quantity (3.66) (or, in the case of detection, compares this quantity to a threshold—refer to Section 2.3.2).

This receiver can be interpreted as follows:

1. Integral with respect to x (3.67): For each direction θ, the signals received at the different abscissa points x are summed in phase, assuming they originate from direction θ; this operation is conventionally carried out by a parabolic reflector steered in direction θ, which phases the signals on a reception horn located at its focal point. The angular scanning in θ is then performed by the simple rotation of the antenna around a vertical axis. In the case of a fixed array of radiating elements (dipoles), this integral, with respect to x, is equivalent to performing a Fourier transform of the signals received at time t on the sensors (previously mixed with a local oscillator, then digitized), the conjugated variables being x/λ and $\sin \theta$. This operation delivers the value of the function to be maximized simultaneously for all directions θ. It can also be carried

out with controllable phase shifters ensuring—in RF—the multiplication of the signal received on the sensor at abscissa x by $e^{-2\pi j \frac{x}{\lambda} \sin \theta}$, or with variable delays $\tau = (x \sin \theta)/c$. In all cases, the purpose is simply to steer the receiver in the right direction using a certain device—parabolic antenna or phased array, for example.

2. Integral with respect to t: For each direction thus isolated, carry out the optimum range-Doppler processing defined in Sections 3.1 and 3.2 (range-Doppler correlation, then square-law envelope detection).[4]

If reception is carried out by consecutive scans of the different directions—calculation of integral (3.67) by a revolving parabola, or by controllable phase shifters or delays—the receiver takes into consideration, for a given steering direction, only the signals coming from this direction [integral (3.67) is equivalent to carrying out "angular filtering" on the received signals, around direction θ]. It is clear that to use the radar power as effectively as possible, it is preferable to "transmit only in the direction we are looking in," that is, to replace the omnidirectional transmitter by a directional transmission on the same antenna (or on an antenna of the same type as the reception antenna). The transmitted signal is then:

$$u(t) = u_0(t) \int e^{2\pi j \frac{x}{\lambda} \sin \theta} dx \tag{3.68}$$

which is an operation symmetrical to the one carried out in reception by (3.67). The processing in reception is obviously not affected by this directivity of transmission, which simply aims to save the necessary power by steering "the torch and the eyes in the same direction."

This type of operation by directive transmission-reception antenna consecutively examining the various directions, by mechanical scan (parabola) or electronic scan (phase shifters), followed by the Doppler-range receiver described above, is very widely used in radar. The procedure of using omnidirectional transmission and simultaneous reception in the different directions by calculation of the integral (3.67) by Fourier transform is more often used in sonar. It is also used in radar in the form of multibeam antennas, in which

4. These two linear operations—spatial processing and time processing—can be implemented in reverse order for modern sonars and digital beamforming radars; time processing is usually performed first for different reasons (such as ease of implementation, necessity of clutter echoes filtering, and necessity of spatial oversampling).

the different channels corresponding to the different angles are generated in RF using waveguides, couplers, and phase shifters.

Finally, it is important to note that the decoupling between the angular analysis and the range-Doppler analysis disappears if the transmission depends both on x and t; that is, if each transmitter of abscissa x transmits a different signal $u_0(t, x)$. Such systems are now being considered, with, for instance, each transmitter of abscissa x transmitting a different frequency as in the RIAS system [9] or the same pulse being successively transmitted by each transmitter, thus generating an artificial Doppler effect depending on the direction of the target [10–12]. In these situations, the angular analysis cannot be differentiated from the range-Doppler analysis, and the measurement accuracies of the three parameters depend on each other. However, this case of sensors differentiated at transmission will not be considered in what follows, and hence the optimum processing will break down into an angular scanning as in (3.67) followed by the optimum range-Doppler processing as in (3.66).

3.5.2 Angular Tracking: Sum Channel, Difference Channel

Section 3.5.1 considered the case of the estimation of $\hat{\theta}$ or of the detection of a target of unknown angular location θ. Turning now to the case of a tracking radar, whose purpose is to become slaved to the target direction, and taking into account the decoupling described in Section 3.5.1 between space and time variables, we can now restrict our reasoning to spatial processing and assume that the transmitted signal is a pure wave:

$$u(t) = e^{2\pi j f_0 t} \tag{3.69}$$

The received signal is then written, in accordance with (3.62):

$$s(x, t) = \alpha e^{2\pi j \frac{x}{\lambda} \sin\theta}, \text{ since } a(t, \tau, \nu) = 1 \tag{3.70}$$

(still for the same reasons of decoupling between angle and range-Doppler, the target Doppler can be assumed to be zero).

Under these conditions, the optimum receiver determines $\hat{\theta}$, according to (3.66) and (3.67), which maximizes $|\Sigma(\theta)|^2$ with:

$$\Sigma(\theta) = \int z(x, t) e^{-2\pi j \frac{x}{\lambda} \sin\theta} dx \tag{3.71}$$

The arguments presented in Sections 2.3.2 and 2.3.4 then apply directly to this optimum receiver with respect to θ. It is particularly useful to introduce the angular diagram $D_\Sigma(\theta, \theta_0)$, representing the optimum receiver response to a target located at θ_0.

$$z(x, t) = e^{2\pi j \frac{x}{\lambda} \sin\theta_0} \tag{3.72}$$

$$\Rightarrow D_\Sigma(\theta, \theta_0) = \frac{1}{L} \int_{-L/2}^{+L/2} e^{2\pi j \frac{x}{\lambda}(\sin\theta_0 - \sin\theta)} dx$$

(L: antenna dimension)

For $\theta_0 = 0$ (broadside observation), the receiver ambiguity function is thus:

$$|D_\Sigma(\theta)^2| = \left| \frac{1}{L} \int_{-L/2}^{+L/2} e^{-2\pi j \frac{x}{\lambda} \sin\theta} dx \right|^2 = \left| \operatorname{sinc}\left(\frac{L}{\lambda} \sin\theta \right) \right|^2 \tag{3.73}$$

where

$$\operatorname{sinc} u = \frac{\sin \pi u}{\pi u}$$

The angular resolution of the antenna is usually defined as the 3-dB width of the main lobe of this diagram (corresponding to the angular separation between to targets of equal amplitude and phases): for this linear antenna, the angular resolution is on the order of $\frac{\lambda}{L}$.

This ambiguity function with respect to θ is known as the *antenna pattern* (see Figure 3.18).

The angular measurement variance can be deduced from the second derivative of the antenna diagram at $\theta = 0$, according to (2.35) and (2.36) (Cramer-Rao bounds):

$$\operatorname{Var}(\hat{\theta}) \geq -\frac{N_0}{A} \left[\left. \frac{d^2 |D_\Sigma(\theta)|^2}{d\theta^2} \right|_{\theta=0} \right]^{-1} = \frac{N_0}{2A} \frac{3}{\pi^2} \left(\frac{\lambda}{L} \right)^2 = \frac{N_0}{2A} \frac{3}{\pi^2} \theta_a^2 \tag{3.74}$$

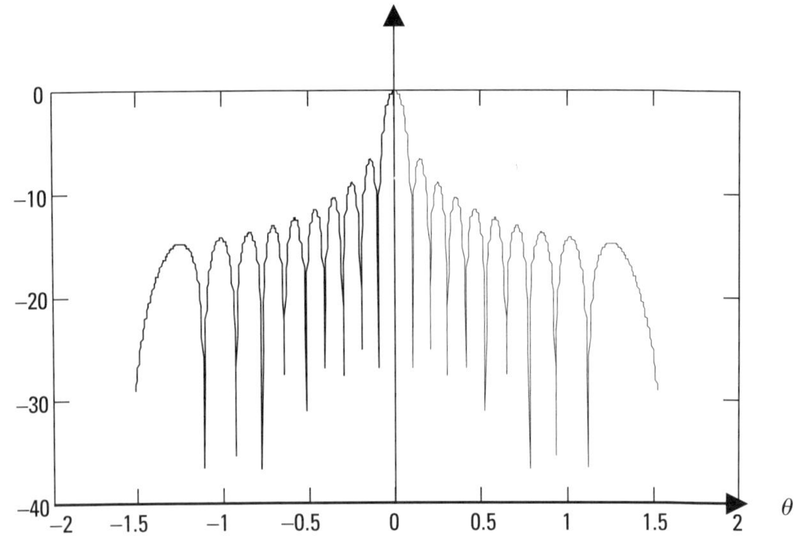

Figure 3.18 Antenna pattern, sum channel (L/λ = 10).

It must be emphasized that in this relation the signal to noise ratio A/N_0 is defined at the output of the signal processing (including the antenna gain), as it was in Chapter 2, as a consequence of the normalization of the received signal.

To perform tracking, $\hat{\theta}$ maximizing $|\Sigma(\theta)|^2$ must be determined at each instant [see (3.71)]. A convenient way of determining the maximum, knowing that the radar is in the tracking phase and is therefore close to the maximum, is to null the derivative:

$$\frac{d|\Sigma(\theta)^2|}{d\theta} = \Leftrightarrow \text{Re}[\Sigma(\theta)\Delta^*(\theta)] = 0, \text{ with } \Delta(\theta) = \frac{d\Sigma(\theta)}{d\theta} \quad (3.75)$$

Δ is conventionally known as the *difference channel* of the antenna, and Σ the *sum channel*. By analogy with the sum channel, a difference channel pattern (see Figure 3.19) can be defined, corresponding to the response to a target of unit amplitude in the direction θ_0 = 0:

$$\left|\frac{dD_\Sigma(\theta)}{d\theta}\right|^2 = \left|\frac{1}{L}\int_{-L/2}^{+L/2} -\frac{2\pi jx}{\lambda}\cos\theta\, e^{-2\pi j\frac{x}{\lambda}\sin\theta}\,dx\right|^2$$

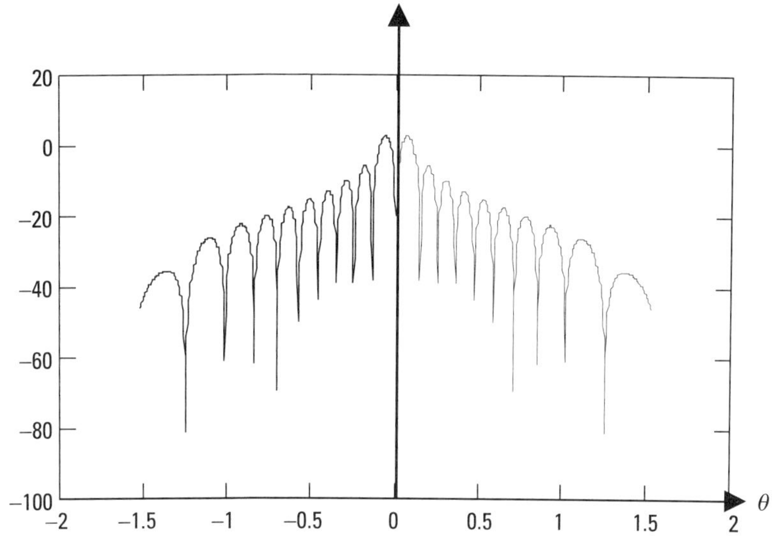

Figure 3.19 Antenna pattern, difference channel ($L/\lambda = 10$).

In the vicinity of the axis, the pattern of the sum channel is essentially parabolic, and that of the difference channel is linear (as a derivative of the former).

In practice, a simple way of generating such a difference channel—in the case of a parabolic antenna—is to place two horns close to the parabola focal point; thus two antennas directed in two close directions differing by an angle of approximately θ_a are built. The signals $s_1(\theta)$ and $s_2(\theta)$ received by these two sources are then summed and differentiated, in RF, so as to synthesize a sum channel of pattern $D_\Sigma(\theta)$ and a difference channel of pattern $D_\Delta(\theta)$:

$$D_\Delta(\theta) = k\theta_a \frac{dD_\Sigma(\theta)}{d\theta} = s_1(\theta) - s_2(\theta) \tag{3.76}$$

where k is a constant depending on the precise implementation of the difference channel.

Noise is then added by the mixers and preamplifiers to these two channels independently, which thus deliver the sum and difference signals:

$$\Sigma = \alpha D_\Sigma(\theta) + n_1 \tag{3.77}$$

$$\Delta = \alpha D_\Delta(\theta) + n_2 = \alpha k \theta_a \frac{dD_\Sigma(\theta)}{d\theta} + n_2 \tag{3.78}$$

Such an antenna, equivalent to two antennas steered in two close directions differing by θ_a, is known as *amplitude monopulse.*

Another way to generate these sum and difference channels consists of arranging two identical antennas, steered in the same direction, but separated by L along the Ox axis (see Figure 3.20). If $d(\theta)$ is the diagram of each antenna, the signals received will be identical on each antenna, except for a phase shift:

$$\phi = 2\pi j \frac{L}{\lambda} \sin\theta \tag{3.79}$$

That is, after mixing and amplification (hence noise addition):

$$\begin{aligned} s_1' &= \alpha d(\theta) e^{\pi j \frac{L}{\lambda} \sin\theta} + n_1 \\ s_2' &= \alpha d(\theta) e^{-\pi j \frac{L}{\lambda} \sin\theta} + n_2 \end{aligned} \tag{3.80}$$

Such an antenna is known as *phase monopulse.* It is merely obtained in the case of an array of radiating elements by separately summing the two halves of the complete antenna.

It is possible to synthesize the sum and difference channels from the s_1' and s_2' channels by obtaining:

$$\begin{cases} \Sigma = (1/2)(s_1' + s_2') \\ \Delta = (1/2j)(s_1' - s_2') \end{cases} \Leftrightarrow \begin{cases} s_1' = \Sigma + j\Delta \\ s_1' = \Sigma + j\Delta \end{cases} \tag{3.81}$$

Figure 3.20 Amplitude and phase monopulse.

The pattern of the sum channel thus constituted is:

$$D_\Sigma(\theta) = d(\theta)\cos\left(\pi\frac{L}{\lambda}\sin\theta\right) \tag{3.82}$$

and that of the difference channel:

$$D_\Delta(\theta) = d(\theta)\sin\left(\pi\frac{L}{\lambda}\sin\theta\right) \tag{3.83}$$

We obtain, in the vicinity of the axis, the approximately parabolic and linear patterns of the sum and difference channels. Those two monopulse types (amplitude or phase) are equivalent via (3.81) relating signals s_1' and s_2' originating from a phase monopulse antenna to Σ and Δ, the sum and difference channels of an amplitude monopulse antenna.

Depending on the problems encountered, it may be preferable to describe the received signals by (3.77) and (3.78), or by (3.80).

3.5.3 Radar Angle-Error Detection

The steering condition to satisfy is therefore, according to (3.75):

$$\text{Re}[\Sigma(\theta)\Delta^*(\theta)] = 0$$

In order to meet this condition with a servo control, it is useful to have, for small deviation angles, an estimation $\hat{\theta}$ of θ in accordance with the signals received on the sum and difference channels. $\hat{\theta}$ is the small correction to apply to the steering direction of the monopulse antenna.

This calculation of $\hat{\theta}$ is the purpose of angle-error detection.

The signals received on the sum and difference channels:

$$\Sigma = \alpha D_\Sigma(\theta) + n_1$$
$$\Delta = \alpha k\theta_a D_\Sigma'(\theta) + n_2$$

(n_1, n_2: independent noise samples of identical variance) thus constitute the vector $\mathbf{r}$ of the received signals: $\mathbf{r} = (\Sigma, \Delta)^t$.

Chapter 2 shows that the optimum processing consists of correlating this signal with a replica of the expected signal (vector $\mathbf{s}$ of the replicas: $\mathbf{s} = (D_\Sigma, k\theta_a D_\Sigma')^t$) and carrying out square-law envelope detection and normalization, that is, determining $\hat{\theta}$ by maximizing:

$$T = \frac{\left| \int \mathbf{r}^{+}(t)\mathbf{s}(t)dt \right|^2}{\int \mathbf{s}^{+}(t)\mathbf{s}(t)dt}$$

(+ designating the transposed-conjugated vector) which is thus written, as only one time sample is available:

$$T = \frac{|\Sigma D_\Sigma + \Delta k\theta_a D'_\Sigma|^2}{D_\Sigma^2 + k^2\theta_a^2 D'^2_\Sigma}$$

where D_Σ and D'_Σ are real (for symmetrical antennas) functions of θ.

In the approximation of a parabolic lobe $D_\Sigma(\theta) = 1 - \theta^2/\theta_a^2$, and in the case in which $\theta << \theta_a$ (tracking with a good signal-to-noise ratio, hence with the antenna almost steered at the target), by nulling the derivative of T with respect to θ, we obtain:

$$\hat{\theta} = -k\theta_a \frac{\mathrm{Re}\Sigma\Delta^*}{|\Sigma|^2 - |\Delta|^2} = -k\frac{\theta_a}{2}\mathrm{Arg}\frac{\Sigma - j\Delta}{\Sigma + j\Delta} = -\frac{k\theta_a}{2\pi}\mathrm{Arg}\frac{s'_2}{s'_1} \quad (3.84)$$

This expression can be applied directly to the case of phase monopulse, with $\theta_a = \lambda/L$. It is easy to check that in the absence of noise—and for small deviation angles—it results in $\hat{\theta} = \theta$, according to (3.80) of s'_1, and s'_2.

The expression (3.84) can be written, for $\theta/\theta_a << 1$, and hence for $|\Delta/\Sigma| << 1$ [according to (3.77) and (3.78)], in the case where D is parabolic:

$$\hat{\theta} = -k\theta_a \frac{\mathrm{Re}(\Sigma\Delta^*)}{|\Sigma|^2} = -k\theta_a \mathrm{Re}\left(\frac{\Delta}{\Sigma}\right) \quad (3.85)$$

This expression is conventionally used in the case of the amplitude monopulse, and in the absence of noise, does indeed lead to steering the antenna in the correct direction:

$$\Sigma = \alpha D_\Sigma(\theta) \text{ and } \Delta = \alpha k\theta_a D'_\Sigma(\theta) \Rightarrow \hat{\theta} = \theta$$

Relations (3.84) and (3.85) are the fundamental relations of the monopulse angle-error detection, used in various forms by all tracking radars.

3.5.4 Comments

The approach followed in the case of a pure wave can easily be transposed to the case of a modulated wave $u(t)$: For each value of τ and ν the sum and difference channels can be written as:

$$\Sigma(\theta) = \int y_{\Sigma}(t)a^{*}(t, \tau, \nu)dt \tag{3.86}$$

$$\Delta(\theta) = \int y_{\Delta}(t)a^{*}(t, \tau, \nu)dt \tag{3.87}$$

where

$$\begin{cases} a^{*}(t, \tau, \nu) = u(t - \tau)e^{2\pi j\nu(t-\tau)} \\ y_{\Sigma}(t)\text{: received signal on the sum channel} \\ y_{\Delta}(t)\text{: received signal on the difference channel} \end{cases}$$

For each bin (τ, ν), the error-angle detection is calculated according to the expression:

$$\theta = -k\theta_a \frac{\mathrm{Re}\Sigma\Delta^{*}}{|\Sigma|^2} = -\frac{k\theta_a}{2}\mathrm{Arg}\frac{\Sigma - j\Delta}{\Sigma + j\Delta} \tag{3.88}$$

It can be seen that the tracking receiver is composed of two identical receivers (Doppler-range correlation) operating on both sum and difference channels of the antenna.

Similarly, this reasoning applied to the case of a one-dimensional antenna, for tracking of only one angle θ can be generalized to the case of elevation-bearing tracking using a two-dimensional antenna (parabola or antenna array). The monopulse antenna then generates a sum channel, with a nearly parabolic pattern in the vicinity of the axis (in elevation as in bearing), and two difference channels, one in elevation and the other in bearing. The receiver processes those three channels concurrently by *Doppler-range correlation*:

$$\Sigma = \int y_\Sigma(t)a^*(t, \tau, \nu)dt$$

$$\Delta_e = \int y_{\Delta_e}(t)a^*(t, \tau, \nu)dt$$

$$\Delta_b = \int y_{\Delta_b}(t)a^*(t, \tau, \nu)dt$$

where y_Σ, y_{Δ_e}, and y_{Δ_b} denote the signals received respectively on the sum, elevation difference, and bearing difference channels. Two angle-error estimations—elevation and bearing—are then calculated by relations similar to (3.84).

In the tracking phase, it is also necessary to track the target range and Doppler. The optimum procedure will not be described in detail here. It can be constructed in a way completely analogous to that used for the angular tracking. In practice, the range and Doppler tracking are often carried out by simple comparison of the levels received on the sum channel in two adjacent bins (as the requirement in angular tracking accuracy is generally more demanding for radar applications than the range or Doppler accuracy, optimum processing is not always necessary for these last two parameters).

This analysis of the radar processing in tracking will be completed in Section 7.2 by taking into account the nonpoint character of the radar targets.

3.6 Implementation

The volume and speed required for radar or sonar signal processing (pulse compression, Doppler filtering, CFAR detection), together with the real-time constraints involved, generally lead to multiprocessor implementations of the algorithms. In this book, which is devoted to processing principles, we will not discuss the specific processors, architectures, and implementations that are subject to rapid evolutions, but instead we will focus on the general approach for hardware selection and programmation. The appropriate methodology for development of those complex architectures and softwares will be described in this section, and criteria for multiprocessor parallel architecture selection (pipeline and data parallelism, for example) will be proposed.

3.6.1 General Approach

Applications of radar/sonar signal processing are implemented through three successive steps, shown in Figure 3.21:

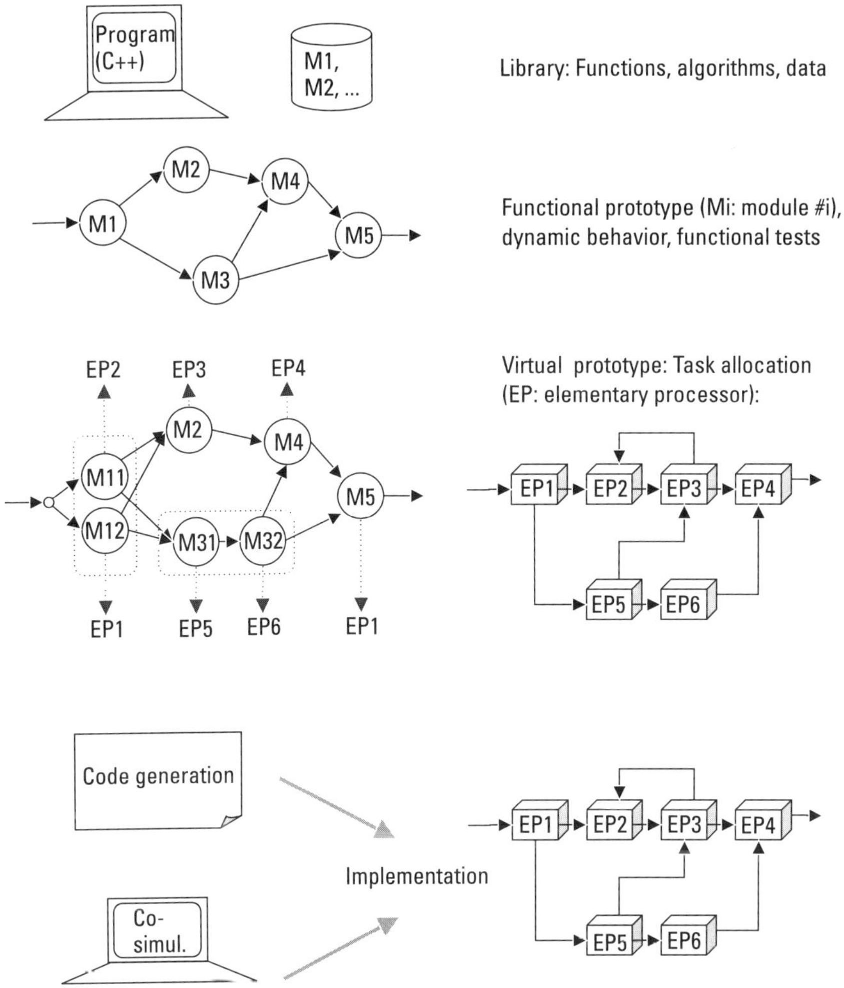

Figure 3.21 Signal processing design and implementation process.

1. Functional processing design, which aims to define the algorithms to be implemented, their timing constraints, and their mutual relations (succession, inputs, and outputs);
2. Architecture design, defining the organization and the operating modes of the selected elementary processors and their interconnections;
3. Hardware and software development, realizing the whole processing system in a multiprocessor machine with appropriate and validated codes.

3.6.2 Functional Design

Functional design is the result of a succession of—or, more realistically, an interaction between—two distinct operations: functional modeling and functional validation.

3.6.2.1 Functional Modeling

From a given library of functions and data, this operation organizes the interactions between the elementary algorithms (such as correlations, Fourier transforms, thresholdings) and data (complex time samples received, replicas), according to a predefined functional schedule and using appropriate links defined by the designer during the preliminary studies. This operation requires a dynamic behavior of the global system to have been previously defined and modeled (e.g., discrete events, synchronous data flow) so that the progress of the algorithms and the radar (or platform) system operation can be described along the same timescale.

During this step a set of functional tests, based on real and/or synthetic data, which will be the basis of the validations for the following steps, is also defined. That set of functional tests plays a key role in the global system design and validation, since it will also be used during the field trials and thus will guarantee the adequation between functional modeling and real system operation.

3.6.2.2 Functional Validation

The goal of this task is to test and evaluate the processing structure previously defined, on the basis of the functional tests defined in the previous step. An iterative operation of these functional modeling and validation steps will generally be required, in order to provide a stabilized version of the processing structure, which will then be used as a reference version by the hardware/software development teams and by the operational test and trial teams.

3.6.3 Architecture Design

Architecture design also involves two successive—or iterative—steps: architecture determination and architecture validation. The selection or combination of data parallelism and program parallelism (see Section 3.7), associated with the appropriate communication network, is the main objective of this central task.

3.6.3.1 Architecture Determination

During this step, for each possible class of multiprocessor architecture (for example, ring structure, central memory structure, use of FPGAs, or ASIC processors) the following are defined:

- The dynamic allocation of the different tasks to the different elementary processors;
- Memory allocations and data partitioning for distribution to the elementary processors;
- The required communications between the processors;
- Sequential operation and latency delays.

This selection takes into account both temporal criteria (real-time operation, latency constraints, constraints of reactivity to external requests and controls) and nonfunctional criteria (cost, volume, and development schedule, for example). It is usually performed with the help of specific prototyping tools, describing the basic components and their association rules, such as elementary processors, communication network, and memories.

3.6.3.2 Architecture Validation

The selected architecture is then validated through analysis of its operation on the previously defined functional tests. A major objective of this validation is to analyze the dynamic behavior of the application; the accuracy of this behavior modeling depends on the available data regarding processors' performances. Some of these components may be available at that time, and directly used in this evaluation.

3.6.4 Hardware and Software Development

On the basis of the selected architecture, and using the functional simulation previously defined, programs are generated, as automatically as possible (especially concerning the communications between the processors). In parallel, a simulation of the target machine is implemented (usually on a workstation), in order to validate the whole system before final production and components programming. These simulations are again based and evaluated on the previously defined functional tests. In the situation where the target machine (or a preliminary version) already exists, this simulation is run on the real machine.

3.7 Systematic Signal Processing

The most intensive part of radar or sonar signal processing is *systematic,* that is, the process is perfectly deterministic, it is known at the design stages, and generally made of matrix operations (e.g., FFT, pulse compression, digital beamforming—this is also true for adaptive beamforming, when block-processing is used, see Chapter 5). It is therefore useful to determine signal processing architectures appropriate for that kind of processing.

A second important characteristic of radar or sonar signal processing applications is the need for real-time reconfiguration, allowing the radar to use the appropriate mode in every situation: from surveillance to tracking modes, but also, as will be seen in the next chapters, from conventional to jammed situations or interleaving surveillance, tracking, and target analysis modes with different processing requirements and dynamic operational constraints. Special attention must therefore be given to dynamic behavior of the computing resources, with explicit description of the overall system behavior.

Finally, a driving factor in processing applications development is the requirement for *evolutivity,* that is, the capability of changing the elementary processors, when obsolescence requires, without redrawing the whole application and testing: Besides being a strong incentive to develop the functional model and the architecture separately, this also leads to cleaner definitions of the architecture, with clear distinction between data and communication networks, for example.

Taking into account these constraints, a widely adopted architecture is *pipeline,* or task—or program—parallelism, which means that a sequence of operations or programs is executed in parallel, with data flowing from one task to the other (see Figure 3.22). This architecture, which clearly dissociates communication and computing functions, guarantees straightforward and explicit programming and a simple communication network. Compared to an ideal architecture using all the processors in parallel (data parallelism, also shown in Figure 3.22), this simplicity is obtained at the cost of some performance degradation:

- Higher latency (delay before arrival of the first results), since the processors not involved in the first calculations are not active at the early stages;
- Need for higher memory and communication capacities, compared to data parallelism.

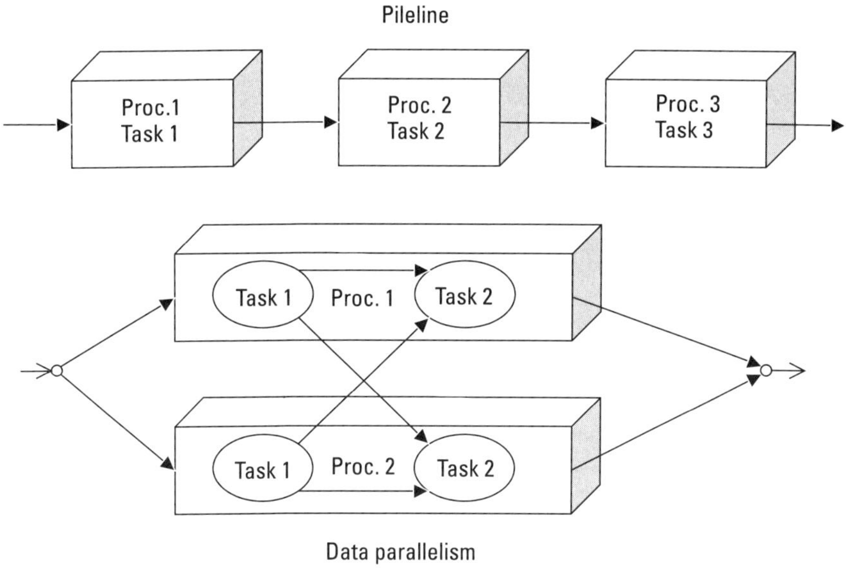

Figure 3.22 Multiprocessor architectures.

For these reasons, and taking advantage of the development tools that become available (and largely alleviate the higher programming complexity), *data parallelism* tends to be preferred, with different processors operating simultaneously on the same data, each one being in charge of a part of the required computations.

Depending on the available resources of communication and computing, it is sometimes preferable to combine those two basic structures with data parallel structures arranged in an overall pipeline structure.

That simple argument shows that a higher automation of the design process is a condition for a better adequation between algorithms and hardware architecture, allowing a maximal use of processors and communications capacities, while maintaining clear programming structures, at each stage of the product life [13, 14].

References

[1] Ville, J., "Théorie et Applications de la Notion de Signal Analytique," *Câbles et Transmission,* No. l, 1948, p. 6074.

[2] Van Trees, H., *Detection, Estimation and Modulation Theory,* New York: Wiley, 1971.

[3] Rihaczek, A. W., *Principles of High Resolution Radar*, New York: McGraw Hill, 1969, Norwood, MA: Artech House, 1996.

[4] Debuisser, J. C., French patent no. 76 38 006, dated Dec. 16, 1976.

[5] Cook, C. E., and M. Bernfeld, *Radar Signals*, New York: Academic Press, 1967, Norwood, MA: Artech House, 1993.

[6] Bellanger, M., *Traitement Numérique du Signal*, Paris: Masson, 1981.

[7] Rohling, H. "Radar CFAR Thresholding in Clutter and Multiple Situations," *IEEE Transactions on AES*, AES-19, 60862 1, July 1983.

[8] Wilson, S. L., "Two CFAR Algorithms for Interfering Targets and Nonhomogeneous Clutter," *IEEE Transactions on AES*, AES-29, 5772, Jan. 1993.

[9] Dorey, J., Y. Blanchard, and F. Christophe, "Le Projet RIAS, une Approche Nouvelle du Radar de Surveillance Aérienne," *L'Onde Electrique*, 64, No. 4, 1978, p. 1520.

[10] Drabowitch, S., and C. Aubry, "Pattern Compression by Space-Time Binary Coding of an Array Antenna," *AGARD CP 66*, Advanced Radar Systems, 1969.

[11] Bienvenu, G., D. Pillon, and P. Tournois, "Ambiguity Functions of an Imaging System Using Simultaneous Multiple Transmissions," *8ème GRETSI*, Nice, June 1981.

[12] Le Chevalier, F., "Future Concepts for Electromagnetic Detection," *IEEE Systems Magazine*, Oct. 99.

[13] Pridmore, J. et al., "Model-Year Architectures for Rapid Prototyping," *Journal of VLSI Signal Processing Systems*, Kluwer Academic Publishers, Vol. 15, 83, Feb. 1997.

[14] Mdahar, B. et al., "ESPADON, Environment for Signal Processing Application Development and Prototyping: Future," *IEEE*, Oct. 99.

Appendix 3A: Relation Between the Duration and Bandwidth of a Signal

Let us consider the case where $\bar{f} = 0$ and $\bar{t} = 0$:

$$\overline{\Delta t}^2 = \int_{-\infty}^{+\infty} t^2 |a(t)|^2 dt$$

$$\overline{\Delta f}^2 = \int_{-\infty}^{+\infty} f^2 |\tilde{a}(f)|^2 df$$

Applying Schwartz inequality: $[\int \text{Re}[uv^*]]^2 \leq \int |u|^2 \int |v|^2$, where $u = ta(t)$ and $v = a'(t)$

$$\left[\int \mathrm{Re}[ta(t)a'^*(t)]dt\right]^2 \leq \int t^2|a(t)|^2 dt \int |a'(t)|^2 dt$$

Integrating by parts:

$$\int \mathrm{Re}[ta(t)a'^*(t)]dt = \left[\frac{t}{2}|a(t)|^2\right]_{-\infty}^{+\infty} - \frac{1}{2}\int_{-\infty}^{+\infty} |a(t)|^2 dt = -\frac{1}{2}$$

Moreover

$$\int |a'(t)|^2 dt = (2\pi)^2 \int f^2 |\tilde{a}(f)|^2 df = (2\pi)^2 \overline{\Delta f^2}$$

$$\Rightarrow \overline{\Delta t} \cdot \overline{\Delta f} \geq \frac{1}{4\pi}$$

Appendix 3B: Radar Range Equation

According to Chapter 2, radar maximum range is limited by the signal-to-noise ratio, which determines both the detection performance (Section 2.3.5) and the location accuracy (Sections 3.2.3 and 3.5.2).

3B.1 Thermal Noise

The thermal noise, essentially generated by the first amplifier of the reception chain, depends only on the noise factor F of this first amplifier and on the bandwidth B of the received signal (since the receiver obviously filters any noise outside the band of this signal), according to the conventional equation:

$$N = kT_0BF \text{ (in watts)}$$

where k is Boltzmann's constant and T_0 the temperature of the receiver (ambient temperature if it is not cooled) expressed in kelvins. If B is expressed in decibels/hertz (for instance, for pulses of 1 μs, and thus of bandwidth 1 MHz, B = 60 dB), kT_0 is −204 dB. The noise factor amounts to a few decibels, for modern low noise amplifiers. For F = 5 dB, the noise power received is therefore approximately equal to:

$$N = -204 + 60 + 5 = -139 \text{ dBW, that is, } 0.01 \text{ pW}$$

(This power should be compared to the transmitted power, usually equal to some kilowatts, which explains why—special cases excluded—it is not possible to receive and transmit at the same time: The receiver is shut down during transmission.)

3B.2 Signal-to-Noise Ratio

The power of the received signal depends on the power transmitted, on the gain of the transmission antenna, on the range, on the target reflectivity, and on the reception antenna gain. Before estimating this power, it is necessary to define the reflectivity of the target (see Chapter 6).

Reflectivity is measured in radar by the radar cross section (RCS), which will be examined in detail in Chapter 6: It corresponds to the area of a target which, if backscattering were isotropic, would return an echo of power identical to the one actually received by the receiver.

In practice, RCSs are on the order of 0.1m^2 for a small aircraft or a missile, 1m^2 for an aircraft, 10m^2 for a vehicle or a helicopter, and over 100m^2 for a ship. This quantity is highly variable, depending on the type of target, but also on its aspect angle, or on the wavelength used. A 20-dB fluctuation for a change in aspect angle smaller than one degree is not abnormal for an aircraft observed at a wavelength of 3 cm.

With this definition it is possible to calculate the power received by the radar, using the diagram in Figure 3B.1.

The received peak power is given by:

$$P_r = P_t G_t \frac{\sigma}{4\pi R_t^2} \frac{A_r}{4\pi R_r^2}$$
$$\uparrow \quad \uparrow \quad \uparrow$$
$$A \quad B \quad C$$

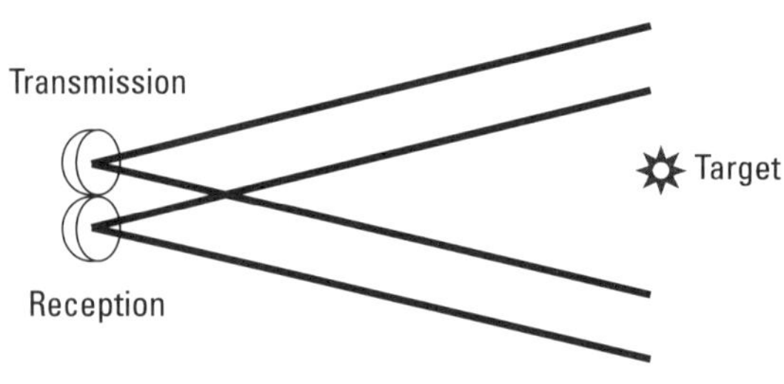

Figure 3B.1 Power received by the radar (geometry).

In this equation, the first term corresponds to transmission; the power transmitted in the direction of the target is equal to the peak power of the transmitter, multiplied by the gain of the transmission antenna. The second term corresponds to the reflection from the target, at range R_t from the transmission antenna; the target intercepts the fraction $\frac{\sigma}{4\pi R_t^2}$ of the radiated energy (ratio between σ, the RCS of the target, and $4\pi R_t^2$, the area of a sphere of radius R_t). The third term corresponds to the return path to the radar; the reception antenna, by definition of the RCS σ, intercepts the fraction $\frac{A_r}{4\pi R_r^2}$ of the energy reradiated by the target in all directions, where A_r is the effective area of the antenna, and R_r the distance between the target and the receiving antenna.

Moreover, the gain G of an antenna of effective area A is equal to the ratio between the directivity of an isotropic antenna, equal to 4π sr, and the directivity of the antenna under consideration, equal to λ^2/A sr (order of magnitude, according to Section 3.5.2).

$$G = \frac{4\pi A}{\lambda^2}$$

It is now possible to compute the signal-to-noise ratio, S/N, after coherent summation of n pulses and pulse compression[5] of factor ρ:

$$(S/N) = P_t G_t \frac{\sigma}{4\pi R_t^2} \frac{A_r}{4\pi R_r^2} \frac{n\rho}{kT_0 BF}$$

with: $G = \frac{4\pi A}{\lambda^2}$ (for transmitting or receiving antennas).

This equation, known as the *radar equation,* enables, for a desired signal-to-noise ratio at the output, the definition of a good compromise between range, wavelength, transmission power, and antenna (generally, the transmission antenna and the reception antenna are one and the same antenna, fed through a duplexer, so that $A_r = A_t$, $G_r = G_t$, and $R_r = R_t$). For a so-called monostatic device (transmission and reception

5. Pulse compression with bandwidth B and pulse duration T is equivalent to the summation of $BT = \rho$ independent samples, and therefore improves the signal-to-noise factor of this same factor ρ.

antennas combined, in contrast to bistatic, in which case they are located at two different places), which is by far the most frequent in active radar and sonar, the radar equation is thus:

$$(S/N) = P_t G^2 \frac{\sigma \lambda^2}{(4\pi)^3 R^4} \frac{n\rho}{kT_0 BF}$$

with

$$G = \frac{4\pi A}{\lambda^2}$$

In particular, the power of the echo received from the target obviously varies according to the fourth power of the range. This explains the permanent limit imposed by the thermal noise, the transmitted power being nearly always insufficient to counterbalance the attenuation effect due to distance.

This equation can be written in terms of average transmitted power $\overline{P_t}$ (peak power weighted by the duty factor of the radar), and final Doppler (integration) bandwidth $b \approx 1/T_{\text{illumination}}$:

$$(S/N) = \overline{P_t} G^2 \frac{\sigma \lambda^2}{(4\pi)^3 R^4} \frac{1}{kT_0 bF}$$

with

$$G = \frac{4\pi A}{\lambda^2}$$

This expression is easier to use during the design phase of radar definition, since it does not require the precise transmitted waveform (repetition frequency, pulse compression factor) to be known at that stage.

3B.3 Performance

It was shown in Sections 2.2.3 and 2.3.5 that the natural search radar performance criterion is the detection probability at a given false alarm probability (Neyman-Pearson criterion).

In practice, the false alarm rate is fixed by the operator, who does not want to be alerted without reason too often. For instance, he or she will tolerate an average of one false alarm per radar surveillance scan. If the maximum radar range is 15 km, with a range resolution of 150m, the number of "range bins" is equal to 100 in each direction. For an antenna directivity of 3°, the number of directions of observation for a full surveillance scan is 120; thus the radar tests $120 \times 100 = 12{,}000$ angle-range bins.

For each of these bins, 50 different assumed Doppler shifts have to be tested, for instance, if 50 pulses are coherently summed (the Fourier transform of 50 samples gives 50 independent frequencies). Consequently, the total number of hypotheses tested by the radar is $12{,}000 \times 50 = 0.6 \cdot 10^6$. For only one false alarm to occur, on average, over all these tests, a false alarm probability lower than 10^{-6} must be ensured; this example clearly shows that the desired false alarm probabilities are extremely low.

The detection probability is set in such a way that at maximum range it is approximately equal to 0.5 to 0.9 depending on the type of application. This is due to the fact that it is generally accepted that a certain detection of the target need not be obtained at the first pass, knowing that the probability of nondetection over two consecutive scans then becomes very low, equal to $(1 - P_d)^2$. Thus, for $P_d = 0.75$, the nondetection probability over two scans is 0.06 and consequently the probability of detection on at least one of these two scans is 0.94.

This choice of detection probability and false alarm probability thus makes it possible to determine the necessary signal-to-noise ratio, and consequently, using the radar equation, the necessary transmission power.

Appendix 3C: Spectral Purity

In actual practice, radar performance in multitarget or target-plus-clutter situations is limited by the sidelobes of the ambiguity function (in Doppler, range, or angle). To prevent masking of a small target by the sidelobes of a strong return (for instance due to clutter), signal processing is modified by altering the replicas involved in optimum signal processing. This is especially needed for Doppler processing, since in a given range cell, the clutter returns are usually much stronger than the target signal. When the radar signal is a pulse train, this reduction of Doppler sidelobes is obtained through weighting of the successive returns used in (3.49). The effectiveness of such weightings, widely used in spectral analysis [1], is however limited by the spectral purity of the transmitted signal. In this appendix, the effect of this limitation is analyzed, following the approach of [2], and evaluated for typical radar signals.

3C.1 Phase and Amplitude Noise

The nonideal characteristics of the signal can be taken into account by a phase and amplitude noise[6] due to the transmitter and to the synthesizer—the latter being present in the transmitted signal and in the local oscillator used on reception.

Transmitter phase noise: $\varphi_t(t)$;
Transmitter amplitude noise: $a(t)$;
Synthesizer phase noise: $\varphi_s(t)$;
Synthesizer amplitude noise: $a_s(t)$ (generally negligible).

Assuming that the radar signal is $u(t)$ with carrier frequency f_0, and neglecting the amplitude noise, the transmitted signal is written:

$$[1 + a(t)]u(t)\exp[2\pi j f_0 t + j\varphi_t(t) + j\varphi_s(t)]$$
$$\approx [1 + a(t) + j\varphi_t(t) + j\varphi_s(t)]u(t)\exp[2\pi j f_0 t]$$
(for small phase noise)

3C.2 Reception

The received signal is written:

$$Au(t - t_0)[1 + A(t - t_0) + j\varphi_t(t - t_0) + j\varphi_s(t - t_0)]\exp[2\pi j(f_0 + f_d)t]$$

After mixing with the local oscillator:

$$Au(t - t_0)\{1 + A(t - t_0) + j[\varphi_t(t - t_0) + \varphi_s(t)]\}\exp[2\pi j f_d t]$$
$$= Au(t - t_0)\{1 + b(t)\}\exp[2\pi j f_d t]$$

The resulting noise $b(t)$ is therefore the sum of the transmitter phase noise, the transmitter amplitude noise, and the synthesizer phase noise $\varphi_s(t)$ filtered by a filter with transfer function $H(f)$ [giving $\varphi_s(t - t_0) - \varphi_s(t)$]. This function $H(f)$ is equal, by definition, to the response of the filter to a complex sinusoid $\exp(2\pi j f t)$:

6. The effect of spurious lines in the synthesized signal and in the transmitted signal can be analyzed separately, using a similar approach (see [3] for a detailed analysis of synthesizers purity).

$$H(f) = \frac{e^{2\pi jf(t-t_0)} - e^{2\pi jft}}{e^{2\pi jft}}$$

$$H(f) = -2je^{-\pi jft_0} \sin \pi ft_0$$
$$|H(f)| = 2|\sin \pi ft_0|$$

This transfer function is shown in Figure 3C.1. In particular, phase noise is obviously suppressed for $t_0 = 0$, that is, for zero-range targets.

The receiver correlates this received signal with a weighted replica $u(t)$ $w(t)$ of the transmitted signal ($w(t)$: weighting function, used for clutter Doppler sidelobes reduction):

$$A\int [1 + b(t)]u(t - t_0)u^*(t - t_1)w(t - t_1) \exp[-2\pi j(f - f_d)t]dt$$

$$= AR(f) + A\int b(t)u(t - t_0)u^*(t - t_1)w(t - t_1) \exp[-2\pi j(f - f_d)t]dt$$

- First term $AR(f)$. Ground clutter filtering residue, taking into account the weighting through u' replica, for a pure transmitted signal and perfect mixing.

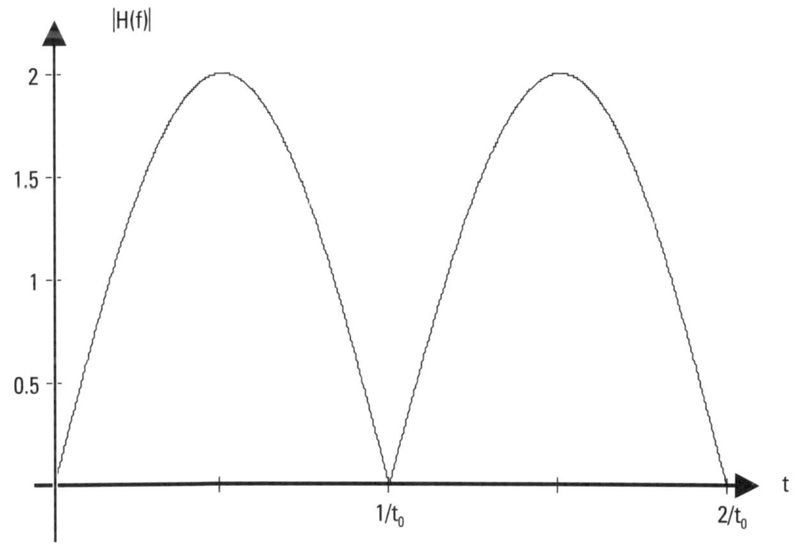

Figure 3C.1 Function $H(f)$.

- Second term. Spurious output resulting from amplitude and phase noises, equal to the Fourier transform of the product of $b(t)$ and $u(t)\, u^*(t)\, w(t)$, and therefore equal to the convolution of the Fourier transforms $\tilde{b}(f)$ and $\tilde{U}(f)$.

Let $P(f)$ be this spurious term:

$$P(f) = Ae^{-2\pi j f t_0} \int \tilde{U}(f - \nu)\tilde{b}^*(\nu)d\nu$$

The average residual energy is therefore:

$$\mathrm{E}|P(f)|^2 = A^2\mathrm{E}\left|\int \tilde{U}(f - \nu)\tilde{b}^*(\nu)d\nu\right|^2$$

But, assuming independent spectral lines for the phase noise:

$$\mathrm{E}[\tilde{b}(\nu)\tilde{b}^*(\nu')] = \delta(\nu - \nu')L(\nu)$$

where $L(\nu)$ is the power spectral density of $b(t)$; hence:

$$\mathrm{E}|P(f)|^2 = A^2 \int |\tilde{U}(f - \nu)|^2 L(\nu)d\nu$$

The cancellation ratio is then defined as:

$$T = \frac{A^2|\tilde{U}(0)|^2}{\mathrm{E}|P(f)|^2} = \frac{|\tilde{U}(0)|^2}{\int |\tilde{U}(f - \nu)|^2 L(\nu)d\nu}$$

with

$$L(\nu) = L_a(\nu) + L_\varphi(\nu)|2\sin \pi t_0 f|^2 + L_\varphi(\nu)$$

Typical examples of spectral noises are shown in Figure 3C.2, from [4], for two types of oscillators at 10 GHz: one using a wave gallery mode, and an estimation of possible future performances for high temperature super conducting oscillators.

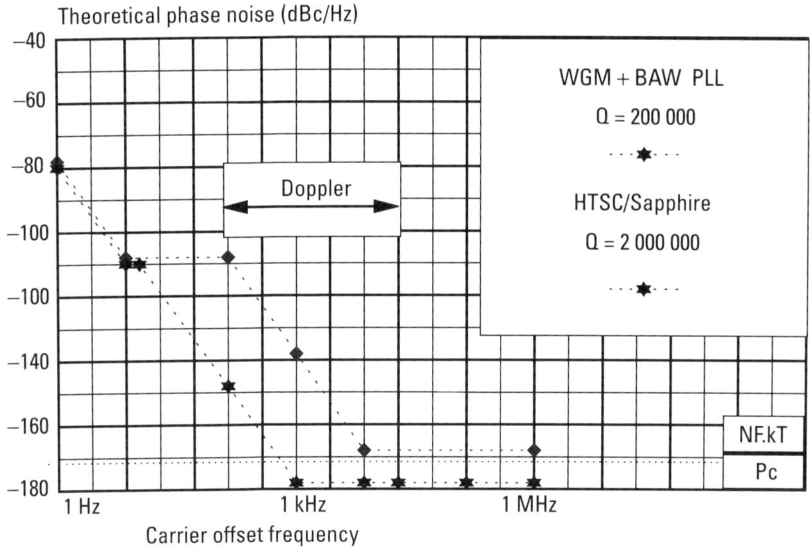

Figure 3C.2 Typical phase noise characteristics at 10 GHz: HTSC Sapphire oscillator 10 GHz. (*From:* [4].)

3C.3 Example: Pulse Radar with High Repetition Frequency

(For further details on high repetition frequency radars, see Chapter 7.)

Instantaneous bandwidth: Δf
Doppler filtering bandwidth: δf

The main lobe ground clutter returns are eliminated by a notch filter (with a few kilohertz bandwidth when the carrier is about 10 GHz). For the other Doppler frequencies, the phase noise can be assumed to be white with (constant) power L_φ.

The filtering of phase noise by $H(f)$ is equivalent, in average, to a 3-dB increase (average of $|2\sin\pi f t_0|^2$, or sum of two independent samples):

$$\mathrm{E}|P(f)|^2 = \int |U(f-\nu)|^2 2L_\varphi d\nu$$

The signal spectrum being made of harmonic lines of width δf (see Section 3.3.5):

$$\mathrm{E}|P(f)|^2 = \sum \delta f\,|\tilde{U}(0)|^2 2L_\varphi$$

(summation of every line of $\tilde{U}(f)$)

$$\mathrm{E}|P(f)|^2 = \frac{\Delta f}{F_r}\delta f\,|\tilde{U}(0)|^2 2L_\varphi$$

$$T = \frac{\tau}{T_r}\frac{1}{2\delta f L_\varphi} = \frac{1}{\frac{B_{\text{inst}}}{F_r}2\delta f L_\varphi}$$

3C.4 Application

Duty factor $\frac{F_r}{B_{\text{inst}}} = 0.1$; $\delta f = 500$ Hz; $L_\varphi = -100$ dBc/Hz $\Rightarrow T = 60$ dB

References

[1] Harris, F. J., "On the Use of Windows for Harmonic Analysis with the Discrete Fourier Transform," *Proceedings IEEE*, Vol. 66, No. l, Jan. 1978.

[2] Lacomme, Ph., et al., *Air and Spaceborne Radar Systems*, Norwich, NY: William Andrew Scitech Publishing, Inc., 2001.

[3] Carpentier, M. H., and B. L. Smith, *The Microwave Engineering Handbook*, London: Chapman & Hall, 1993.

[4] Mage, J. C. et al., "Low Phase Noise Oscillator for Stealth Target Detection," *International Conference on Radar*, Paris, 1994.

4

Optimum Reception in Colored Noise

4.1 Summary

Various other spurious signals are, in the real situation, added to the amplification thermal noise (or to ambient sea noise in sonar), hereto modeled by the white and Gaussian additive noise in the frequency band considered. These are: (1) external spurious signals that, intentionally or not, jam reception; (2) echoes from various reflectors in the landscape; and (3), in sonar, reverberation (see Chapter 6). Although these spurious signals can often also be treated as noise, the fact that this noise is not white—and sometimes not Gaussian, although this deviation from the model is generally less of a problem—must usually be taken into account. It is therefore necessary to extend the preceding arguments to the case of colored noise. Initially, the characteristics (power and spectral density) of the noise are assumed to be known; that is, we have a *noise-only reference,* either by a priori knowledge of the phenomena or by a supplementary measurement (the optimum processing in the absence of a noise-only reference will be determined in Chapter 5).

An intuitive approach to the transition from the optimum receiver in white noise to the optimum receiver in colored noise is to use the concept of *whitening.* Suppose that a filter exists that is capable of converting the received colored noise into white noise. If the received noisy signal is first passed through this filter, then at the filter output we are faced with the previously discussed problem of detection or estimation of a signal of known form (because it is distorted by the filter in a known manner) in the presence of Gaussian white additive noise (because a filtered Gaussian noise remains Gaussian). If such a whitening filter exists, then the optimum processing consists of passing the received noisy signal through this whitening filter and then correlating the obtained signal with a set of replicas, themselves whitened by the same filter (to repeat on the replica the distortion introduced into the signal by the whitening filter); performing quadratic detection; normalizing with respect to the power of the whitened replica; and finally determining the maximum—or comparing with a threshold for detection.

According to another approach, the contribution of the colored part of the noise to the received noisy signal is first estimated, and this estimate is then subtracted from the received signal. Assuming that the signal has not been affected by this operation, it must then be correlated with the replica set, detected, and normalized. Theoretical analysis then yields two results that were not a priori obvious:

1. This approach, by subtraction of the colored part of the noise, is strictly equivalent to the whitening approach.
2. The normalization to be performed is still a normalization with respect to the power of the whitened replica, although the correlation is made with the original—unwhitened—replica.

A third intuitively reasonable approach is to correlate the received noisy signal with the replica, then subtract from the result of this correlation the estimate of the correlation of the replica with the colored part of the noise. This third approach is again equivalent to the first two, as can be shown by considering an example in which the colored noise is formed by a spurious signal of the same form as the transmitted signal (jammer at position θ_I) and of random amplitude.

The optimum processing consists of first estimating the amplitude and the phase of the jammer (its position is assumed to be known, since it is assumed that the noise characteristics are known in this chapter) by correlating the received noisy signal with the replica corresponding to the position θ_I. Once these quantities have been estimated, it is possible to calculate, for each potential target position, how much the jammer would falsify the result of the conventional processing by correlation (optimum in white noise) for that position. The estimated jammer contribution to that position can then be subtracted. This procedure of subtracting the estimated contribution of colored noise is just a third version of the optimum processing, after whitening and subtraction of the colored part of the noise.

The equivalence between these three different receiver structures is a consequence of the two following properties:

1. The result of the correlation of a signal with a filtered signal does not change if the filtering "branch" is changed (filtering of the signal is equivalent to filtering of the replica).
2. The whitening filter has a specific structure, as a consequence of the specific structure of the noise (white noise added to colored noise).

This equivalence is shown in this chapter in the very general case where the signal depends on a set of unspecified parameters to be estimated (e.g., range, speed, bearing), the observation period is finite, and the colored noise is nonstationary.

4.2 Introduction

4.2.1 Background

Although generally confirmed in radar, the assumption of white noise is far from being universally acceptable. In particular, if the receiver noise can generally be considered white to a good approximation, the same cannot be said of the external noises, such as echoes from various landscape elements or clouds. Similarly, in sonar, sound reverberation phenomena in water can be described as colored noise added to the signal. Similarly again, spurious signals, whether from intentional jammers or interference, cannot be assumed to be white noise, whereas colored noise can provide a valid model for them.

4.2.2 Method

The following two cases should be considered:

1. The characteristics of the noise are known (known covariance); the optimum receiver can then be constructed, as described in this chapter.
2. The characteristics of the noise are unknown and possibly variable; the generalized maximum likelihood criterion must then be applied, in which the noise characteristics are first estimated and then used in

the optimum processing in known colored noise. These procedures, leading to processing known as adaptive, will be described in Chapter 5.

Throughout this chapter the theory of optimum estimation in Gaussian colored noise will be presented without specifying the nature of the estimated parameters (e.g., range, Doppler shift, angle) so that it can also be used in the processing of sensor arrays as well as in sonar situations, where the representation of the Doppler effect by a simple frequency shift is not always legitimate (see Chapter 1).

In order to clarify the physical interpretation of the different optimum receiver structures, the theory will first be presented in the time domain in the case of a nonstationary additive noise (covariance dependent on time) and of a finite observation period, using the approach in (3.1), then in the spectral domain in the case of stationary noise with an infinite observation period. Although less general, the latter situation enables the acquisition of a more intuitive understanding of the operation of the receiver, useful in practice when designing real systems.

4.3 Receiver Structure

4.3.1 Reception by Whitening

Keeping the discussion as general as possible, we shall again consider the case in which the parameters θ on which the received signal depends are not specified (as in Chapter 2). The received signal is then given by:

$$r(t) = s(t, \theta) + n(t) = z(t)e^{j\omega_0 t} \tag{4.1}$$

where

$$s(t, \theta) = \alpha a(t, \theta)e^{j\omega_0 t} \tag{4.2}$$
$$n(t) = b(t)e^{j\omega_0 t}$$

We have previously seen (Figure 2.1) that in the presence of white noise the optimum receiver performs, for each possible value θ_i of the vector parameter θ, matched filtering, quadratic detection, and normalization, and then chooses the parameter $\hat{\theta}$ corresponding to the maximum output. The corresponding diagram consists of a set of M parallel channels (M is the number of possible values of the vector parameter θ) of the type shown in Figure 4.1, followed by the search for a maximum (or by a comparison with a threshold, in the case of detection).

Let us consider the case in which the duration of observation is limited to the period $[T_i, T_f]$ and where the noise $b(t)$ consists of two uncorrelated components: a white part, $b_0(t)$, of covariance $N_0\delta(t-u)$, and a colored part, $b_c(t)$, nonstationary, of covariance $K_c(t, u)$. (The presence of the white part is necessary to eliminate the singular—and unrealistic—situations in which a spectral component of the signal is noise-free.)

$$K_b(t, u) = \mathrm{E}[b(t)b^*(u)] = K_c(t, u) + N_0\delta(t-u) \tag{4.3}$$

Let us assume that there is a filter, with impulse response $h_w(t, u)$, capable of whitening the noise, that is, such that:

$$b_w(t) = \int_{T_i}^{T_f} h_w(t, u)b(u)\,du \tag{4.4}$$

is a white noise.

If the response of this filter can be calculated (as a function of the input noise characteristics), then the problem can be reduced to that of reception in the presence of white noise, examined in the preceding chapters, by building the receiver according to Figure 4.2. This filter, placed on the reception channel, whitens the noise, but also distorts the received signal $a(t, \theta)$, which becomes:

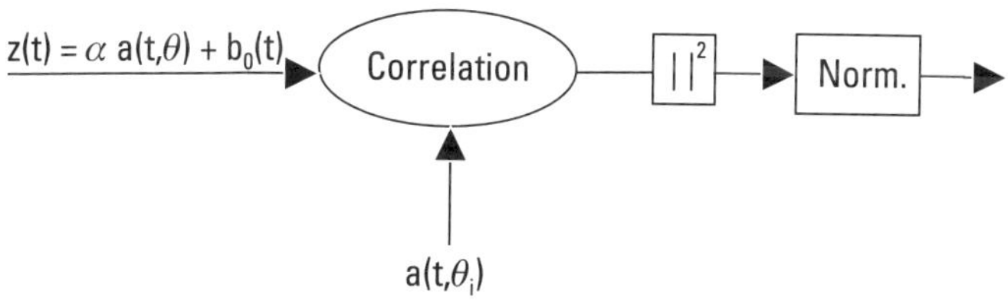

Figure 4.1 Optimum receiver in white noise.

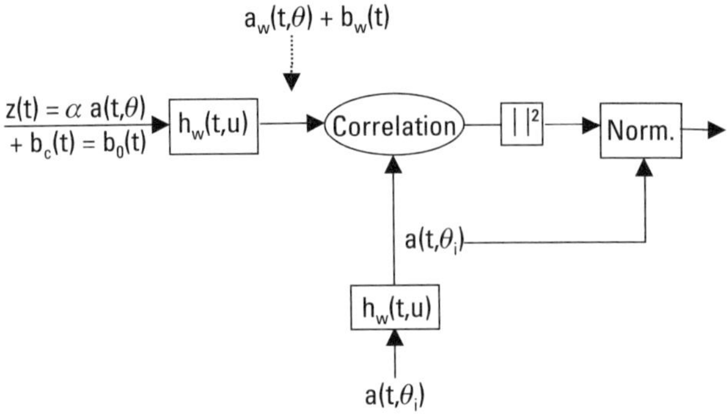

Figure 4.2 Optimum receiver in colored noise (whitening).

$$a_w(t, \theta) = \int_{T_i}^{T_f} h_w(t, u) a(u, \theta)\, du \tag{4.5}$$

To take this distortion into account, the same filter is inserted on the "replica" channel, transforming $a(t, \theta_i)$ into $a_w(t, \theta_i)$. We then face the problem of estimating the parameter θ in the presence of white noise $n_w(t)$, solved as in the preceding chapters by correlation, quadratic detection, normalization with respect to the power of the whitened replica $a_w(t, \theta_i)$, and determination of the maximum with respect to θ_i (Figure 4.2).

The noise whitening condition is given by:

$$\mathrm{E}[b_w(t) b_w^*(u)] = \delta(t, u) \text{ on interval } [T_i, T_f] \tag{4.6}$$

$$\Rightarrow \delta(t, u) = \mathrm{E}\left[\int_{T_i}^{T_f} h_w(t, x) b(x) b^*(y) h_w^*(u, y)\, dx\, dy\right]$$

$$\delta(t, u) = \int_{T_i}^{T_f} h_w(t, x) K_b(x, y) h_w^*(u, y)\, dx\, dy \tag{4.7}$$

The optimum processing therefore calculates:

$$r_i = \frac{\left| \iint\limits_{T_i}^{T_f} z(x) h_w(u, x) h_w^*(u, y) a^*(y, \theta_i)\, dx\, dy\, du \right|^2}{\int\limits_{T_i}^{T_f} a(x, \theta_i) h_w(u, x) h_w^*(u, y) a^*(y, \theta_i)\, dx\, dy\, du} \tag{4.8}$$

and chooses the value $\hat{\theta} = \theta_i$ of the parameter θ which corresponds to the highest value of r_i, knowing that h_w satisfies the whitening relation (4.7). To make these relationships clear, let:

$$Q_b(x, y) = \int\limits_{T_i}^{T_f} h_w(u, x) h_w^*(u, y)\, du \tag{4.9}$$

then r_i is given by:

$$r_i = \frac{\left| \iint\limits_{T_i}^{T_f} z(x) Q_b(x, y) a^*(y, \theta_i)\, dx\, dy \right|^2}{\int\limits_{T_i}^{T_f} a(x, \theta_i) Q_b(x, y) a^*(y, \theta_i)\, dx\, dy} : \hat{\theta}/r_i \text{ Max} \tag{4.10}$$

and the whitening condition (4.7) can be multiplied by $h_w^*(t, v)$ and integrated with respect to t:

$$h_w^*(u, v) = \int\limits_{T_i}^{T_f} h_w^*(t, v) h_w(t, x) h_w^*(u, y) K_b(x, y)\, dx\, dy\, dt$$

$$h_w^*(u, v) = \int\limits_{T_i}^{T_f} Q_b^*(v, x) K_b(x, y) h_w^*(u, y)\, dx\, dy$$

$$\text{over }]T_i, T_f[\; \delta(y - v) = \int\limits_{T_i}^{T_f} Q_b^*(v, x) K_b(x, y)\, dx \tag{4.11}$$

The same whitening argument can be applied to the estimator of α, denoted by $\hat{\alpha}$. We thus obtain:

$$\hat{\alpha} = \frac{\int_{T_i}^{T_f} z(x) Q_b(x, y) a^*(y, \hat{\theta})\, dx\, dy}{\int_{T_i}^{T_f} a(x, \hat{\theta}) Q_b(x, y) a^*(y, \hat{\theta})\, dx\, dy} \tag{4.12}$$

Equations (4.10), (4.11), and (4.12) define the optimum whitening receiver.

These three equations are demonstrated more precisely in Appendix 4B using the Karhunen-Loève expansion, whose main properties are summarized in Appendix 4A.

4.3.2 Reception by Subtraction of the Colored Part of the Noise

For a better understanding of this optimum receiver, it can be generated in a different way by simply manipulating (4.10), (4.11), and (4.12).

The form of (4.11) defining Q_b suggests that Q_b can be written in the same form as K_b, that is, from (4.3):

$$K_b(t, u) = N_0 \delta(t - u) + K_c(t, u)$$

Therefore let

$$Q_b^*(t, x) = \frac{1}{N_0} [\delta(t - x) - h_0(t, x)] \tag{4.13}$$

h_0 to be determined.

Substituting (4.13) and (4.3) in (4.11), an equation defining h_0 over $]T_i, T_f[$ is obtained:

$$N_0 h_0(t, u) + \int_{T_i}^{T_f} h_0(x, u) K_c(t, x)\, dx = K_c(t, u) \tag{4.14}$$

This equation is none other than the one defining the optimum filter for estimating $b_c(t)$, the colored part of the noise, when the total noise $b_0(t) + b_c(t)$ is observed; this result, conventional in estimation theory (Wiener-Hopf equation), is demonstrated in Appendix 4C.

The diagram of the resulting receiver is shown in Figure 4.3. The noisy signal received is first passed through a filter whose purpose is to subtract the best estimate of the colored part of the noise, then fed to the optimum receiver in white noise (correlation, detection, normalization), except that the normalization is not performed with respect to the power of the replica but with respect to the power of the whitened replica.

We shall see in Chapter 5 that this structure for the optimum receiver leads naturally to structures for adaptive receivers, for dealing with situations in which the noise characteristics are unknown.

4.4 Application to Spurious Echoes (or Jammers)

To illustrate the results of (4.10) and (4.11), let us apply them to the case in which the colored part of the noise is caused by a spurious point target, characterized by the value θ_I of the parameter, whose level I fluctuates according to a complex Gaussian distribution of standard deviation θ:

$$b_c(t) = I(t, \theta_I);\ \mathrm{E}|I|^2 = \sigma_I^2;\ \text{jammer energy} \tag{4.15}$$

$$K_c(t, u) = \mathrm{E}[b_c(t)b_c^*(u)] = \sigma_I^2 a(t, \theta_I)a^*(u, \theta_I) \tag{4.16}$$

The above results are applicable if the parameters θ_I and σ_I are assumed to be known. This is generally acceptable, since a spurious target is a problem

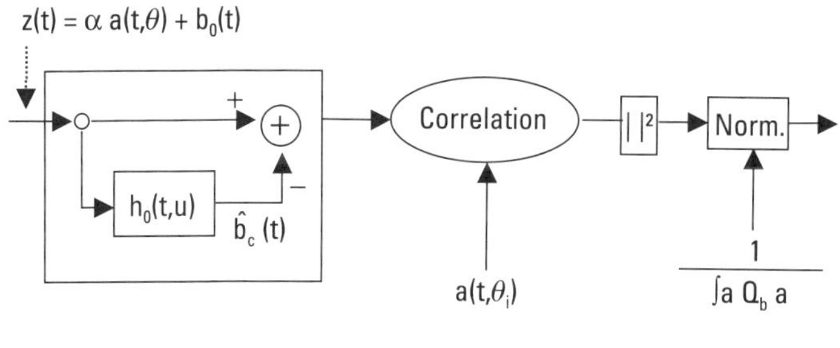

Figure 4.3 Optimum receiver by subtraction of the colored part of the noise.

only if it is relatively powerful, in which case its position and power can be measured with satisfactory accuracy.

The problem is therefore to estimate θ, the received signal being of the form:

$$z(t) = \alpha a(t, \theta) + b(t) \text{ (after demodulation by the carrier } e^{j\omega_0 t})$$

$$\text{with} \begin{cases} b(t) = Ia(t, \theta_I) + b_0(t) \\ K_b(t, u) = K_c(t, u) + N_0 \delta(t-u)\text{: } N_0\text{: noise spectral density} \\ K_c(t, u) = \sigma_I^2 a(t, \theta_I) a^*(u, \theta_I) \end{cases} \tag{4.17}$$

The signal is assumed to be normalized in energy: $\int |a(t, \theta)|^2 dt = 1$ Following (4.10), and letting:

$$g(x, \theta) = \int_{T_i}^{T_f} Q_b(x, y) a(y, \theta)\, dy \tag{4.18}$$

the optimum receiver calculates:

$$H = \frac{\left| \int z(x) g^*(x, \theta)\, dx \right|^2}{\int a(x, \theta) g^*(x, \theta)\, dx} \tag{4.19}$$

Let us determine $g(x, \theta)$ a priori assumed of the form:

$$g(t, \theta) = g_0 a(t, \theta) - g_I a(t, \theta_I) \tag{4.20}$$

where g_0 and g_I are to be determined.

From (4.11), we obtain:

$$\delta(t-u) = \int Q_b^*(u, y) K_b(y, t)\, dy$$

and

$$\delta(t-u) = \int K_b(t, y) Q_b(y, u)\, dy$$

since

$$K_b^*(y, t) = K_b(t, y) \Rightarrow Q_b^*(u, y) = Q_b(y, u)$$

Hence

$$a(u, \theta)\delta(t-u) = \int K_b(t, y) Q_b(y, u) a(u, \theta)\, dy$$

$$a(t, \theta) = \int K_b(t, y) g(y, \theta)\, dy \tag{4.21}$$

Substituting (4.20) in (4.21), we obtain:

$$\begin{aligned} a(t, \theta) = \sigma_I^2 \int a^*(y, \theta_I) a(t, \theta_I)[g_0 a(y, \theta) - g_I a(y, \theta)]\, dy \\ + N_0[g_0 a(t, \theta) - g_I a(t, \theta_I)] \end{aligned}$$

Multiplying by $a^*(t, \theta)$ and integrating leads to:

$$\begin{aligned} 1 = \sigma_i^2 g_0 |\chi(\theta, \theta_I)|^2 - \sigma_I^2 g_I \chi(\theta_I, \theta_I) \chi(\theta, \theta_I) \\ + N_0 g_0 - N_0 g_I \chi(\theta, \theta_I) \end{aligned}$$

where $|\chi(0, 0_I)|^2$ is the ambiguity function associated with the signal:

$$\chi(\theta, \theta_I) = \int a(t, \theta_I) a^*(t, \theta)\, dt$$

We can thus take:

$$\begin{cases} g_0 = \dfrac{1}{N_0} \\ g_I = \dfrac{\sigma_I^2}{N_0} \dfrac{\chi^*(\theta, \theta_I)}{N_0 + \sigma_I^2} \end{cases} \tag{4.22}$$

Consequently, the optimum receiver calculates:

$$H = \frac{\left| \int z(x) g^*(x, \theta)\, dx \right|^2}{\int a(x, \theta) g^*(x, \theta)\, dx} \tag{4.23}$$

where

$$g(x, \theta) = \frac{1}{N_0} a(x, \theta) - \frac{\chi^*(\theta, \theta_I)}{N_0 + \sigma_I^2} a(x, \theta_I) \frac{\sigma_I^2}{N_0} \tag{4.24}$$

Compared with the optimum receiver in white noise, the operation of correlation with the replica $a(t, \theta)$ that was performed in that case is supplemented by the subtraction of an estimate of the contribution of the interference to the result of this correlation (in amplitude and phase), and the subsequent normalization is similarly modified.

The preceding demonstration provides a theoretical basis for introducing, in a sensible way, the "cleaning" methods used in antenna processing, or more generally, in image processing. These methods consist of subtracting, usually by successive approximations, the estimated contribution of the spurious signals from each pixel in an image with interference. It appears that these methods, usually described as heuristic, are in fact a reasonable practical implementation of the optimum receiver in colored noise.

This jammer cancellation example will be examined in detail in Section 5.4, regarding the case of sampled signals. The performance of the optimum receiver will then be examined.

4.5 Stationary Colored Noise and Infinite Observation Time

For a better understanding of the principle of reception in the presence of colored noise, it is worth analyzing the operation of the receiver in the case of a stationary noise and infinite observation time, a situation in which discussion can be based on power spectra. We shall first examine the case of reception by whitening, then by subtraction of the colored part of the noise. Finally, the gain in signal-to-noise ratio provided by this processing will be evaluated.

4.5.1 Whitening

Let $N(f)$ be the power spectral density of the noise:

$$N(f) = N_c(f) + N_0 \tag{4.25}$$

The diagram of processing by whitening is shown in Figure 4.4 (equivalent to Figure 4.2 in the nonstationary case). The whitening condition is written:

$$|h_w(f)|^2 N(f) = 1 \tag{4.26}$$

A schematic interpretation is given in Figure 4.5, which illustrates the way in which the filters affect the spectra of the signal, the noise, and the replica, respectively. The operation carried out by the receiver is written as:

$$\text{Max}/\theta_i = \frac{\left|\int z(f)h_w(f)h_w^*(f)a^*(f, \theta_i)\,df\right|^2}{\int a(f, \theta_i)h_w(f)h_w^*(f)a^*(f, \theta_i)\,df} \tag{4.27}$$

This operation can also be written:

$$\text{Max}/\theta_i = \frac{\left|\int z(f)q_n(f)a^*(f, \theta_i)\,df\right|^2}{\int a(f, \theta_i)q_n(f)a^*(f, \theta_i)\,df} \quad \text{where } q_n(f) = |h_w(f)|^2 \tag{4.28}$$

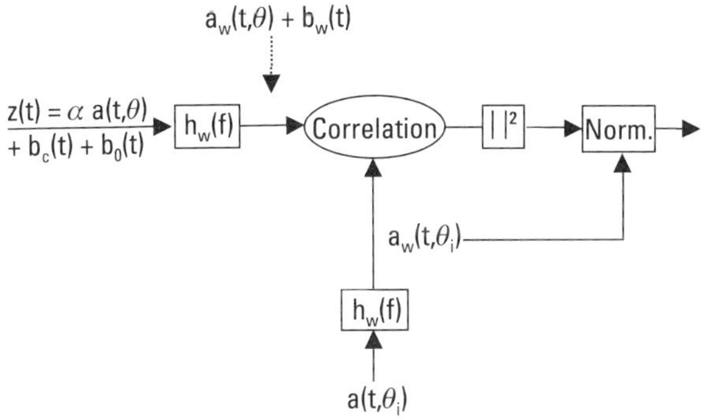

Figure 4.4 Processing by whitening (stationary case).

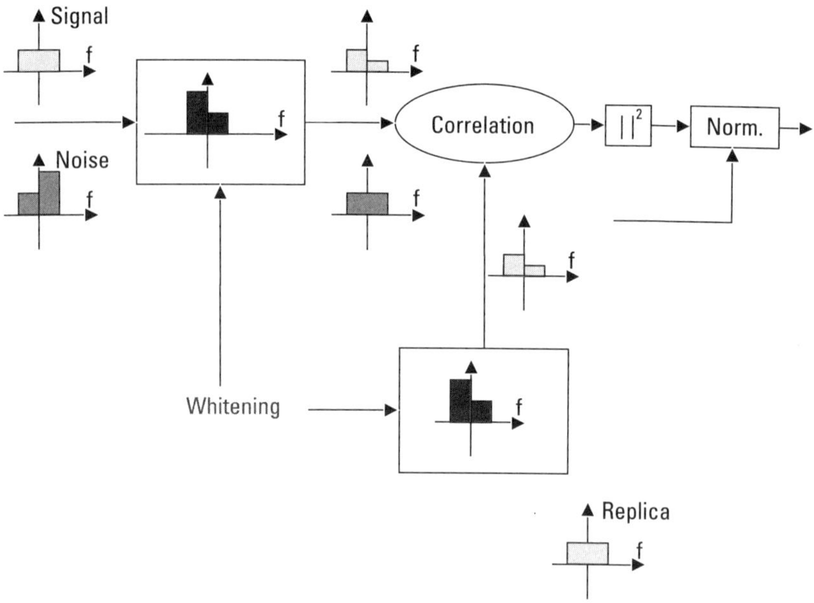

Figure 4.5 Interpretation of processing by whitening.

The diagram of the corresponding processing is shown in Figure 4.6 and illustrated by Figure 4.7. It can be seen that the filter $q_n(f)$ overwhitens the noise. The interpretation of this filter as an estimator-subtractor will lead to the subtraction approach in Section 4.5.2.

4.5.2 Subtraction of the Colored Part of the Noise

As in Section 4.2.2, let us now look for a filter of the same form as $N(f)$:

$$N(f) = N_0 + N_c(f)$$

$$q_n(f) = \frac{1}{N_0}[1 - h_0(f)] \tag{4.29}$$

$h_0(f)$ to be determined.

From (4.28):

$$h_0(f)[N_c(f) + N_0] = N_c(f) \tag{4.30}$$

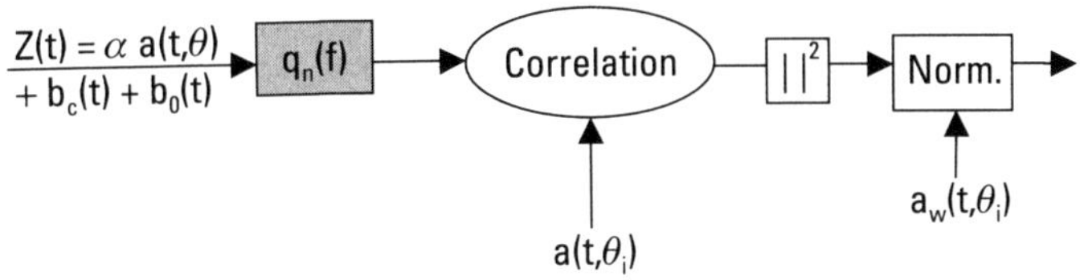

Figure 4.6 Processing by subtraction of colored noise (stationary case).

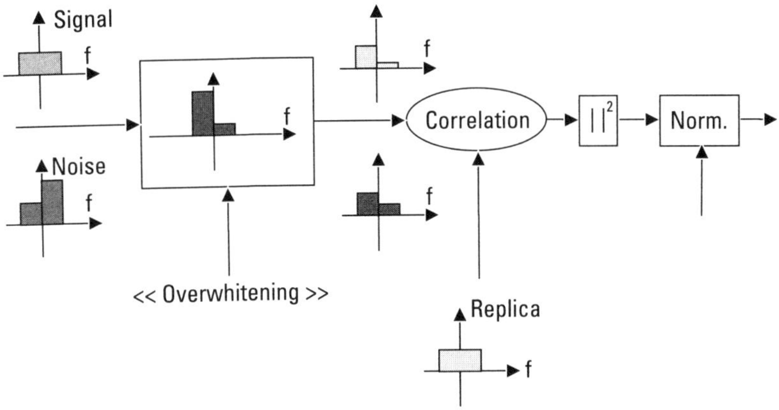

Figure 4.7 Interpretation of processing by subtraction of colored noise.

It is easy to check that h_0 is the optimum filter for estimating $n_c(t)$ in the presence of $n_0(t)$. In order to shorten the calculations, this is confirmed by using the minimum variance criterion.

Given $b(t) = b_c(t) + b_0(t)$, the problem consists of determining h_0 so that:

$$\hat{n}_c(t) = \int h_0(t-u)n(u)\,du$$

where

$$\mathrm{E}\left[\int |\hat{b}_c(t) - b_c(t)|^2 dt\right] \min/h_0$$

$$\Rightarrow \mathrm{E}\left[\int |\hat{b}_c(f) - b_c(f)|^2 df\right] \min/h_0$$

$$\Rightarrow \mathrm{E}[|\hat{b}_c(f) - b_c(f)|^2] \min/h_0$$

$$\Rightarrow \mathrm{E}[|h_0(f)b(f) - b_c(f)|^2] \min/h_0$$

$$\Rightarrow |h_0(f)|^2 N(f) + N_c(f) - 2\,\mathrm{Re}\{h_0(f)\mathrm{E}[b(f)b_c^*(f)]\} \min/h_0$$

$$\Rightarrow |h_0(f)|^2 N(f) + N_c(f) - 2\,\mathrm{Re}\{h_0(f)N_c(f)\} \min/h_0$$

$$\Rightarrow h_0(f)N(f) = N_c(f)$$

(by nulling of the gradient).

The expected result is therefore obtained: h_0 is the optimum filter for estimating the colored part of the noise in the presence of the white part.

It is worth noting that this subtraction of the best estimate of the colored part of the noise, performed by the filter of transfer function $q_n(f) = [1 - h_0(f)]\dfrac{1}{N_0}$, does not output a white noise (which is consistent with the fact that, as we have seen, $q_n(f)$ overwhitens the noise), except when the ratio N_0/N_c tends to infinity. The spectral density of the noise at the filter output is in fact:

$$[N_c(f) + N_0][q_n(f)]^2 = [N_c(f) + N_0]\frac{1}{[N_c(f) + N_0]^2}$$

$$= \frac{1}{N_c(f) + N_0}$$

This is explained by the fact that the spectral density of $\hat{b}_c(t)$ is itself different from $N_c(f)$, since it is equal to:

$$[N_0 + N_c(f)]h_0^2(f) = [N_0 + N_c(f)]\frac{N_c(f)^2}{[N_0 + N_c(f)]^2} \neq N_c(f)$$

4.5.3 Optimum Processing Gain

The gain provided by optimum processing, in comparison with conventional processing, can be estimated in this stationary situation. Let:

- $(S/N)_0$ be the signal-to-noise ratio after correlation, without whitening;
- $(S/N)_w$ be the signal-to-noise ratio after correlation, with whitening.

The calculations are straightforward, since the spectral density of a noise at the output of a filter is equal to the input spectral density multiplied by the transfer function of the filter. Letting $A(f)$ be the power spectrum of the signal $\alpha a(t, \theta_i)$, the output signal-to-noise ratio is written:

$$(S/N)_0 = \frac{\left[\int A(f)\,df\right]^2}{\int A(f)N(f)\,df}$$

and

$$(S/N)_w = \int \frac{A(f)}{N(f)}\,df$$

The gain provided by the adaptation to the noise color is therefore:

$$G = \frac{\int A(f)N(f)\,df \int \frac{A(f)}{N(f)}\,df}{\left[\int A(f)\,df\right]^2}$$

From Schwarz's inequality, we have $G \geq 1$, equality being obtained for $N(f) = \text{Const.}$ (white noise).

This formula enables easy evaluation of the value of the processing in practical cases for noise of given spectral density $N(f)$. The same type of argument can be used to study the loss experienced when, because of inadequate knowledge of $N_c(f)$, the filter $h_0(f)$ is not the optimum filter.

4.6 Discussion

The preceding demonstrations have enabled the building of different receivers optimized in terms of maximum likelihood. The demonstration given in

Appendix 4B alone is sufficient to define the optimum receiver in the presence of Gaussian additive noise. However, the practical and theoretical importance of the whitening concept justifies the description of other forms of receiver that are open to more intuitive interpretations and are therefore easier to extrapolate to practical cases, as we shall see in Chapter 5.

At this stage, it is probably useful to assess the tools used so far. This chapter has shown that it is possible, using relatively simple intuitive arguments, to build optimum receivers (nevertheless based on the optimum receiver in white noise). This can be considered to be satisfactory, since it shows that the theoretical assumptions made (maximum likelihood as criterion, for example) lead to results that are confirmed by common sense. This could also be considered embarrassing, in that relatively general assumptions and relatively sophisticated demonstrations give the same results as those obtained by the straightforward reasoning of an electronics engineer adjusting filters to the received power spectra.

In fact, the main limitation of the above results appears to be a consequence of the Gaussian assumption, which limits signal exploitation to its second-order characteristics (correlations and spectra), to which intuitive arguments can be applied. It is also the Gaussian assumption that makes the results obtained by the maximum likelihood and by least squares identical (typically power-based argument). Remember that it is also this Gaussian assumption that explains the central role of correlation, as noted in Section 3.1. However, the preceding demonstrations have enabled the optimum receivers to be constructed properly in the Gaussian case and have the advantage of being directly transposable to the case of other noise distributions. They have also shown the equivalence between the different possible approaches (whitening or subtraction) and enabled the normalizations, whose role is essential, to be specified.

Moreover, it should be noted that the optimum receivers in the presence of colored noise that we have constructed assume that the covariance function of the noise alone is known. As this assumption is not always valid in practice, Chapter 5 will show how adaptive processing can overcome this limitation.

Appendix 4A: Karhunen-Loève Expansion

4A.1 The Problem

We are looking for an expansion of $x(t)$, a random process over $[0, T]$, on a basis of orthogonal functions $\phi_i(t)$ [see (2.1) and (2.2)]:

$$x(t) = \lim_{N \to \infty} \sum_{i=1}^{+\infty} x_i \phi_i(t) \quad t \in [0, T]$$

satisfying the following conditions:

1. $\displaystyle\int_0^T \left| x(t) - \sum_{i=1}^{N} x_i \phi_i(t) \right|^2 dt$ minimum
2. The x_i are not intercorrelated
3. $\displaystyle\lim_{N \to \infty} \mathrm{E}\left[\left| x(t) - \sum_{i=1}^{N} x_i \phi_i(t) \right|^2 \right] = 0, \ t \in [0, T]$

Condition 1: $I = \displaystyle\int_0^T \left| x(t) - \sum x_i \phi_i(t) \right|^2 dt$ minimum

$$\frac{\delta I}{\delta x_i} = 0 \Rightarrow x_i = \int_0^T x(t) \phi_i^*(t)\, dt$$

Condition 2: Let us assume that $\mathrm{E}[x_i] = 0$ (generalization is straightforward)

$$\mathrm{E}(x_i x_j^*) = \lambda_i \delta_{ij} = \int_0^T \phi_i^*(t) \int_0^T K_x(t, u) \phi_j(u)\, dt\, du \ \forall \ i, j$$

$$\Rightarrow \int_0^T K_x(t, u) \phi_j(u)\, du$$

orthogonal to any other vector of the orthonormal base

$$\Rightarrow \lambda_j \phi_j(t) = \int_0^T K_x(t, u) \phi_j(u)\, du$$

The functions $\phi_j(t)$ are thus the solutions of an integral equation (or of a search for eigenfunctions).

Condition 3: See properties of the integral equation.

4A.2 Properties of the Integral Equation

- $\exists$ at least one solution $\phi(t)$ with $\lambda \neq 0$.
- The eigenfunctions corresponding to two distinct eigenvalues are orthogonal.
- $\exists$ at most a countable infinity of eigenvalues and they are all bounded.
- We have $K_x(t, u) = \sum \lambda_i \phi_i(t)\phi_i^*(u)$, $t, u \in [0, T]$ with uniform convergence.
- If $K_x(t, u)$ is definite positive, $\{\phi_i\}$ = orthonormal base; if $K_x(t, u)$ is definite nonnegative, the ϕ_i can be supplemented to obtain an orthonormal base.
- $\mathrm{E}\left[\int_0^T |x(t)|^2 dt\right] = \int_0^T K_x(t, t)\, dt = \sum_i \lambda_i.$
- Let $\zeta_N(t) \underline{\Delta} \mathrm{E}\left\{\left|x(t) - \sum_{i=1}^{N} x_i \phi_i(t)\right|^2\right\}$: average truncation error.

One can show: $\zeta_N = K_x(t, t) - \sum \lambda_i |\phi_i(t)|^2$ converges uniformly toward 0 $\Rightarrow$ satisfies condition 3.

Value: transforms the study of a random *process* into that of a set of independent random *variables*.

Appendix 4B: Optimum Receiver in Colored Noise

Consider the problem of the optimum estimation (in terms of maximum likelihood) of the parameter vector θ from the received signal:

$$z(t) = \alpha a(t, \theta) + b(t) \quad t \in [T_i, T_f]$$

$b(t)$: centered Gaussian noise of covariance $K(t, u)$

$$\mathrm{E}[b(t)b^*(u)] = K(t, u)$$

α: complex amplitude of the signal.

The Karhunen-Loève expansion (see Appendix 4A) associated with the noise $b(t)$ has base function $\phi_i(t)$ verifying:

$$\lambda_i \phi_i(t) = \int_{T_i}^{T_f} K_x(t, u)\theta_i(u)\,du$$

with

$$K_x(t, u) = \sum \lambda_i \phi_i^*(t)\phi_i(u)$$

$$z_i = \int_{T_i}^{T_f} \phi_i^*(t)z(t)\,dt$$

$$a_i(\theta) = \int_{T_i}^{T_f} \phi_i^*(t)a(t, \theta)\,dt$$

$$b_i = \int_{T_i}^{T_f} \phi_i^*(t)b(t)\,dt$$

where b_i is Gaussian (as linear form of a Gaussian process) of variance λ_i (straightforward verification).

Calculate the likelihood function for the set of independent observations z_i (since the b_i are independent because they result from a Karhunen-Loève decomposition), for $i = 1, \ldots, K$

$$\begin{aligned}\log\Lambda_K &= \log\prod_{i=1}^{K} p_{z_i/\theta}(z_i)\\ &= -\sum_{i=1}^{K}(z_i - \alpha a_i(\theta))^* \frac{1}{\lambda_i}(z_i - \alpha a_i(\theta)) + \text{Const. }(\theta)\end{aligned}$$

that is

$$\log\Lambda_K = -\sum_{i=1}^{K}\frac{z_i^* z_i}{\lambda_i} + 2\ \mathrm{Re}\sum_{i=1}^{K}\alpha\frac{z_i^* a_i}{\lambda_i} - \sum_{i=1}^{K}|\alpha|^2\frac{a_i^* a_i}{2\lambda_i} + 2\ \mathrm{Const.}\ (\theta)$$

As $\frac{z_i^* z_i}{\lambda_i}$ does not depend on θ, we obtain:

$$\log\Lambda_K = 2\ \mathrm{Re}\sum_{i=1}^{K}\alpha\frac{z_i^* a_i}{\lambda_i} - \sum_{i=1}^{K}|\alpha|^2\frac{a_i^* a_i}{2\lambda_i} + \mathrm{Const.}\ (\theta)$$

Returning to functions of time and letting K tend to infinity, the function to maximize $/\theta,\alpha$ is written:

$$2\ \mathrm{Re}\ \alpha\iint z^*(x)Q(x, y)a(y, \theta)\,dx\,dy \qquad (4B.1)$$
$$-|\alpha|^2\iint a^*(x, \theta)Q(x, y)a(y, \theta)\,dx\,dy$$

where

$$Q(x, y) = \sum_i \frac{\phi_i^*(x)\phi_i(y)}{\lambda_i}$$

It is easy to verify that:

$$\int_{T_i}^{T_f} Q^*(x, y)K(y, t)\,dy = \delta(x - t)$$

Considering (4B.1) as the beginning of a squared modulus, it can be written that the following expression must be minimized with respect to α and θ:

$$\left\{\iint a^*Qa\right\}\left\{\left|\alpha^* - \frac{\iint z^*Qa}{\iint a^*Qa}\right|^2 - \left|\frac{\iint z^*Qa}{\iint a^*Qa}\right|^2\right\}$$

where the integration variables have been omitted for simplicity.

Hence, the optimal receiver is obtained (the integrals being calculated between T_i and T_f):

$$\hat{\alpha}^* = \frac{\iint z^*(x)Q(x, y)a(y, \hat{\theta})\,dx\,dy}{\iint a^*(x, \hat{\theta})Q(x, y)a(y, \hat{\theta})\,dx\,dy}$$

$$\hat{\theta}: \mathrm{Max}_{/\theta} = \frac{\left|\iint z^*(x)Q(x, y)a(y, \theta)\,dx\,dy\right|^2}{\iint a^*(x, \theta)Q(x, y)a(y, \theta)\,dx\,dy}$$

where again

$$\int Q^*(x, y)K(y, t)\,dy = \delta(x - t)$$

The optimum receiver structure, defined in Section 4.3.1, is thus obtained by a more precise proof.

Appendix 4C: Estimation of a Random Signal in the Presence of White Noise—Wiener-Hopf Equation

Let $r(t) = n_c(t) + n_0(t)$, $t \in [T_i, T_f]$ where $n_0(t)$ Gaussian noise, $\mathrm{E}[n_0(t)n_0^*(u)] = K_0(t, u) = N_0\,\delta(t - u)$ and $n_c(t)$ Gaussian, independent of $n_0(t)$: $\mathrm{E}[n_c(t)n_c^*(u)] = K_c(t, u)$.

Let $\hat{n}_c(t)$ be the estimator of $n_c(t)$ in terms of the *maximum a posteriori* criterion (which, for estimation of random parameters or signals, plays the role of the maximum likelihood for nondeterministic unknown signals or parameters):

$$\hat{n}_c(t) \text{ maximizing } p[n_c(t)/r(t)] \text{ with respect to } n_c(t).$$

Let $\phi_i(t)$ be the base functions of the Karhunen-Loève expansion associated with $K_c(t, u)$ (and also with $K_0(t, u)$ since the equation defining the Karhunen-Loève functions associated with white noise is verified for any orthonormal function base):

$$\lambda_i \phi_i(t) = \int_{T_i}^{T_f} K_c(t, u)\phi_i(u)\,du$$

We obtain

$$r_i = \int_{T_i}^{T_f} r(u)\phi_i^*(u)\,du$$

$$n_{ci} = \int_{T_i}^{T_f} n_c(u)\phi_i^*(u)\,du$$

$$n_{0i} = \int_{T_i}^{T_f} n_0(u)\phi_i^*(u)\,du$$

where

$$\begin{aligned} \mathrm{E}(n_{ci}n_{cj}^*) &= \lambda_i \delta_{ij} \\ \mathrm{E}(n_{oi}n_{oj}^*) &= N_0 \delta_{ij} \\ \mathrm{E}(n_{oi}n_{cj}^*) &= 0 \end{aligned}$$

and the function to maximize is then written (independent samples):

$$\prod_{i=1}^{\infty} p(n_{ci}/r_i) = \prod_{i=1}^{\infty} p(r_i/n_{ci})p(n_{ci})/p(r_i)$$

Maximizing this function with respect to n_{ci} is equivalent to maximizing the following with respect to n_{ci}:

$$\sum_i \frac{|r_i - n_{ci}|^2}{N_0} - \sum_i \frac{|n_{ci}|^2}{\lambda_i}$$

By differentiating with respect to n_{ci} we obtain:

$$-\frac{r_i}{N_0} = \frac{n_{ci}}{N_0} + \frac{n_{ci}}{\lambda_i}$$

hence the estimator of n_{ci}:

$$\hat{n}_{ci} = \frac{\lambda_i r_i}{N_0 + \lambda_i}$$

which can also be written:

$$\hat{n}_c(t) = \int_{T_i}^{T_f} h(t, \tau) r(\tau)\, d\tau$$

with

$$h(t, \tau) = \sum_i \frac{\lambda_i}{\lambda_i + N_0} \phi_i(t) \phi_i^*(\tau)$$

It is easy to verify that $h(t, \tau)$ then verifies the equation:

$$\int_{T_i}^{T_f} h(t, \tau)[K_c(\tau, u) + K_0(\tau, u)]\, d\tau = K_c(\tau, u)$$

5

Adaptive Processing

5.1 Summary

In the real world, external noise and interference have characteristics that are often difficult to determine accurately and, furthermore, that change over time. It is then no longer possible to assume that the characteristics of the colored noise that is used to model them are known. Two solutions can then be considered.

1. Try to build the receiver that most effectively extracts the signal, based on noise characteristics that are unknown (or modeled with a high degree of tolerance: zero-average noise, for example). This is the so-called *nonparametric* approach, which states that "all that is not signal is noise." Subject to various restrictions, this approach can sometimes be used—and is in fact used for some radar or sonar processing—after conventional optimum processing. Due to its specificity, this approach will not be discussed here (see [1]).
2. Try to build the receiver that, based on the received signals, measures the noise characteristics and uses the optimum processing for these noise characteristics in order to extract the useful signal.

The latter approach leads directly to adaptive processing, as the noise characteristics are continually assessed, which enables the receiver to be adapted to a variable environment (or at least to an environment that varies slowly). Moreover, we shall see that this adaptivity is provided by a closed-loop system, which also enables the receiver to be adapted to various imperfections of the antenna or the transmitter, therefore resulting in better performance than that achieved when these elements are assumed to be perfect.

The noise characteristics can be easily measured if an auxiliary reception channel is available that only receives external noise (noise-only reference). This is the case, for instance, in radar or in active sonar while transmission is cut off (pulse system); the signals received at this moment obviously contain only noise. This is also the case if an auxiliary antenna with very low directivity is available. As a result, such an antenna picks up all the interference, with an interference/useful signal power ratio much greater than that at the output of the directional main antenna. Under these circumstances, as an estimate of the colored part of the noise is available, it is natural to adopt the second approach to optimal processing described in Chapter 4, which consists of subtracting this colored part of the noise. However, as the noise reference channel is distinct from the signal channel, the differences in parameters such as antenna gain and sensitivity between the two channels must be compensated for by adequately filtering this reference channel before subtracting it from the signal channel.

The problem is then to calculate this filter—whose characteristics change over time according to the positions and power levels of the noise sources—in real time. We then show that this filter must in fact minimize the average power that remains after subtraction; this can be explained intuitively, because the reference noise, once filtered, should ideally be equal to the colored noise present on the useful signal channel. As it is further assumed that the useful signal and the noise are decorrelated, the subtraction cannot reduce the signal power and can therefore only—ideally—suppress the colored noise, while leaving the signal unchanged. This equivalence between estimation-subtraction of the colored part and minimization of the average power of the signal obtained after subtraction serves as the basis for adaptive processing, as it provides a closed-loop system, which adapts the filter coefficients as a function of the observed power. The power after subtraction is a direct quality criterion for the filter inserted in the reference channel.

Then all that is needed is to specify the architecture of the filter—which, for the sake of simplicity, is generally a transversal filter—in order to deduce the algorithm for computing its coefficients. An adaptive algorithm can then be defined, by using a gradient method to minimize not the average energy but the instantaneous energy. This algorithm (which is extremely simple as it requires only p multiplications, where p is the filter order, at each received signal sample) converges asymptotically, in the case of stationary noise, toward the optimum processing.

In parallel, it is interesting to assess the influence of the presence of signal on the noise reference channel. In this processing by subtraction, it is clear that when a signal is present on this reference channel (e.g., in the case where this channel is provided by a low-directivity auxiliary antenna), the consequence will be an attenuation of the output signal. The theoretical analysis confirms this tendency, as it shows that the signal-to-noise ratio after subtraction is simply equal to the reciprocal of the signal-to-noise ratio on the noise reference channel. This result clearly illustrates the importance of the quality of the noise reference for the final performance.

Another receiver architecture can also be used, based on the optimum receiver by whitening described in Chapter 4; the signal channel is filtered so that the noise at the filter output is white. The filter characteristics are then calculated, assuming that the useful targets are sufficiently scattered for the signal received from bearings or ranges close to the observed bearing and range to consist essentially of noise. This approach is being adopted in Doppler processing, using a "lattice" filter structure that has the advantage, when adaptation is performed by a gradient algorithm, of converging more rapidly toward the optimum filter, owing to a lower correlation between the individual coefficients.

The analysis of the optimum spatial processing of a sensor array—that is, of the processing of the N complex values recorded, at a given time, by N sensors—provides a better understanding of the operation and limitations of colored noise processing. As we have seen in Chapter 4, jammer suppression is performed by subtraction of its contribution to the target. The analysis further shows that the signal-to-noise ratio (jammer included) at processing output is zero when the target is in the jammer direction—which is normal, as nothing then distinguishes the target from the jammer—and that the presence of the jammer significantly degrades target location accuracy when the angular deviation between them is less than the intrinsic directivity of the sensor array.

Up to now, our arguments assumed that the noise characteristics were known a priori, or more or less accurately measured by various processes (auxiliary antenna). We can also state directly the problem of estimating the parameters of the target in the absence of noise-only reference. The problem then consists of estimating the correlation matrix of the noise alone, which is necessary to build the whitening filter, from the signal + noise correlation matrix, which is the only one accessible in practice. A relatively natural approach consists, for each value of the parameters, in:

- Assuming that a target with these parameters is present;
- Under this assumption, estimating the amplitude and phase of this target;
- Deducing the value of the noise-only correlation matrix, by subtracting the signal correlation matrix (estimated above) from the measured signal + noise matrix;
- Implementing the optimum parameter estimation processing, described in the previous chapter.

The detailed analysis of this procedure shows that it is sufficient to implement the first two steps of this approach. As it is initially assumed that there is a target in each direction, the subsequent estimation of the optimum direction can only lead to indeterminacy. Consequently, this approach simply results in delivering, for each position, the estimated amplitude and phase of the target, if any.

This approach leads to a target location procedure with a simple expression [see (5.87)], generally referred to as *maximum likelihood procedure* (neglecting the generalization made to take into account the absence of noise-only reference). Applied to the problem of angle-error measurement in the presence of external noise, this procedure leads to a simple but effective modification of the relation giving the estimated direction (described in Chapter 3).

To define an adaptive processing, still in the absence of a noise-only reference, we first show that the maximum likelihood processing in the absence of a noise-only reference is equivalent to a filtering process seeking to minimize the output power, under the constraint of letting any useful signal pass. This equivalence then enables us, by specifying the filter architecture, to determine the rule for calculating these coefficients in an adaptive form by taking an approach very similar to that which had been adopted when using an auxiliary noise reference. Further, we also show that these two approaches—minimization under the constraint of letting the signal pass and minimization after forming a reference channel and then subtracting this channel from the received signal—are strictly equivalent when processing an array of sensors.

This proof enables us to "close the loop" in the absence of a noise-only reference by establishing the equivalence between the following various approaches considered in this chapter:

- Subtraction of a correctly filtered noise reference (made up of the received noisy signals);

- Whitening;
- Maximum likelihood (generalized);
- Constrained minimization.

As shown by the examples discussed in this chapter, this equivalence enables us to select, in each special case, the approach that allows the best use to be made of the knowledge of the colored noise characteristics and minimum complexity processing to be defined, which is obviously a priority goal.

Implicit in the definition of these algorithms was the assumption of a perfect knowledge of the characteristics of the receiving channels (elementary antennas, amplifiers, and mixers) in amplitude and phase. Even when special care is taken in the design and calibration of the receiving channels, this assumption is not usually true to much better than within 1° in phase and 0.5 dB in amplitude. Such errors are sufficient to annihilate much of the signal-to-jammer improvement expected from adaptive processing, since the subtraction of the estimated jamming interference will be incorrect when the channel gain is not precisely known. Careful analysis shows that the effect of calibration errors mainly consists of a signal attenuation, leaving the jammer suppression efficiency mainly unchanged, by virtue of the closed-loop adaptation attempting to minimize the total output power.

Another limitation of the adaptive algorithms occurs in the presence of coherent signals coming from different directions; in this situation (common, for instance, in the presence of convoys or raids, or due to multipath effect, as described in Chapter 6), the subtraction of estimated unwanted signals would cancel the useful signals. Moreover, this coherence actually prevents an efficient estimation of the interfering signals to be performed, since it can be shown that the correlation matrix becomes singular. An efficient method of operation in such situations is called *spatial smoothing* and can be thought of as a spatial dithering of the array, decorrelating the signals from different directions by superposing a kind of artificial Doppler effect, depending on the direction of arrival.

The comparison between the approach used in reception by active sonars and radars and those used by passive listening systems (where, except for multipath effects, the sources can be assumed to be incoherent) is instructive, as it gives a better grasp of the influence of the assumption made (in active mode) according to which we have a priori knowledge of the form of the transmitted signal. It amounts, in a way, to moving from a parametric situation (known form, unknown parameters) to a nonparametric situation (unknown form). Although the extreme diversity, already emphasized above, of the situations experienced in passive listening prevents us from dealing with this point in entirely general terms, it is nevertheless useful to compare various commonly proposed procedures for locating the sources.

The usual approach consists of operating within the frequency domain, that is, determining the direction of origin of each spectral component of the signals picked up by the sensor array. A heuristic approach, tailored to each special situation, then enables the useful received signal to be reconstructed through summation of its frequency components, for each direction of arrival.

Three types of problems must be distinguished in passive listening:

1. Detection of the presence of a signal in each direction;
2. Estimation of the directions of arrival of useful signals;
3. Extraction of the signal coming from each direction, for the purpose of later identification of the transmission source.

The maximum likelihood procedure previously described—whether in adaptive form or not—enables the latter problem to be solved by minimizing, for each observed direction, the power at the output of a filter, under the constraint of letting the signal from this direction pass. When the situation is known to consist of multiple sources, however, this procedure does not make use of this "multiple source" information, as it considers the received signals to be representing one useful source and colored noise.

A more appropriate approach then determines, based on the received signals, which is the most probable point source distribution. This approach leads to the so-called spectral decomposition procedure [such as MUltiple SIgnal Classification (MUSIC)], which, although its performance in angular resolution is indeed better, does not, however, allow reconstruction of the useful signal from each source. Through an analysis of the eigenvalues of the correlation matrix of the signals recorded at each frequency on the sensors, this procedure enables the direction and power of the sources, at each frequency, to be obtained.

Lastly, in the case where only two sensors are used, a so-called generalized correlation procedure—based on the maximum likelihood principle, with the signals assumed random and of unknown shape—enables the sources to be located by making the best use of their spectral characteristics. This procedure consists of obtaining the delay applied to the signal received by one of the sensors, which maximizes its correlation with the signal received by the other sensor after filtering, whose purpose is to minimize the influence of the spurious noise.

This chapter shows how, by taking into account the poorly known characteristics of the noise (adaptive receivers) and then of the signals (passive listening), the structure of the receivers can be modified or even redesigned. The following chapters will expand this analysis even further, by attempting to take into account the specific nature of the received signals.

5.2 Introduction

The previous chapters have enabled us to build optimum receivers based on the assumption that the correlation function of the noise was known. In practice, this assumption is rarely valid, either because no measurement of this function has been made, or because the noise is not strictly stationary (motion of the radar or noise sources).

Three different approaches can then be considered:

1. Define the optimum receiver for an unknown noise correlation function; with no additional assumption, this approach is not possible, as the amount of information available does not enable the signal to be distinguished from the noise.
2. Define a receiver whose architecture is independent of the noise correlation function; this is the nonparametric approach, which is beyond the scope of this discussion but has a number of practical and theoretical advantages.
3. Define the optimum receiver after an estimation of the correlation function; this approach, which leads to adaptive processing, is the one we shall now describe.

5.3 Adaptive Noise Suppression

5.3.1 Receiver Structure

The problem therefore consists of estimating θ from the received signal $z(t)$ over a time interval $[T_i, T_f]$

$$z(t) = \alpha a(t, \theta) + b(t) \tag{5.1}$$

(after demodulation by the carrier $e^{j\omega_0 t}$) where $b(t)$ is a colored noise with zero mean:

$$b(t) = b_0(t) + b_c(t) \tag{5.2}$$

The respective covariances of $b_0(t)$ and $b_c(t)$ are:

$$\begin{aligned} E[b_0(t)b_0^*(t+\tau)] &= K_0(t, \tau) = N_0\delta(t-\tau) \\ E[b_c(t)b_c^*(t+\tau)] &= K_c(t, \tau) \end{aligned} \tag{5.3}$$

The optimum receiver by subtraction of the colored part of the noise was built in Section 4.3.2. The diagram of this receiver is shown in Figure 5.1. In this diagram, $h_0(t, u)$ is the impulse response of the filter yielding optimum estimation of the colored part $b_c(t)$ in the presence of white noise $b_0(t)$. Note that consequently, $h_0(t, u)$ provides an estimation $b_c(t)$ that is incorrect due to the presence of the signal $a(t, \theta)$: this influence will be examined later. The so-called Wiener-Hopf equation, (4.14), is satisfied by $h_0(t, u)$:

$$K_c(t, \tau) = \int_{T_i}^{T_f} K_b(t, x)h_0(x, u)\,dx \tag{5.4}$$

and the normalization indicated in the diagram is a division by:

$$S_i = \frac{1}{N_0}\int_{T_i}^{T_f} a(t, \theta_i)[\delta(t-x) - h_0^*(t, x)]a^*(x, \theta_i)\,dt\,dx \tag{5.5}$$

(S_i can also be interpreted as the energy of the whitened replica, as indicated in Section 4.3.1.)

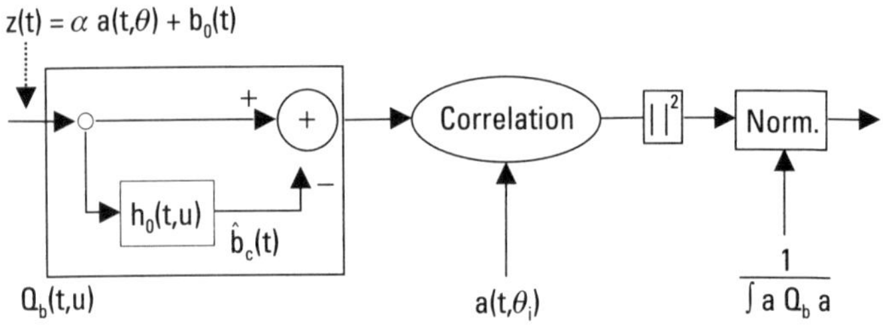

Figure 5.1 Optimum receiver by subtraction of the colored part of the noise.

It is clear that the better the estimate of the colored part of the noise, the better the receiver performance. Let us then consider the case of colored noise sources external to the system, such as interference, reverberation, and environment clutter. A simple way of estimating these types of noise consists of using a sensor specially designed for this purpose. For example, in radar, a low-directivity auxiliary antenna is placed next to the main antenna so that the former, known as the reference antenna, collects all the external noise and only a fraction of the useful signal. This results in Figure 5.2, which would replace only the box $Q_b(t, u)$ on the diagram in Figure 5.1, and would therefore be followed by the operation of correlation with the replicas $a(t, \theta_i)$ and normalization with respect to S_i.

In this new receiver, the filter h_0 remains to be determined. Let us consider the stationary case, with an infinite observation time:

$$\begin{cases} K_b(t, u) = K_b(t, u) = N_0 \delta(t, u) + K_c(t, u) \\ T_f = -T_i = +\infty \end{cases} \tag{5.6}$$

Let $x(t)$ be the signal received on the reference channel, and $y(t)$ the output of filter h_0 (best estimate of the colored part of the noise, b_c). Adopting the variance minimization criterion, we determine h_0 such that:

$$\mathrm{E}\int |y(t) - b_c(t)|^2 dt \text{ min} \tag{5.7}$$

giving, as $b_0(t)$ and $b_c(t)$ are independent:

$$\mathrm{E}\int |y(t) - b(t)|^2 dt \text{ min} \tag{5.8}$$

giving, as $a(t, \theta)$ and $b(t)$ are independent:

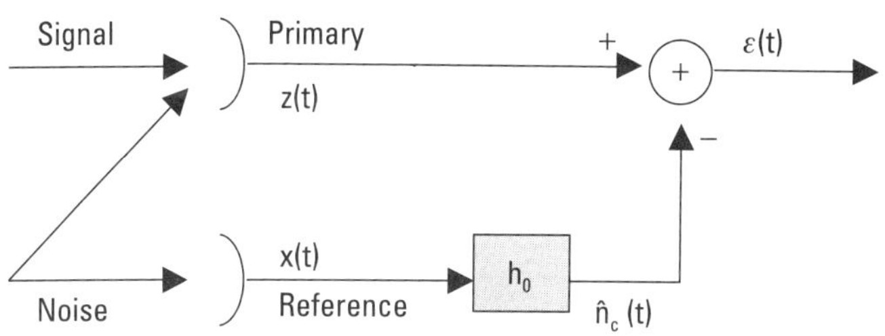

Figure 5.2 Noise reference provided by an auxiliary sensor.

$$\mathrm{E}\int |y(t) - z(t)|^2 dt \min \tag{5.9}$$

Converting to Fourier transforms:

$$\mathrm{E}\int |\tilde{z}(f) - \tilde{x}(f)\tilde{h}(f)|^2 df \min \tag{5.10}$$

$$\Rightarrow \mathrm{E}|\tilde{z}(f) - \tilde{x}(f)\tilde{h}_0(f)|^2 df \min$$

By annulation of the gradient with respect to h_0, we obtain:

$$\tilde{h}_0(f) = \frac{S_{ZX}(f)}{S_{XX}(f)} \tag{5.11}$$

where

$$\begin{cases} S_{ZX}(f) = \mathrm{E}[\tilde{z}(f)\tilde{x}^*(f)] \\ S_{XX}(f) = \mathrm{E}[\tilde{x}(f)\tilde{x}^*(f)] \end{cases} \tag{5.12}$$

The transfer function $h_0(f)$ is thus obtained from the spectral density of the reference channel, $S_{XX}(f)$, and the cross-spectral density between reference and primary channels, $S_{ZX}(f)$.

It is interesting to measure the influence on this processing of the presence of the signal on the noise reference channel. Let us therefore consider the case in which the signal arrives on the signal channel but also on the reference channel through a transfer function $I(f)$ [2]. Similarly, the noise arrives on the signal channel and on the reference channel through a transfer function $J(f)$ (Figure 5.3), so that the following holds:

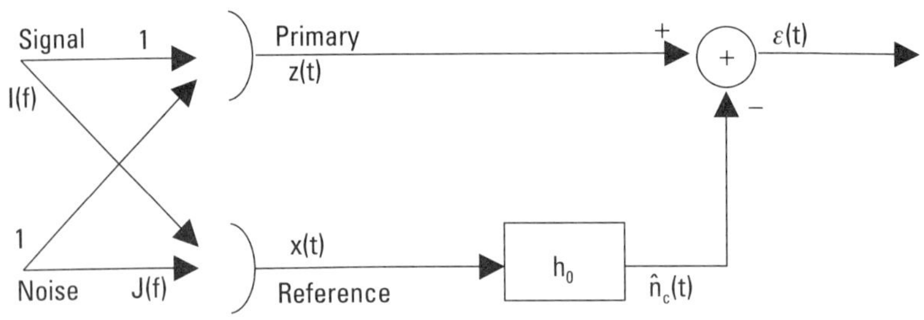

Figure 5.3 Presence of noise on the reference channel.

$$\tilde{z}(f) = \tilde{a}(f)\tilde{b}(f)$$
$$\tilde{x}(f) = I(f)\tilde{a}(f) + J(f)\tilde{b}(f)$$

From (5.11), assuming independent noise and signal, the transfer function can then be written:

$$\tilde{h}_0(f) = \frac{S_{aa}(f)I^*(f) + S_{bb}(f)J^*(f)}{S_{aa}(f)|I(f)|^2 + S_{bb}(f)|J(f)|^2} \tag{5.13}$$

where

$S_{aa}(f)$: spectral density of the signal $a(t, \theta)$
$S_{bb}(f)$: spectral density of the noise $b(t)$.

The spectral density of the signal at the output of the device is then:

$$S_{\text{signal}}(f) = S_{aa}(f)|1 - I(f)\tilde{h}_0(f)|^2 \tag{5.14}$$

and that of the noise:

$$S_{\text{noise}}(f) = S_{bb}(f)|1 - J(f)\tilde{h}_0(f)|^2 \tag{5.15}$$

Substituting (5.13) for $h_0(f)$ into (5.14) and (5.15), we obtain the signal-to-noise ratio (at each frequency) at the output of the device:

$$\rho_{\text{out}}(f) = \frac{S_{\text{signal}}(f)}{S_{\text{noise}}(f)} = \frac{S_{bb}(f)}{S_{aa}(f)} \frac{|J(f)|^2}{|I(f)|^2} \tag{5.16}$$

Now, the signal-to-noise on the reference channel is:

$$\rho_{\text{ref}}(f) = \frac{S_{aa}(f)}{S_{bb}(f)} \frac{|I(f)|^2}{|J(f)|^2} \tag{5.17}$$

hence the relation:

$$\rho_{\text{out}}(f) = 1/\rho_{\text{ref}}(f) \tag{5.18}$$

The signal-to-noise ratio at the output of the device is the reciprocal of the signal-to-noise ratio on the reference channel. This very simple relation confirms the remarks made previously regarding the requirement for a good noise reference. In practice, the signal-to-noise ratio is also limited by the thermal noise added independently on each channel x and z, which prohibits a perfect estimation of the noise, and therefore an infinite signal-to-noise ratio.

5.3.2 Adaptive Filtering Using a Transversal Filter

The use of the above receiver comes up against the problem of calculating the filter h_0, which requires the correlation functions to be estimated as in (5.11). This estimation is both costly in terms of computing time and of doubtful value in a real situation with nonstationary noise.

This leads us to define a receiver in closed-loop form, continuously optimizing the filter as a function of the received signals.

To adapt the filter, its structure must be specified; consider, for example, a moving average filter or transversal filter, whose production in digital technology is quite simple (Figure 5.4). This filter includes a number of delays τ whose outputs are weighted by the gains w_i and summed to deliver y:

$$y = \sum_{i=1}^{p} w_i^* x_{n-i} = \mathbf{W}^t \mathbf{X}_n \tag{5.19}$$

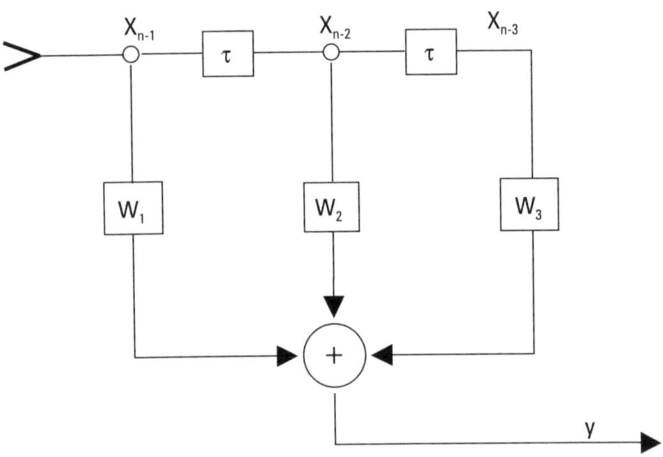

Figure 5.4 Implementation of h_0 as a transversal filter.

where $\mathbf{W}$ is the vector of the gains, $\mathbf{W}^t$ is the transposed conjugate of vector $\mathbf{W}$, and $\mathbf{X}_n$ is the vector of the samples $x_{n-1}, \ldots, x_{n-p}$. The impulse response of this digital filter is the series of samples $w_1^*, \ldots, w_p^*$, with τ the sampling period of this response (generally chosen to be identical with the sampling period of signals x and y).

From the above paragraph, the optimum filter minimizes:

$$E|z - \mathbf{W}^t\mathbf{X}_n|^2 = E|\epsilon|^2 \tag{5.20}$$

where ϵ is the output signal of the device.

The vector $\mathbf{W}$ is obtained by nulling the gradient of this quadratic form (single minimum):

$$\mathbf{W}_{\text{opt}} = \mathbf{R}_{XX}^{-1}\mathbf{R}_{ZX} \tag{5.21}$$

if $\mathbf{R}_{XX}$ and $\mathbf{R}_{ZX}$ are the correlation matrices of $\mathbf{X}$ with itself and of z with $\mathbf{X}$. For example, for Doppler adaptive processing, the noise is made of clutter returns whose spectral characteristics may be considered constant on a few adjacent range gates (especially for atmospheric clutter). In this situation, the primary channel is constituted with the time samples z_i collected on one pulse train, in the range gate under test. The reference channel uses the samples x_i received in an adjacent range gate, as shown in Figure 5.5. The averaging that has to be done in order to estimate the matrices $\mathbf{R}_{XX}$ and $\mathbf{R}_{ZX}$ can be conducted on different range gates surrounding the gate under test. The number of range gates used for estimation results from a trade-off

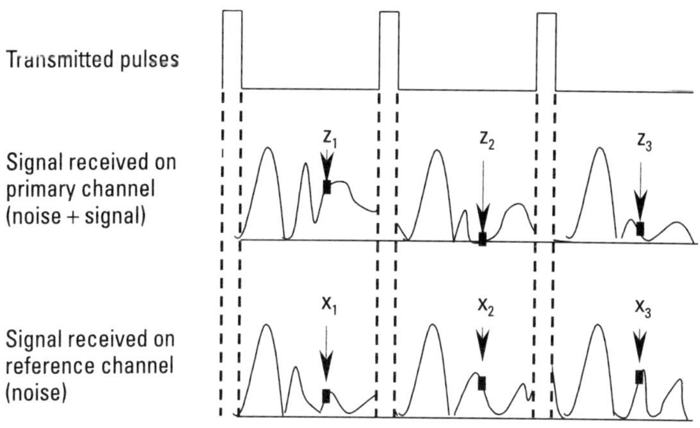

Figure 5.5 Definition of the samples.

between the quality of correlation matrices estimation, which requires many range gates to be used, and the nonstationarity of clutter from one range gate to the next, which limits the extent of the reference area.[1] These limitations are in some aspects comparable to the situation of CFAR detection (see Section 3.3.8).

The algorithm defined by (5.21) is called *block-adaptive,* because it relies on an estimation of the correlation matrices from successive blocks of data.

As these matrices are unknown and slowly variable, it can be appropriate to look for an adaptive solution where $\mathbf{W}$ is modified at each new received sample (which is another way of estimating the correlation matrices $\mathbf{R}_{XX}$ and $\mathbf{R}_{XZ}$).

If $\mathbf{W}(j)$ is the vector of the filter coefficients at iteration j, we choose to go from $\mathbf{W}(j)$ to $\mathbf{W}(j+1)$ in order to follow the line of greatest slope of the function $|\epsilon|^2$ to be minimized:

$$\mathbf{W}(j+1) = \mathbf{W}(j) - \mu\nabla(j), \text{ with } \nabla(j) = \frac{d(\mathrm{E}[|\epsilon|^2])}{d\mathbf{W}} \tag{5.22}$$

This approach is particularly appropriate here, as the function to be minimized is a quadratic function of the coefficients $\mathbf{W}$, which guarantees the uniqueness of the minimum. Let $\mathbf{X}(j)$ be the value of vector $\mathbf{X}$ at iteration j, and $z(j)$ that of z at the same time. Adaptivity is achieved by replacing $\nabla(j)$ by its instantaneous estimate:

$$\begin{cases} \mathrm{Re}\,\hat{\nabla}(j) = \nabla_{\mathrm{Re}\,W(j)}|z(j) - \mathbf{W}^{\mathrm{t}}(j)\mathbf{X}(j)|^2 \\ \mathrm{Im}\,\hat{\nabla}(j) = \nabla_{\mathrm{Im}\,W(j)}|z(j) - \mathbf{W}^{\mathrm{t}}(j)\mathbf{X}(j)|^2 \end{cases} \tag{5.23}$$

which means that we replace the gradient of the average of $|\epsilon|^2$ by that of its instantaneous value.

Elementary calculations then lead to:

$$\hat{\nabla}(j) = -2\epsilon^*(j)\mathbf{X}(j) \tag{5.24}$$

1. Another limitation arises from the possible targets in these reference range gates, whose presence would have the effect of canceling any target with the same velocity in the range gate under test.

which leads to the filter coefficient calculation algorithm:

$$\mathbf{W}(j+1) = \mathbf{W}(j) + 2\mu\epsilon^*(j)\mathbf{X}(j) \quad (5.25)$$

where: $\epsilon(j) = z(j) - \mathbf{W}^t(j)\mathbf{X}(j)$

where μ is a parameter on which the algorithm convergence rate depends (see the following paragraph).

The diagram of the receiver finally obtained is shown in Figure 5.6. In this figure, two adaptation procedures are shown: the gradient adaptive algorithm, (5.25), and block adaptive algorithm, (5.21). The equivalence between the two algorithms, after adaptation in the stationary situation, is shown in the next section.

It should be noted that the definition of this receiver is incomplete, as the normalization factor is not specified. This factor, which was equal to the energy of the whitened replica in (5.5), the noise reference being taken on the signal channel (known noise covariance), can no longer be specified now, as the transfer function of the "equivalent filter" applied to the signal depends on the transfer functions of the signal and the noise, from the signal channel to the reference channel [called $I(f)$ and $J(f)$ above], which are unknown; the receiver thus becomes suboptimum, by an amount that is difficult to assess.

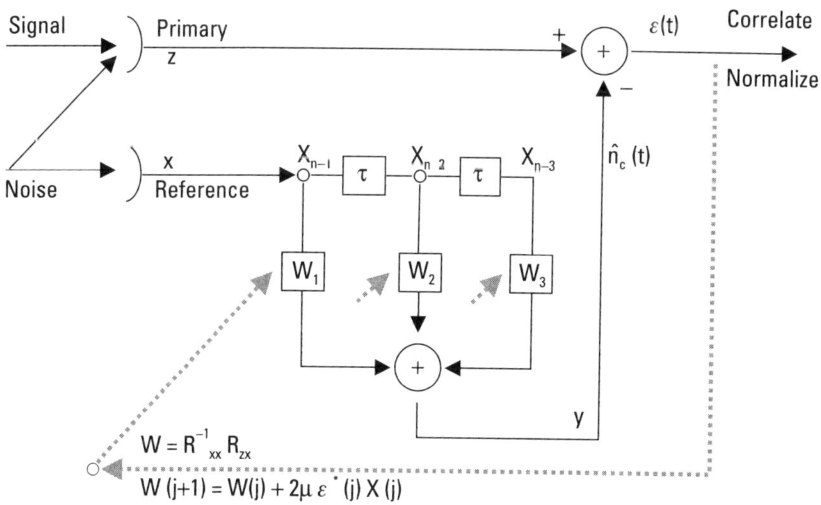

Figure 5.6 Adaptive noise suppression.

The best one can say is that if whitening by subtraction of the filtered reference does not change the power of the useful signal significantly, the receiver remains quasi-optimum (because the energy of the whitened replica is then close to that of the original replica). This critical aspect of the normalization coefficient—which is a priori different for each replica $a(t, \theta)$, and hence for each value of the parameter θ, and therefore directly impacts the estimate of θ—will be examined in further detail in Section 5.5.

5.3.3 Convergence

The adaptive filter obtained here is therefore characterized by the vector $\mathbf{W}$ of the filter coefficients [see (5.25)].

In the stationary case, the optimum filter is

$$\mathbf{W}_{\text{opt}} = \mathbf{R}_{XX}^{-1}\mathbf{R}_{ZX}$$

Given the approximation made in (5.23) to build the adaptive receiver (replacement of the average gradient by the instantaneous gradient), it is necessary to check the convergence of $\mathbf{W}(j)$ toward $\mathbf{W}_{\text{opt}}$ in the stationary case (see [2]).

$$\mathrm{E}[\mathbf{W}(j+1)] = \mathrm{E}[\mathbf{W}(j)] + 2\mu\mathrm{E}\{[z^* - \mathbf{X}_n^t\mathbf{W}(j)]\mathbf{X}_n\}$$
$$\mathrm{E}[\mathbf{W}(j+1)] = \mathrm{E}[\mathbf{W}(j)][\mathbf{I} - 2\mu\mathbf{R}_{XX}] + 2\mu\mathbf{R}_{ZX}$$

hence by recurrence:

$$\mathrm{E}[\mathbf{W}(j+1)] = [\mathbf{I} - 2\mu\mathbf{R}_{XX}]^{j+1}\mathbf{W}(0) + 2\mu\sum_{i=0}^{j}[\mathbf{I} - 2\mu\mathbf{R}_{XX}]^i\mathbf{R}_{ZX} \tag{5.26}$$

$\mathbf{R}_{XX}$, the autocorrelation matrix of the signal received on the reference channel, is diagonalizable, with positive eigenvalues:

$$\mathbf{R}_{XX} = \mathbf{M}^{-1}\mathbf{\Lambda}\mathbf{M} \tag{5.27}$$

where

$$\mathbf{\Lambda} = \begin{bmatrix} \lambda_1 & & 0 \\ & \vdots & \\ 0 & & \lambda_n \end{bmatrix} \tag{5.28}$$

$$E[\mathbf{W}(j+1)] = \mathbf{M}^{-1}(\mathbf{I} - 2\mu\mathbf{\Lambda})^{j+1}\mathbf{M}\mathbf{W}(0) + 2\mu\mathbf{M}^{-1}\sum_{i=0}^{j}(\mathbf{I} - 2\mu\mathbf{\Lambda})^{i}\mathbf{M}\mathbf{R}_{ZX}$$

If μ is chosen such that:

$$|1 - 2\mu\lambda_i| < 1 \quad \forall \mathrm{i} \in (1, \ldots p) \tag{5.29}$$

then the first term tends to 0 with increasing j. The second term is a geometric series that tends to:

$$2\mu\mathbf{M}^{-1}\mathbf{L}\mathbf{M}\mathbf{R}_{ZX} \tag{5.30}$$

where

$$\mathbf{L} = \begin{bmatrix} \nu_1 & & 0 \\ & \vdots & \\ 0 & & \nu_n \end{bmatrix}$$

and

$$\nu_i = \frac{1}{1 - (1 - 2\mu\lambda_i)} = \frac{1}{2\mu\lambda_i} \Rightarrow \mathbf{M}^{-1}\mathbf{L}\mathbf{M} = \mathbf{R}_{XX}^{-1}\frac{1}{2\mu}$$

$$E[\mathbf{W}(j+1)] = \mathbf{R}_{XX}^{-1}\mathbf{R}_{ZX} = \mathbf{W}_{\mathrm{opt}} \tag{5.31}$$

The adaptive filter therefore does, in the stationary case, converge toward the optimum filter, under the condition in (5.29), and this condition will be verified if:

$$0 < \mu < \frac{1}{\lambda_{\max}} \tag{5.32}$$

The rate of convergence of mode i associated with the modal decomposition in (5.27) is characterized by time constant τ_i:

$$1 - 2\mu\lambda_i = e^{-1/\tau_i} \sim 1 - \frac{1}{\tau_i} \tag{5.33}$$

$$\tau_i \sim \frac{1}{2\mu\lambda_i} \tag{5.34}$$

The maximum time constant is therefore set by the minimum eigenvalue λ_{min}:

$$\tau_{\text{max}} = \frac{1}{2\mu\lambda_{\text{min}}} \tag{5.35}$$

that is, given the convergence condition in (5.32):

$$\tau_{\text{max}} = \frac{\lambda_{\text{max}}}{2\lambda_{\text{min}}} \tag{5.36}$$

Therefore, in summary, convergence is achieved in the stationary case if the condition in (5.32) is valid, at a speed limited by the dispersion of the eigenvalues of the reference channel correlation matrix. Equation (5.32) is easy to verify in practice, as $\lambda_{\text{max}} < \text{tr}\,(\mathbf{R}_{xx})$, and therefore a sufficient convergence condition is:

$$0 < \mu < \frac{1}{\sigma^2},\ \sigma^2 = \text{tr}\mathbf{R}_{XX} \tag{5.37}$$

σ^2 being the power of the reference channel. This condition guarantees that the correction term applied for adaptation does not increase with increasing power at the reference sensor.

The main advantage of gradient adaptive least-square noise canceling is that it requires minimum calculation load. The trade-off is a relatively slow convergence, which restricts the scope of application to cases in which the noise sources are relatively stationary. In most radar and sonar applications, the block adaptive algorithm in (5.21) is now usually preferred.

5.3.4 Application

This algorithm can be applied to the processing of sensor arrays, as indicated in the diagram shown in Figure 5.7. The phase shifts $\Delta_1, \Delta_2, \ldots, \Delta_N$ are

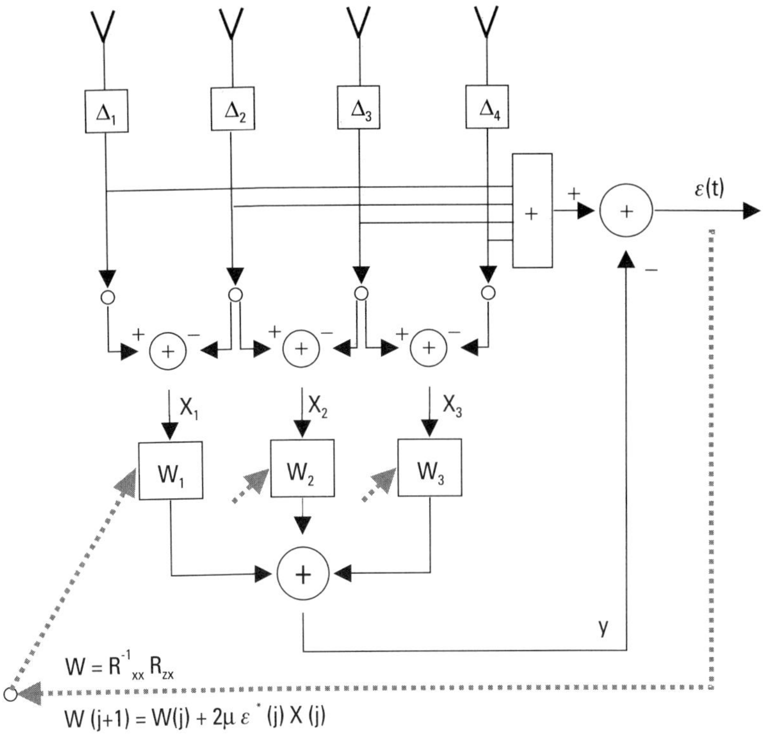

Figure 5.7 Adaptive suppression of jammers in sensor array processing.

applied to the signals received by the N sensors, assumed to be aligned along axis $O\xi$, to "steer the antenna" in the desired observation direction θ, in accordance with the processing described in Section 3.5.1, with:

$$\Delta_i = \exp\left(-2\pi j \frac{\xi_i}{\lambda} \sin\theta\right)$$

where ξ_i is the abscissa of sensor number i and θ the angle between the steering direction and the normal to the sensor axis.

The phase shifted signals are then:

- On the one hand, summed to make up the usual sum channel (optimum processing in white noise) z;
- On the other hand, subtracted two by two to make up a noise reference $(X_1, X_2, \ldots, X_{N-1})$, as indicated in the diagram shown in Figure 5.7.

This subtraction eliminates the useful signal, as this signal was, after the phase shifts Δ_i, in phase on all channels; the signal resulting from this subtraction, $(X_1, X_2, \ldots, X_{N-1})$ is therefore a noise reference, which is processed by the algorithm described in Section 5.3.3: adaptive weighting, and subtraction from the signal of the sum channel [both adaptive procedures are shown here, as in Figure 5.6—block adaptive, as in (5.21), or gradient adaptive, as in (5.25)]. The power of the residual noise at the output of the array processed in this way will then decrease smoothly, at a rate that depends on the characteristics of the noise, as was seen in Section 5.3.3. The number of iterations necessary to obtain good convergence generally amounts to several hundred, and therefore allows effective action to be taken only against relatively stationary interference.

The signal from this array is then processed conventionally (see Section 3.5.1) by correlation and quadrature detection, then comparison with a threshold (for a detection system) or determination of a maximum (for a location system). We can see that this processing is no longer optimum, because the normalization operation (with respect to the energy S_i of the whitened replica) is omitted (because the whitening filter, equivalent to the operation performed by subtraction of the filtered reference, is not explicitly known, since the noise transfer function between signal channel and reference channel is not known). The corresponding loss of optimality is to be assessed in each particular case, depending on the nature of the signals and the noise. Generally, it remains negligible when the target is not located close to the interference source, as shown by the example discussed in Section 5.5.2. It must also be emphasized that this normalization is not a problem for detection situations in which the threshold is usually determined through comparison with adjacent cells in range or Doppler (CFAR, see Section 3.3.8).

5.4 Adaptive Whitening

5.4.1 Receiver Structure

Considering the same problem of estimating θ from the received signal $z(t)$ (after demodulation by the carrier $e^{j\omega_0 t}$), let us first examine the stationary case, with an infinite observation time:

$$\begin{cases} z(t) = \alpha a(t, \theta) + b(t) \\ b(t) = b_0(t) + b_c(t) \text{ (white noise + colored noise)} \\ E[b_0(t) b_0^*(t+\tau)] = N_0 \delta(\tau) \\ E[b_c(t) b_c^*(t+\tau)] = K_c(t) \end{cases} \tag{5.38}$$

The optimum whitening processing was described in Section 4.3.1. It consists of N parallel branches (each adapted to one value θ_i of the parameter), of the type shown in Figure 5.8. In this diagram, filter h_w is a whitening filter, that is, the noise at the output of this filter must be white [see (4.6)]—the signal plus noise combination is generally not white at the output of this filter, but of variable color according to the value of the signal-to-noise ratio.

To achieve whitening, the signal must somehow be distinguished from the noise, which means that a noise-only reference must be available.

This reference can be obtained, in the case of a pulse radar, if it is assumed that the targets are sufficiently dispersed. If T is the duration of the pulse or pulse train, we can then use samples received before $t_0 - T$, or after $t_0 + T$ to detect a target at $t = t_0$. In the sampled situation, if N samples are taken during the duration of the pulse or pulse train, the problem therefore consists of determining the whitening filter to be applied to this series of N samples, based on the previous samples that make up the noise-only reference.

It is worth noting that this way of producing the noise-only reference enables this reference to contain, in addition to the noise due to sources internal or external to the radar, echoes from the various landscape elements located in front of (or behind) the potential target. In fact, in many applications (see Section 5.4.3), the echoes from these elements are considered as noise (echoes from clouds, rain, landscape background in radar, reverberation in sonar), and they make it necessary to take noise color into account.

Let $z(t)$, with discrete t, be this series of noise-only reference samples; the whitening filter h_w will be a filter built from these reference samples in

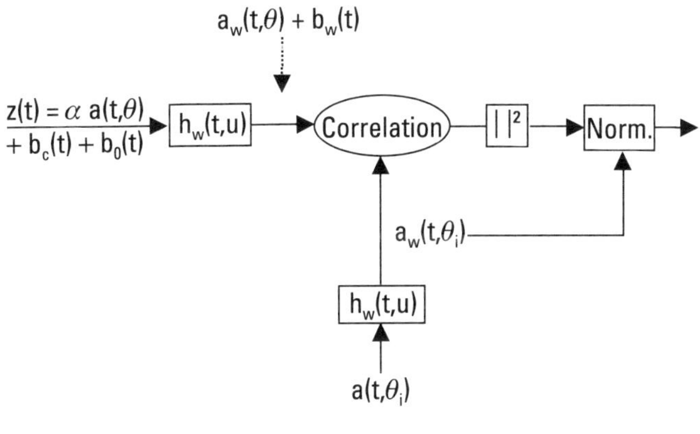

Figure 5.8 Reception by whitening.

order to whiten them. The filter will then be applied to the series of samples to be tested (neighborhood of $t = t_0$).

To carry out an order p filtering, we assume that $z(t)$ is, in the absence of a useful signal, a linear combination of the p preceding samples added to white noise (which will be the result of the whitening):

$$z(t) + a_1 z(t-1) + \ldots + a_p z(t-p) = f_p(t) \tag{5.39}$$

"forward" prediction error difference between actual and predicted $z(t)$.

If this relation is verified, a relation of the same type, but reversed, will also be verified in the stationary case:

$$z(t-p-1) + c_1 z(t-p) + \ldots + c_p z(t-1) = b_p(t-1) \tag{5.40}$$

"backward" prediction error.

For the prediction error to be minimum, it must no longer contain information about the signal z; consequently, $f_p(t)$ must be orthogonal to the p preceding samples:

$$\mathrm{E}[f_p(t) z^*(t-j)] = 0 \qquad 1 \le j \le p \tag{5.41}$$

similarly

$$\mathrm{E}[b_p(t-1) z^*(t-j)] = 0 \qquad 1 \le j \le p \tag{5.42}$$

To carry out filtering of higher order $p + 1$, we must predict $z(t)$ from the previous samples up to $z(t-p-1)$. The information contained in the latter sample is already present in $b_p(t-1)$, according to (5.40). It is therefore reasonable to define:

$$f_{p+1}(t) = f_p(t) - k^f_{p+1} b_p(t-1) \tag{5.43}$$

still with the same orthogonality conditions:

$$\mathrm{E}[f_{p+1}(t) z^*(t-j)] = 0 \qquad 1 \le j \le p+1 \tag{5.44}$$

The same argument applied to the backward prediction error—prediction of $z(t-p-1)$ from the following samples up to $z(t)$—leads to:

$$b_{p+1}(t) = b_p(t-1) - k^b_{p+1} f_p(t) \tag{5.45}$$

still with the condition:

$$E[b_{p+1}(t-1)z^*(t-j)] = 0 \qquad 1 \le j \le p+1$$

hence again

$$E[b_{p+1}(t)z^*(t-j)] = 0 \qquad 0 \le j \le p \tag{5.46}$$

The two new orthogonality conditions, (5.44) and (5.46), imply in particular:

$$E[f_{p+1}(t)z^*(t-p-1)] = 0 \tag{5.47}$$

$$E[b_{p+1}(t)z^*(t)] = 0 \tag{5.48}$$

Given (5.43) and (5.45), these two conditions determine coefficients k^f_{p+1} and k^b_{p+1}:

$$k^f_{p+1} = \frac{E[f_p(t)b^*_p(t-1)]}{E[|b_p(t-1)|^2]} \tag{5.49}$$

$$k^b_{p+1} = \frac{E[f^*_p(t)b_p(t-1)]}{E[|f_p(t)|^2]} \tag{5.50}$$

[These coefficients are referred to as PARtial CORrelation (PARCOR) coefficients in the literature.]

We thus build, recursively on the order, a lattice filter whose structure is shown in Figure 5.9.

This construction by consecutive orthogonalization steps enables the coefficients k_i to be defined independently of one another, which, in a changing situation, results in a faster adaptation than that of coefficients a_i and c_i of the autoregressive models (5.39) and (5.40), when this adaptation is carried out by a gradient algorithm.

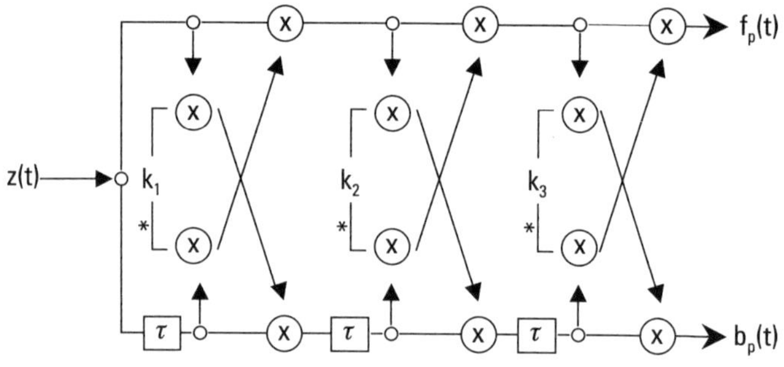

Figure 5.9 Lattice filter (* conjugation).

When the signal is stationary, for reasons of symmetry, we should have:

$$k_p^f = (k_p^b)^* \tag{5.51}$$

In practice, these coefficients are estimated from the measurements. Various estimates have been proposed. Burg's algorithm [3] has various advantages; it is obtained by determining $k_p^f = (k_p^b)^*$, such that the quadratic sum of the forward and backward prediction errors is minimum:

$$\mathrm{E}[|f_{p+1}|^2 + |b_{p+1}|^2] \text{ minimum}$$

Substituting (5.43) and (5.45) and determining the minimum by gradient annulation, we obtain:

$$k_{p+1}^f = (k_{p+1}^b)^* = -2\frac{\mathrm{E}[f_p(t)b_p^*(t-1)]}{\mathrm{E}[|f_p(t)|^2 + |b_p(t-1)|^2]} \tag{5.52}$$

This estimation of k guarantees that $|k_p| < 1$, and therefore that the lattice filter is stable.

From a set of N samples of $z(t)$, measured in the absence of useful signal, and assuming that coefficients k_j, $j \leq p$ have been calculated, it is necessary, to calculate $k_j + 1$, to filter the N samples at order j, in order to obtain N values of $f_j(t)$ and $b_j(t)$ and deduce the mean values involved in (5.52); the required computation load is therefore significant. There are, however, algorithms that enable these coefficients to be calculated directly from the estimate of the signal correlation function [3]. The adaptive configuration described in Section 5.4.2 is undoubtedly the most economical way of calculating these coefficients.

5.4.2 Adaptive Whitening by Lattice Filter

Various algorithms have been proposed to calculate the coefficients $k_p(t)$ adaptively. A conventional technique consists of applying exponential weighting for the calculation of the means in (5.52).

$$k_{p+1}(t) = \frac{2\sum_{i+1}^{n} \mu^{i-1} f_p(t) b_p^*(t-1)}{\sum_{i=1}^{n} \mu^{i-1} [|f_p(t)|^2] + |b_p(t-1)|^2]} \tag{5.53}$$

The adaptation is thus performed according to the following relations:

$$\begin{cases} k_{p+1}(t) = \dfrac{u_{p+1}(t)}{v_{p+1}(t)} \\ u_{p+1}(t) = \mu u_{p+1}(t-1) + 2f_p(t)b_p^*(t-1) \\ v_{p+1}(t) = \mu v_{p+1}(t-1) + |f_p(t)|^2 + |b_p(t-1)|^2 \end{cases} \tag{5.54}$$

which enable the prediction errors at order $p + 1$ to be calculated.

- The forward prediction error $f_{p+1}(t)$, which is also the output of the whitening filter of order $p + 1$:

$$f_{p+1}(t) = f_p(t) - k_{p+1} b_p(t-1) \tag{5.55}$$

- The backward prediction error $b_{p+1}(t)$,

$$b_{p+1}(t) = b_p(t-1) - k_{p+1}^* f_p(t) \tag{5.56}$$

taking as initial conditions: $u_{p+1}(0) = v_{p+1}(0) = 0 \quad \forall p$

5.4.3 Application

As an illustration, let us consider the application of this processing by a whitening filter to a pulse Doppler radar, that is, a radar transmitting elementary pulses $u_e(t)$ at a repetition period T_r and processing the received signals in trains of $2n + 1$ pulses (see Section 3.3.6). We have seen that the optimum processing in white noise of this type of signal includes three consecutive operations:

1. Matched filtering, or correlation by the elementary pulse $u_e(t)$, which delivers a compressed signal;
2. Sampling at times $\tau_i + kT_r$, where τ_i characterizes the delay—hence the range—and where k varies from $-n$ to $+n$; this sampling therefore delivers, in each range bin characterized by τ_i, a series of $2n + 1$ complex samples;
3. Discrete Fourier transform of the samples, making up the Doppler analysis.

Then, one way of taking into account the colored nature of the external noise (or of the spurious echoes on the diffuse elements of the landscape) consists of keeping the matched filtering at the start of the processing, while assuming that the noise is colored within each range gate, its characteristics varying slowly from one range gate to the next. This is the case, for example, when this noise is made up of echoes from a rain cloud; the extent of the cloud is greater than the depth of a range bin, and the Doppler spectrum of the raindrops depends on various factors (e.g., wind and turbulence in the cloud) which result in a shape varying slowly from one range gate to the next. Similar situations can be observed in the presence of ground clutter or, for sonar, in the presence of reverberation (surface or volume reverberation).

Under these conditions, the signal to be whitened consists of the $2n + 1$ complex samples of each range bin. The noise reference, which is used for calculating the whitening filter, consists of the samples received from the previous range bins. Doppler filtering is then adaptive in range, which means that, for a given antenna steering direction, we use (5.54)–(5.56) to calculate the coefficients k_p and the values of the whitened signal f_p, $p = 1, 2 \ldots, p_0$, on the basis of the $2n + 1$ samples of range bin number i. This calculation is then iterated on the basis of the $2n + 1$ samples in bin number $i + 1$, and so on. Given the memory provided by the weighting coefficient μ, the whitening filter applied in a given range bin is a function of the noise present in a few preceding range bins.

Another way of performing this filtering is to adopt angle adaptivity [4]. In each range bin, the whitening filter is then calculated according to the same expressions, but by iterating from one bearing to another and no longer from one range to another. Obviously, any combination of these techniques may be used, depending on the radar characteristics, the interference characteristics, and the computing power available [5]. According to the situation, a direct calculation (nonadaptive) by blocks of samples may be preferred, in particular for better processing of the nonstationary aspects of the noise (boundary between different types of landscape, for example). The order p_o of the whitening filtering is generally less than 5.

Once the samples in each range bin have been filtered, the Doppler filtering, previously performed by Fourier transform, should theoretically be modified. The same whitening filter should be inserted in the replica channel, and therefore the Fourier transform (correlation with sinusoids) should be replaced by a correlation with the sine waves whitened by this filter.

Alternatively, and to retain the simplicity provided by the FFT, the whitened signal $f_{p_o}(t)$ should pass through the transfer function filter $h_w^*(t)$—if $h_w(f)$ is the whitening filter transfer function—in accordance with (4.27) and (4.28). In practice, this filter is generally omitted, and detection is carried out directly on the whitened signal, or after Doppler filtering by Fourier transform.

As an illustration, Figure 5.10—taken from [5]—shows the outputs of the Doppler filter with or without a whitening filter inserted. The fifth order lattice filter is calculated by averaging 32 adjacent resolution bins (angle and range). The noise consists of real clutter echoes (sea clutter + hail, clearly visible on the top left picture showing the 32 spectra of the signal received in adjacent cells). The lower left picture shows the average

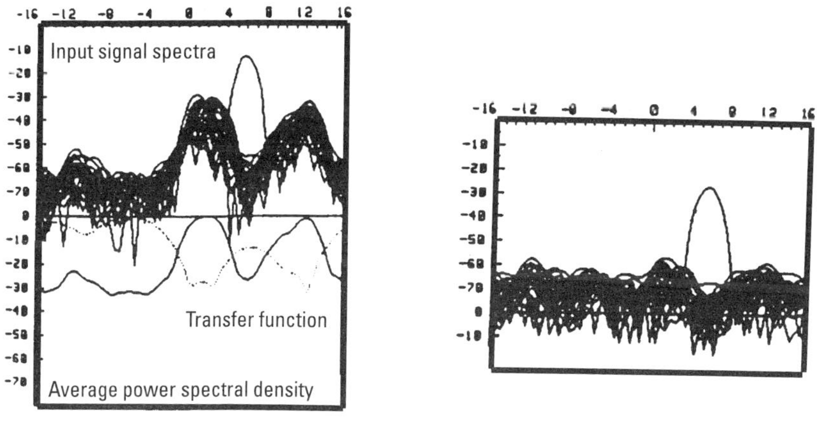

Figure 5.10 Doppler filtering output [5].

power spectral density (average of the 32 former spectra), and the transfer function obtained for the adaptive filter. The right picture shows the whitening result, after adaptation, for each of the 32 reference cells and for the cell under test. The clutter noise has been indeed normalized, allowing a standard CFAR detection procedure to perform detection.

5.5 Sensor Arrays

5.5.1 Problem Statement

The optimum receiver of a signal, a known function of a set of parameters θ, buried in nonstationary colored noise, has been built in Section 4.2. The received signal being:

$$\begin{cases} r(t) = z(t)e^{j\omega_0 t} \\ z(t) = \alpha a(t, \theta) + b(t) \quad t \in [T_i, T_f] \\ b(t) = b_0(t) + b_c(t) \\ E[b(t)b^*(u)] = K(t, u) = N_0\delta(t-u) + K_c(t, u) \end{cases} \tag{5.57}$$

The optimum receiver estimates α, an unknown complex coefficient, and θ, the vector of the unknown parameters (i.e., angles, range, and Doppler) according to (4.10), (4.11), and (4.12), by:

$$\begin{cases} \hat{\theta} / = \dfrac{\left| \displaystyle\iint a^*(x, \theta)Q(x, y)z(y)\,dx\,dy \right|^2}{\displaystyle\iint a^*(x, \theta)Q(x, y)a(y, \theta)\,dx\,dy} \text{ maximum} \\ \hat{\alpha} = \dfrac{\displaystyle\iint a^*(x, \hat{\theta})Q(x, y)z(y)\,dx\,dy}{\displaystyle\iint a^*(x, \hat{\theta})Q(x, y)a(y, \hat{\theta})\,dx\,dy} \end{cases} \tag{5.58}$$

where

$$\int Q^*(x, y)K(y, t)\,dt = \delta(x-t)$$

Let us consider the sampled case where the received signal is a complex vector **z** with N components. By discretization of the preceding relations:

$$\begin{cases} \mathbf{z} = \alpha \mathbf{a}(\theta) + \mathbf{b} \\ \mathbf{b} = \mathbf{b}_0 + \mathbf{b}_c \\ \mathrm{E}[\mathbf{b}\mathbf{b}^t] = \mathbf{K} = N_0\mathbf{I} + \mathbf{K}_c \end{cases} \tag{5.59}$$

where ^{t} denotes the transposed, conjugated vector, and:

α: unknown complex scalar;
z, **a**, **b**, $\mathbf{b}_0$, $\mathbf{b}_c$: vectors, with N components;
K, $\mathbf{K}_c$: Nth order correlation matrices (**I**: identity matrix).

The optimum receiver estimates θ and α through:

$$\begin{cases} \hat{\theta} / = \dfrac{|\mathbf{a}^t(\theta)\mathbf{Q}\mathbf{z}|^2}{\mathbf{a}^t(\theta)\mathbf{Q}\mathbf{a}(\theta)} \mathrm{Max}/\theta \\ \hat{\alpha} = \dfrac{\mathbf{a}^t(\hat{\theta})\mathbf{Q}\mathbf{z}}{\mathbf{a}^t(\hat{\theta})\mathbf{Q}\mathbf{a}(\hat{\theta})} \end{cases} \tag{5.60}$$

where: $\mathbf{Q}^t\mathbf{K} = \mathbf{I} \Rightarrow \mathbf{Q} = \mathbf{K}^{-1}$

where the optimum receiver:

$$\begin{cases} \hat{\theta} / = \dfrac{|\mathbf{a}^t(\theta)\mathbf{K}^{-1}\mathbf{z}|^2}{\mathbf{a}^t(\theta)\mathbf{K}^{-1}\mathbf{a}(\theta)} \mathrm{Max}/\theta \\ \hat{\alpha} = \dfrac{\mathbf{a}^t(\hat{\theta})\mathbf{K}^{-1}\mathbf{z}}{\mathbf{a}^t(\hat{\theta})\mathbf{K}^{-1}\mathbf{a}(\hat{\theta})} \end{cases} \tag{5.61}$$

K: noise correlation matrix.

Remember that in these expressions $\mathbf{a}^t(\theta)\mathbf{K}^{-1}\mathbf{a}(\theta)$ represents the power of the whitened replica, and $\mathbf{a}^t(\theta)\mathbf{K}^{-1}\mathbf{z}$ represents the result of the correlation between the whitened received signal and the whitened replica (whitening always assuming that the noise correlation matrix is known).

Alternatively, these expressions can also be obtained by looking for the whitening filter $\mathbf{H}_w$ (matrix) defined by the whitening condition:

$$E[\mathbf{n}_w \mathbf{n}_w^t] = \mathbf{I} = E[\mathbf{H}_w \mathbf{n}\mathbf{n}^t \mathbf{H}_w^t] = \mathbf{H}_w \mathbf{K} \mathbf{H}_w^t \Rightarrow \mathbf{H}_w \mathbf{H}_w^t \underline{\Delta} \mathbf{Q} = \mathbf{K}^{-1}$$

In these expressions, no assumption is made as to the domain, whether temporal or spatial, in which the samples are taken. Thus, $\mathbf{z}$ may consist of the series of samples taken consecutively on various sensors and at different times, the first N values corresponding to the outputs of the N sensors at one time, the next N values to the outputs of the same sensors at the next time, and so on. Equations (5.61) define the optimum "space-time" processing, in the nonstationary situation, providing the following are known:

- The vector function $\mathbf{a}(\theta)$ of the unknown parameters θ (such as angle, Doppler);
- The correlation matrix $\mathbf{K}$ of the noise sampled on the various sensors at the different times.

This processing is therefore directly applicable to the case in which the noise correlation matrix can be measured, that is, where a noise-only reference is available. For example, when the colored noise consists of external jammers—whether intentional or not—it is sufficient to cut off the transmitter for a short time to record the noise arriving on the sensors and to measure their correlation matrix. When this is not possible, either because the noise is not stationary or because it is not possible to cut off the radar transmission (or again because passive mode is being used, that is, reception only), it becomes necessary to use another processing whose performance will obviously be worse, since less information will be available.

In what follows, this processing without noise-only reference will first be deduced from the previous processing by estimation of the noise + signal correlation matrix, which can be measured. Then a different approach, by constrained minimization, will be used to build the corresponding adaptive receiver (as in Sections 5.3 and 5.4).

However, before examining the case in which no noise-only reference is available, it is worth examining in detail the operation of the optimum processing in the case of an array of sensors, to better understand how a jammer can be suppressed.

5.5.2 Application to Jammer Cancellation

To evaluate the performance characteristics of a receiver with a noise-only reference, let us examine in detail the case of a linear sensor array in the presence of a single jammer, in the case of narrowband, or one-dimensional processing (which is also the case of time processing in the case of a single sensor).

According to (5.59) and (3.71), the received vector signal can be written as:

$$\mathbf{z} = \alpha \mathbf{a}(\theta) + \mathbf{b} \tag{5.62}$$

with

$$\mathbf{a}(\theta) = \begin{bmatrix} a_1(\theta) \\ \vdots \\ a_N(\theta) \end{bmatrix}$$

$$a_i(\theta) = \exp\left(2\pi j \frac{d_i}{\lambda} \sin\theta\right)$$

where $\mathbf{a}(\theta)$ is the vector of the samples received on the sensors for a target in direction θ, and $\mathbf{b}$ the colored noise.

$$E[\mathbf{b}\mathbf{b}^t] = \mathbf{K} \tag{5.63}$$

If θ_J is the jammer direction and N_0 the power of the isotropic white noise, the noise correlation matrix can be written:

$$\mathbf{K} = N_0\mathbf{I} + J\mathbf{a}(\theta_J)\mathbf{a}^t(\theta_J) \tag{5.64}$$

where J is the power of the jammer.

When the matrix $\mathbf{K}$ is known (either estimated before the measurement, or otherwise known), the previously defined optimum processing is given by:

$$\hat{\theta}/ = \frac{|\mathbf{a}^t(\theta)\mathbf{K}^{-1}\mathbf{z}|^2}{\mathbf{a}^t(\theta)\mathbf{K}^{-1}\mathbf{a}(\theta)} \text{Max}/\theta \tag{5.65}$$

$$\hat{\alpha} = \frac{\mathbf{a}^t(\hat{\theta})\mathbf{K}^{-1}\mathbf{z}}{\mathbf{a}^t(\hat{\theta})\mathbf{K}^{-1}\mathbf{a}(\hat{\theta})} \tag{5.66}$$

Let us examine the operation of the optimum receiver in this single-jammer configuration. It is useful to define the antenna radiation pattern (see Section 3.5.2):

$$\chi(\theta,\ \theta_0) = \frac{1}{N}\,\mathbf{a}^t(\theta)\mathbf{a}(\theta_0) \tag{5.67}$$

Normalization by N ensures that $\chi(\theta_0,\ \theta_0) = 1$; to simplify the notation, we shall further assume that the sensors are symmetrically distributed about the origin, so that $\chi(\theta,\ \theta_0)$ is a real quantity.

Let us calculate the output of the optimum receiver (with noise-only reference) in the presence of a target in direction θ_0, a jammer of power J in direction θ_J, and thermal white noise (Gaussian) of power N_0. The receiver calculates, from (5.65):

$$Q(\theta,\ \theta_0) = \frac{\left|\mathbf{a}^t(\theta)\mathbf{K}^{-1}\mathbf{z}\right|^2}{\mathbf{a}^t(\theta)\mathbf{K}^{-1}\mathbf{a}(\theta)} \tag{5.68}$$

where

$$\mathbf{K} = N_0\mathbf{I} + J\mathbf{a}(\theta_J)\mathbf{a}^t(\theta_J) \tag{5.69}$$

Using the lemma of the inverse matrix (Appendix 5A), $\mathbf{K}^{-1}$ is written:

$$\mathbf{K}^{-1} = \frac{1}{N_0}\left(\mathbf{I} - \frac{\mathbf{a}(\theta_J)\mathbf{a}^t(\theta_J)}{N + \dfrac{N_0}{J}}\right) \quad (\mathbf{I}\text{: identity matrix}) \tag{5.70}$$

In most interesting cases, the jammer is strong relative to the thermal noise: $NJ >> N_0$

$$\mathbf{K}^{-1} \sim \frac{1}{N_0}\left(\mathbf{I} - \frac{1}{N}\,\mathbf{a}(\theta_J)\mathbf{a}^t(\theta_J)\right) \tag{5.71}$$

The received signal is:

$$\mathbf{z} = \alpha \mathbf{a}(\theta_0) + \beta \mathbf{a}(\theta_J) + \mathbf{b}_0 \tag{5.72}$$

where b_0 is a white noise, and we obtain, by substituting (5.72) and (5.71) into (5.68) and neglecting the terms due to thermal noise:

$$Q(\theta, \theta_0) = \frac{N}{N_0} \frac{|\alpha\chi(\theta, \theta_0) + \beta\chi(\theta, \theta_J) - \alpha\chi(\theta, \theta_J)\chi(\theta_J, \theta_0) - \beta\chi(\theta, \theta_J)|^2}{1 - \chi^2(\theta, \theta_J)} \tag{5.73}$$

$$Q(\theta, \theta_0) = \frac{N|\alpha|^2}{N_0} \frac{[\chi(\theta, \theta_0) - \chi(\theta, \theta_J)\chi(\theta_J, \theta_0)]^2}{1 - \chi^2(\theta, \theta_J)} \tag{5.74}$$

To understand the action of the receiver, it is interesting to rewrite $Q(\theta, \theta_0)$ as a function of the output $C(\theta)$ of the conventional receiver $C(\theta) = \mathbf{a}^t(\theta)\mathbf{z}$ (usually referred to as the conventional beamformer):

$$Q(\theta, \theta_0) = \frac{|C(\theta) - C(\theta_J)\chi(\theta, \theta_J)|^2}{1 - \chi^2(\theta, \theta_J)} \frac{1}{NN_0} \tag{5.75}$$

The operation carried out by the optimum receiver can thus be interpreted as follows:

- Calculation of the conventional beamformer output for the steering direction θ;
- Calculation of the conventional beamformer output for the steering direction θ_J, and subtraction of the contribution of this jammer to the output for the steering direction θ, in amplitude and phase;
- Normalization to ensure, at the receiver output, a noise + jammer power independent of the steering direction θ.

This normalization "with respect to the power of the whitened replica" is in fact only necessary when $\chi^2(\theta_J, \theta)$ is not negligible relative to 1. In practice, this means that normalization can be omitted if the jammer is not located within the same antenna beamwidth as the target. Conversely, normalization is necessary when the jammer and the target are located within the same antenna beamwidth, that is, when the processing is required to have better resolution properties than those used conventionally ("super-resolution").

This interpretation of the processing performed by the optimum receiver obviously agrees with the conclusions of Section 4.4 concerning the suppression of jammers in the time domain. It is important to see that the suppression of the jammer is possible through information on its location provided by the noise-only correlation matrix—then through the subtraction of its estimated amplitude and phase contribution, which is enabled by the simultaneous formation of beams $C(\theta)$ and $C(\theta_J)$ in the directions of the target and the jammer.

Let us calculate the signal-to-noise ratio at the receiver output before quadratic detection, as a function of the jammer position:

$$\frac{S}{N}(\theta_0, \theta_J)$$

According to (5.65), the receiver calculates $y = \mathbf{w}^t\mathbf{z}$, where

$$\mathbf{w} = \frac{\mathbf{K}^{-1}\mathbf{a}(\theta)}{[\mathbf{a}^t(\theta)\mathbf{K}^{-1}\mathbf{a}(\theta)]^{1/2}}$$

The output signal power, when the receiver is steered in the signal direction, is then

$$|\alpha|^2 = \frac{|\mathbf{a}^t(\theta_0)\mathbf{K}^{-1}\mathbf{a}(\theta_0)|^2}{\mathbf{a}^t(\theta_0)\mathbf{K}^{-1}\mathbf{a}(\theta_0)}$$

and the output average noise + jammer power is:

$$\frac{\mathbf{a}^t(\theta_0)\mathbf{K}^{-1}\mathbf{K}\mathbf{K}^{-1}\mathbf{a}(\theta_0)}{\mathbf{a}^t(\theta_0)\mathbf{K}^{-1}\mathbf{a}(\theta_0)} = 1$$

which results, by construction, from the fact that the receiver whitens and normalizes the noise (5.60).

The output signal-to-noise ratio can therefore be written as:

$$\frac{S}{N}(\theta_0, \theta_J) = \mathbf{a}^t(\theta_0)\mathbf{K}^{-1}\mathbf{a}(\theta_0)|\alpha|^2$$

Given (5.71) for $\mathbf{K}^{-1}$, we obtain:

$$\frac{S}{N}(\theta_0, \theta_J) = \frac{|\alpha|^2 N}{N_0}[1 - |\chi(\theta_0, \theta_J)|^2]$$

It therefore appears that the signal-to-noise ratio becomes zero in the jammer direction (a more precise analysis shows that: $\frac{S}{N} \to \frac{|\alpha|^2}{J}$).

Calculating the derivative of $Q(\theta, \theta_0)$, (5.74) enables us to verify that $(dQ/d\theta) = 0$ at $\theta = \theta_0$, $\forall\ \theta_J$; as a result, the target is correctly located in average, regardless of the jammer position, assuming that the jammer-to-thermal-noise ratio tends to infinity (the estimation of θ is asymptotically unbiased).

Location accuracy can be estimated by calculating the second derivative of the ambiguity function $Q(\theta, \theta_0)$ at θ_0. We obtain, after all calculations have been completed:

$$\frac{1}{|\alpha|^2}\frac{N_0}{2N}\frac{d^2Q}{d\theta^2}\bigg|_{\theta=\theta_0} = \chi''(\theta_0, \theta_0) + \frac{\chi'^2(\theta_0, \theta_J)}{1 - \chi^2(\theta_0, \theta_J)} \tag{5.76}$$

This expression is to be compared with what would be obtained in the absence of jammer (or for $\chi'(\theta_0, \theta_J) = 0$)

$$\frac{1}{|\alpha|^2}\frac{N_0}{2N}\frac{d^2Q}{d\theta^2}\bigg|_{\theta=\theta_0} = \chi''(\theta_0, \theta_0) \tag{5.77}$$

It appears that the presence of a jammer induces a degradation of angular accuracy, since the term $\frac{\chi'^2(\theta_0, \theta_J)}{1 - \chi^2(\theta_0, \theta_J)}$ is always positive (or null). To evaluate this accuracy degradation, let us consider the case in which the jammer is "in the main lobe" of the array, that is, less than one beamwidth from the useful target, and let us approximate this lobe with an exponential:

$$\chi(\theta, \theta_0) = \exp\left[-\beta\frac{(\theta_0 - \theta_J)^2}{\theta_a^2}\right] \tag{5.78}$$

where θ_a is the 3-dB width of the main lobe, if the coefficient β is equal to 1.4.

$$\chi'(\theta_0, \theta_J) = -2\beta\frac{(\theta_0 - \theta_J)}{\theta_a^2}\exp\left[-\beta\frac{(\theta_0 - \theta_J)^2}{\theta_a^2}\right]$$

$$\chi''(\theta_0, \theta_J) = -2\frac{\beta}{\theta_a^2}\exp\left[-\beta\frac{(\theta - \theta_0)^2}{\theta_a^2}\right] + \frac{4\beta^2}{\theta_a^4}(\theta - \theta_0)^2\exp\left[-\beta\frac{(\theta - \theta_0)^2}{\theta_a^2}\right] \tag{5.79}$$

for

$$\theta = \theta_0 \colon \chi''(\theta_0, \theta_0) = -2\frac{\beta}{\theta_a^2}$$

Expression (5.76) can then be written, for $\theta_0 - \theta_J = \Delta\theta << \theta_a$, as:

$$\frac{N_0}{2|\alpha|^2 N}\frac{d^2Q}{d\theta^2}\bigg|_{\theta=\theta_0} \sim -\frac{2\beta}{\theta_a^2}\left[1 - \frac{2\beta\left(\frac{\Delta\theta}{\theta_a}\right)^2}{\exp\left(2\beta\left(\frac{\Delta\theta}{\theta_a}\right)^2\right) - 1}\right] \tag{5.80}$$

to be compared, from (5.77), to $-\frac{2\beta}{\theta_a^2}$ in the absence of the jammer. For small values of $\Delta\theta/\theta_a$, the ratio of the standard deviations with and without jamming is approximately equal to:

$$\frac{\sigma_{\text{without jamming}}}{\sigma_{\text{with jamming}}} = \sqrt{1 - \frac{2\beta\left(\frac{\Delta\theta}{\theta_a}\right)^2}{\exp\left(2\beta\left(\frac{\Delta\theta}{\theta_a}\right)^2\right) - 1}} \sim \sqrt{\beta}\frac{\Delta\theta}{\theta_a}$$

5.5.3 Adaptive Digital Beamforming with Noise-Only Reference

It was shown in Section 3.5.1 [(3.66) and (3.67)] that the optimal angle-range-Doppler processor could be decomposed into an angular processor followed by a range-Doppler processor.

Similarly, the adaptive angular processing described by (5.65) and (5.66) can be combined with the usual Doppler processing in each range gate. The angular processing amounts to spatial filtering with the filter $\mathbf{W}(\theta)$ shown in Figure 5.11:

$$\mathbf{w}(\theta) = \frac{\mathbf{K}^{-1}\mathbf{a}(\theta)}{\sqrt{\mathbf{a}^{t}(\theta)\mathbf{K}^{-1}\mathbf{a}(\theta)}} \tag{5.81}$$

This angular processing is followed, for each direction θ and in each range cell, by the time correlation (Doppler analysis, which usually amounts to Fourier transform in radar applications), quadratic detection, and thresholding, for target detection—or maximization relative to θ, for angular location.

A useful measure of the efficiency of the adaptive processing is the output signal to interference (jammers + noise) ratio S/I:

$$\left(\frac{S}{I}\right)_{\text{OUT}} = \frac{|\alpha|^2 |\mathbf{W}^{t}\mathbf{a}(\theta)|^2}{|\mathbf{W}^{t}\mathbf{b}|^2} = |\alpha|^2 |\mathbf{a}^{t}(\theta)\mathbf{K}^{-1}\mathbf{a}(\theta)| \tag{5.82}$$

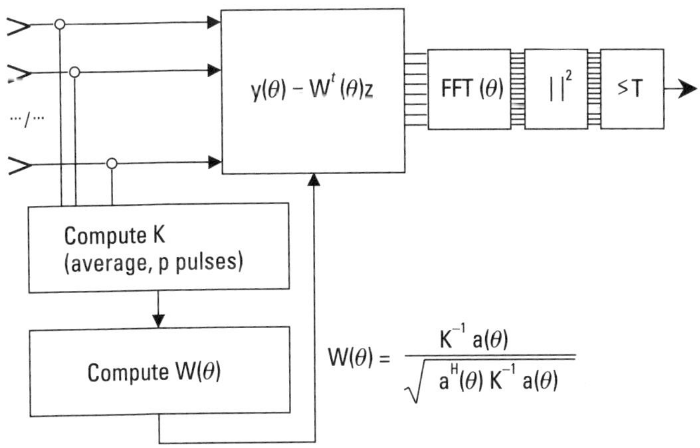

Figure 5.11 Digital beamforming with noise-only reference.

This expression—already used in Section 5.5.2—allows a prediction of performances to be conducted for each specific interference situation. However, in this implementation, the main issue is the estimation of the correlation matrix **K**. For rejection of continuous noise jammers, a good estimate can be obtained by averaging the estimation on successive range gates where few targets can be present (e.g., the farthest range gates where targets' contribution should be minimal).

5.5.4 Processing Without Noise-Only Reference

In the absence of a noise-only reference, the processing defined by (5.61) is no longer applicable, as the sensor noise correlation matrix **K** is no longer available. It is then interesting to examine what becomes of this processing when we use the correlation matrix of the noisy signals instead of matrix **K**:

$$\mathbf{R} = \mathbf{K} + |\alpha|^2 \mathbf{a}(\theta)\mathbf{a}^t(\theta) \text{ (noise and signals are assumed independent)} \tag{5.83}$$

Using the lemma of the inverse matrix (Appendix 5A), and substituting into (5.61), we obtain:

$$\frac{|\mathbf{a}^t(\theta)\mathbf{R}^{-1}\mathbf{z}|^2}{\mathbf{a}^t(\theta)\mathbf{R}^{-1}\mathbf{a}(\theta)} = \frac{|\mathbf{a}^t(\theta)\mathbf{K}^{-1}\mathbf{z}|^2}{\mathbf{a}^t(\theta)\mathbf{K}^{-1}\mathbf{a}(\theta)[1 + |\alpha|^2\mathbf{a}^t(\theta)\mathbf{K}^{-1}\mathbf{a}(\theta)]} \tag{5.84}$$

$$\frac{\mathbf{a}^t(\hat{\theta})\mathbf{R}^{-1}\mathbf{z}}{\mathbf{a}^t(\hat{\theta})\mathbf{R}^{-1}\mathbf{a}(\hat{\theta})} = \frac{\mathbf{a}^t(\hat{\theta})\mathbf{K}^{-1}\mathbf{z}}{\mathbf{a}^t(\hat{\theta})\mathbf{K}^{-1}\mathbf{a}(\hat{\theta})} = \hat{\alpha}$$

Consequently, if $|\alpha|^2\mathbf{a}^t(\theta)\mathbf{K}^{-1}\mathbf{a}(\theta)$ is much lower than 1, that is, if the signal-to-noise ratio at the processing output is small, then this processing remains very close to the optimum processing. This means that this processing is only close to the optimum processing if it is followed by other processes designed to extract the signal from the remaining noise, for example, if this so-called spatial processing of the sensor network is followed by a Doppler analysis, which will eliminate, by filtering, a significant part of the remaining noise.

When the signal-to-noise ratio at the processing output is not low, another processing suitable for the situation must be defined. This is what we are going to do now by defining the optimum processing without noise-only reference.

Let **R** be the correlation matrix of the noisy signals recorded on the sensors. The problem then consists of estimating **K**, α, and θ from these noisy signals (in fact, the only quantity we want to determine is generally θ, but the estimation of θ requires that α and **K** be known). A reasonable way of estimating **K** consists of writing:

$$\hat{\mathbf{K}} = \mathbf{R} - |\hat{\alpha}|^2 \mathbf{a}(\hat{\theta})\mathbf{a}^t(\hat{\theta}) \tag{5.85}$$

This amounts to saying that the autocorrelation **R** of the noisy signal is equal to the sum of the signal autocorrelation (for its estimated parameters $\hat{\alpha}$ and $\hat{\theta}$) and the noise autocorrelation **K**, with signal and noise assumed uncorrelated.

If we then apply the lemma of the inverse matrix (Appendix 5A), we obtain, by replacing in (5.61) **K** by its estimate $\hat{\mathbf{K}}$:

$$\hat{\theta}/ = \frac{|\mathbf{a}^t(\theta)\mathbf{R}^{-1}\mathbf{z}|^2}{\mathbf{a}^t(\theta)\mathbf{R}^{-1}\mathbf{a}(\theta)} \text{Max}/\theta \tag{5.86}$$

$$\hat{\alpha} = \frac{\mathbf{a}^t(\hat{\theta})\mathbf{R}^{-1}\mathbf{z}}{\mathbf{a}^t(\hat{\theta})\mathbf{R}^{-1}\mathbf{a}(\hat{\theta})}$$

It then appears that the structure of the receiver is the same as in the case where a noise-only reference is available, the correlation matrix of the noise being replaced by that of the noisy signals. In contrast, although the structure of the receiver is the same, the performance will obviously be worse, as the information concerning the noise is mixed with that concerning the signal.

For a better understanding of the operation of the receiver thus defined, it is interesting to calculate the average values of quantities (5.86) by statistical averaging of **z** for a given situation; that is, for a given value of the parameters θ, we obtain:

$$\text{E}\left[\frac{|\mathbf{a}^t(\theta)\mathbf{R}^{-1}\mathbf{z}|^2}{\mathbf{a}^t(\theta)\mathbf{R}^{-1}\mathbf{a}(\theta)}\right] = 1 \tag{5.87}$$

independent of θ and **z**

$$E[|\hat{\alpha}|^2] = \frac{1}{\mathbf{a}^t(\theta)\mathbf{R}^{-1}\mathbf{a}(\theta)} \tag{5.88}$$

This result simply expresses the fact that the procedure used to estimate **K** in (5.85) amounts to assuming, for each direction θ, that there is a target in this direction and to determine its level. This result is not surprising inasmuch as, in the absence of noise-only reference, there is nothing to differentiate a target from a colored noise component.

Therefore, in practice, reception consists of estimating **R**, the correlation matrix of the signals received on the sensors, then calculating, for each value of θ, the expression:

$$\frac{1}{\mathbf{a}^t(\theta)\mathbf{R}^{-1}\mathbf{a}(\theta)} \tag{5.89}$$

The maximum of this expression indicates the direction $\hat{\theta}$ of the target, and the level of this maximum is the estimated power $|\alpha|^2$ of this target.

Note that this processing is also used for spectral analysis of time signals; the components of the vector $\mathbf{a}(\theta)$ are then the series of samples of the replica. For spectral analysis, $\mathbf{a}(\theta)$ is the series of samples of a sine wave with frequency θ. In publications dealing with modern spectral analysis, this processing is known as *maximum likelihood* processing (strictly speaking, this should be termed "generalized maximum likelihood in the absence of noise-only reference").

5.5.5 Radar Angle-Error Measurement in the Presence of External Interference

Equation (5.89) for estimation in the presence of colored noise applies directly to the case of tracking antennas discussed in Section 3.4. Remember that such antennas deliver a vector signal consisting of a sum channel and a difference channel [(see (3.77) and (3.78)]:

$$\mathbf{z}^t = (\Sigma, \Delta) \tag{5.90}$$

where

$$\Sigma = \alpha D_\Sigma(\theta) + n_\Sigma \approx \alpha(1 - \theta^2/\theta_a{}^2) + n_\Sigma \tag{5.91}$$

$$\Delta = \alpha\theta_a D'_\Sigma(\theta) + n_\Delta \approx \alpha(1 - 2\theta/\theta_a) + n_\Delta \tag{5.92}$$

θ_a: antenna beamwidth

$$\mathbf{a}^t(\theta) = [D_\Sigma(\theta),\ \theta_a D'_\Sigma(\theta)] = [1 - \theta^2/\theta_a{}^2,\ -2\theta/\theta_a] \tag{5.93}$$

The problem consists of estimating θ on the basis of consecutive samples of the received signal z, in the absence of noise-only reference. According to the preceding paragraph, when n_Σ and n_Δ are colored noise samples, the processing determines the maximum with respect to θ of the expression:

$$\frac{1}{\mathbf{a}^t(\theta)\mathbf{R}^{-1}\mathbf{a}(\theta)}$$

The correlation matrix $\mathbf{R}$ is here written as:

$$\mathbf{R} = \begin{bmatrix} \mathrm{E}|\Sigma|^2 & \mathrm{E}(\Sigma\Delta^*) \\ \mathrm{E}(\Delta\Sigma^*) & \mathrm{E}|\Delta|^2 \end{bmatrix} \tag{5.94}$$

From which we can deduce:

$$\mathbf{R}^{-1} = \begin{bmatrix} \mathrm{E}|\Delta|^2 & -\mathrm{E}(\Sigma\Delta^*) \\ -\mathrm{E}(\Delta\Sigma^*) & \mathrm{E}|\Sigma|^2 \end{bmatrix} \times \frac{1}{\mathrm{E}|\Sigma|^2\mathrm{E}|\Delta|^2 - |\mathrm{E}(\Delta\Sigma^*)|^2} \tag{5.95}$$

In the tracking phase, we determine the maximum of the expression by canceling its derivative:

$$\frac{d}{d\theta}\left[\frac{1}{\mathbf{a}^t(\theta)\mathbf{R}^{-1}\mathbf{a}(\theta)}\right] = 0 \Rightarrow \mathrm{Re}\left[\frac{d\mathbf{a}^t(\theta)}{d\theta}\mathbf{R}^{-1}\mathbf{a}(\theta)\right] = 0 \tag{5.96}$$

Substituting in this relation the expression of $\mathbf{a}(\theta)$ from (5.93) of its derivative, and of $\mathbf{R}^{-1}$, we obtain:

$$\hat{\theta} = -\theta_a \frac{\mathrm{Re}[\mathrm{E}(\Delta\Sigma^*)]}{2\mathrm{E}|\Sigma|^2 - \mathrm{E}|\Delta|^2} \sim -\theta_a \frac{\mathrm{Re}[\mathrm{E}(\Sigma\Delta^*)]}{2\mathrm{E}|\Sigma|^2} \text{ since } \frac{|\Delta|}{|\Sigma|} \sim \frac{2\theta}{\theta_a} << 1 \tag{5.97}$$

The angle-error measurement expression is therefore similar to the expression obtained for the angle-error measurement in the presence of white noise, except that the numerator $\mathrm{Re}(\Sigma\Delta^*)$ and the denominator $|\Sigma|^2$ are first averaged over all the available samples as long as the angular position of the target has not changed significantly. It will be shown in Section 7.3.3 that tracking radars can use this angle-error measurement procedure for improved location of extended targets or in the presence of spurious reflections from the ground.

5.5.6 Capon's Adaptive Processing by Constrained Minimization

The processing defined by (5.89) can also be obtained by an energy minimization type argument, which will be used, as in Section 5.3, to derive the structure of an adaptive receiver.

In this heuristic approach, we attempt to build, for each value of the parameter θ, the linear receiver (filter) that does not affect the signal while reducing as much as possible the average output power, which amounts to reducing the output noise power as much as possible. This is written, if $\mathbf{z}$ is the vector of the received samples, as:

$$\begin{cases} \text{Determine } \mathbf{W}\text{: } \mathrm{E}|y|^2 \text{ min, with: } y = \mathbf{W}^{\mathrm{t}}\mathbf{z} \\ \text{Constraint: } \mathbf{W}^{\mathrm{t}}\mathbf{a}(\theta) = 1 \end{cases} \tag{5.98}$$

Using Lagrange's multiplier:

$$\mathrm{Min}/_{\mathbf{w}} H(\mathbf{W}) = \frac{1}{2}\mathrm{E}(\mathbf{W}^{\mathrm{t}}\mathbf{Z}\mathbf{Z}^{t}\mathbf{W}) + \lambda(\mathbf{W}^{\mathrm{t}}\mathbf{a} - 1)$$

$$\mathrm{Min}/_{\mathbf{w}} H(\mathbf{W}) = \frac{1}{2}\mathbf{W}^{\mathrm{t}}\mathbf{R}\mathbf{W} + \lambda(\mathbf{W}^{\mathrm{t}}\mathbf{a} - 1)$$

$$\begin{cases} \nabla_w H(\mathbf{W}) = \mathbf{R}\mathbf{W} + \lambda\mathbf{a} = 0 & \Rightarrow \mathbf{W} = \lambda\mathbf{R}^{-1}\mathbf{a} \\ \mathbf{W}^{\mathrm{t}}\mathbf{a} = 1 & \Rightarrow \lambda = \dfrac{1}{\mathbf{a}^{t}\mathbf{R}^{-1}\mathbf{a}} \end{cases} \tag{5.99}$$

where finally

$$\mathbf{W} = \frac{\mathbf{R}^{-1}\mathbf{a}(\theta)}{\mathbf{a}^{t}(\theta)\mathbf{R}^{-1}\mathbf{a}(\theta)} \tag{5.100}$$

We thus obtain a receiver that calculates, for each value of the parameters,

$$y = \frac{\mathbf{a}^t(\theta)\mathbf{R}^{-1}\mathbf{z}}{\mathbf{a}^t(\theta)\mathbf{R}^{-1}\mathbf{a}(\theta)} \tag{5.101}$$

and determines $\hat{\theta}$ by seeking the maximum of $|y(\theta)|^2$ (or performs detection by comparison of $|y|^2$ with a threshold).

This receiver is none other than the estimator of $|\alpha|^2$ obtained previously [see (5.86)] as generalized optimum processing in the absence of noise-only reference—as we have seen that θ estimator became ineffective in this case [see (5.87)].

As a result, in the absence of a noise-only reference, the receiver determines the maximum of:

$$\frac{1}{\mathbf{a}^t(\theta)\mathbf{R}^{-1}\mathbf{a}(\theta)}$$

or, similarly, determines, the maximum of $|y(\theta)|^2$:

$$y = \frac{\mathbf{a}^t(\theta)\mathbf{R}^{-1}\mathbf{z}}{\mathbf{a}^t(\theta)\mathbf{R}^{-1}\mathbf{a}(\theta)} \text{ since } \mathrm{E}|y(\theta)|^2 = \frac{1}{\mathbf{a}^t(\theta)\mathbf{R}^{-1}\mathbf{a}(\theta)}$$

$y(\theta)$ results from the minimization of the output power of filter y: $\mathbf{W}^t\mathbf{Z}$, under the constraint of letting the desired signal $\mathbf{a}(\theta)$ pass.

The adaptive angular receiver can be combined, as in Section 5.5.3, with the usual Doppler processing in each range gate. The angular processing amounts to spatial filtering with the filter $\mathbf{W}(\theta)$ [see (5.100)] shown in Figure 5.12:

$$\mathbf{W}(\theta) = \frac{\mathbf{R}^{-1}\mathbf{a}(\theta)}{\mathbf{a}^t(\theta)\mathbf{R}^{-1}\mathbf{a}(\theta)} \tag{5.102}$$

where $\mathbf{R}$ is estimated in the range gate under test (signal + noise).

This angular processing is followed, for each direction θ, and in each range cell, by the time correlation (Fourier Doppler analysis), quadratic

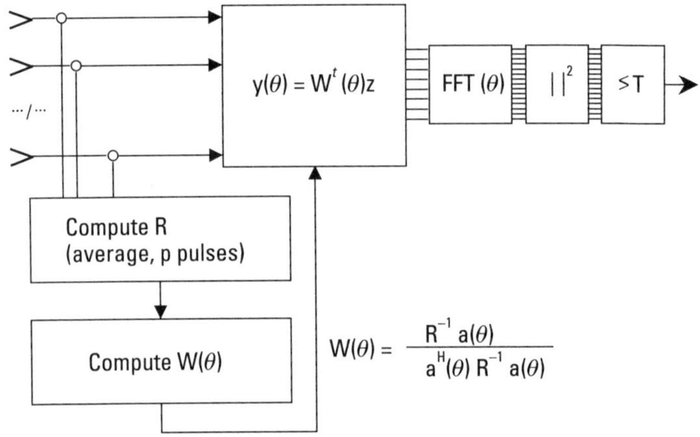

Figure 5.12 Digital beamforming without noise-only reference.

detection, and thresholding, for target detection—or maximization relative to θ for angular location.

A useful measure of efficiency of the adaptive processing is the output signal to interference + noise ratio S/I, in the signal direction:

$$\left(\frac{S}{I}\right)_{\text{OUT}} = \frac{|\alpha|^2 |\mathbf{W}^t \mathbf{a}(\theta)|^2}{|\mathbf{W}^t \mathbf{b}|^2}$$

Using the matrix inversion lemma for $\mathbf{R} = \mathbf{K} + |\alpha|^2 \mathbf{a}(\theta)\mathbf{a}^t(\theta)$, the spatial filter is written:

$$\mathbf{W} = \frac{\mathbf{R}^{-1}\mathbf{a}(\theta)}{\mathbf{a}^t(\theta)\mathbf{R}^{-1}\mathbf{a}(\theta)} = \frac{\mathbf{K}^{-1}\mathbf{a}(\theta)}{\mathbf{a}^t(\theta)\mathbf{K}^{-1}\mathbf{a}(\theta)}$$

Therefore, the expression of the signal to interference + noise is the same as (5.82)

$$\left(\frac{S}{I}\right)_{\text{OUT}} = |\alpha|^2 |\mathbf{a}^t(\theta)\mathbf{K}^{-1}\mathbf{a}(\theta)| \tag{5.103}$$

The main issue in this implementation is again the estimation of matrix **R**, which requires as many samples as possible while staying in the range gate under test. One possible solution is to split up the pulse train in different "subtrains" and average the signals over the different subtrains, at the cost of smaller coherent trains, which means a lower Doppler resolution. Another solution consists of commuting the spatial and Doppler processing; the Doppler processing is implemented first, and followed by angular beamforming where the averaging is made on the different Doppler frequencies. Anyway, this processing structure without noise-only reference is the only solution when target density is too high to allow noise-only estimation.

Figure 5.13 shows the 16-antenna demonstrator SAFRAN, built and experimented by THALES (formerly Thomson-CSF), and a typical result of adaptive processing without noise-only reference. In this situation, the cancellation ratio is larger than 40 dB, so the target at 750 Hz Doppler can be easily detected, and the tracking scenario demonstrates an excellent tracking of two targets in presence of two noise jammers.

5.5.7 Application: Space-Time Processing

The adaptive algorithm thus defined applies directly to the case of space processing of a sensor array (Figure 5.14). Let us assume that the antenna made up by the sensors is looking for the presence of a target in a given direction θ_0, and that the transmitted signal is a pure wave $e^{j\omega_0 t}$. The phase shifts $\Delta_1, \Delta_2, \ldots, \Delta_N$ are first applied to the signals received by the N sensors, to steer the antenna in the desired direction, in accordance with the processing described in Section 3.5.1 (with $\Delta_i = \exp\left[-2\pi j \frac{\xi_i}{\lambda} \sin\theta_0\right]$), where ξ_i is the abscissa of sensor number i in this case of linear array of sensors aligned along axis $O\xi$.

Downstream of each phase shifter is placed a gain w_i, governed by the algorithm, under the constraint:

$$\Sigma w_i = \sqrt{N} \tag{5.104}$$

as in this case, the desired signal is $\mathbf{a}^t = \frac{1}{\sqrt{N}}(1 \ldots . 1)^t$ (the factor $1/\sqrt{N}$ being introduced to comply with: $\mathbf{a}^t\mathbf{a} = 1$). The algorithm then calculates $\mathbf{W}$ according to (5.101):

$$y = \mathbf{W}^t z = \frac{\mathbf{a}^t(\theta)\mathbf{R}^{-1}\mathbf{z}}{\mathbf{a}^t(\theta)\mathbf{R}^{-1}\mathbf{a}(\theta)}$$

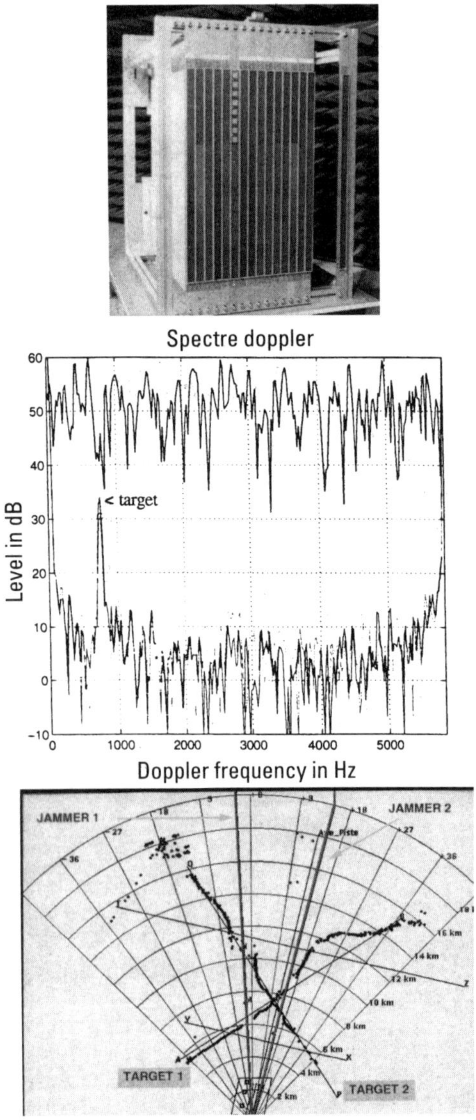

Figure 5.13 SAFRAN demonstrator.

The same algorithm is also applied to "space-time" processing of sensor array of the type shown in Figure 5.15. In the same way that the "spatial" phase shifts Δ_i were applied downstream of each sensor, "time" phase shifts Γ_i are applied between the consecutive signal samples on each sensor, to adapt the antenna to the pulsation ω_0. Consequently:

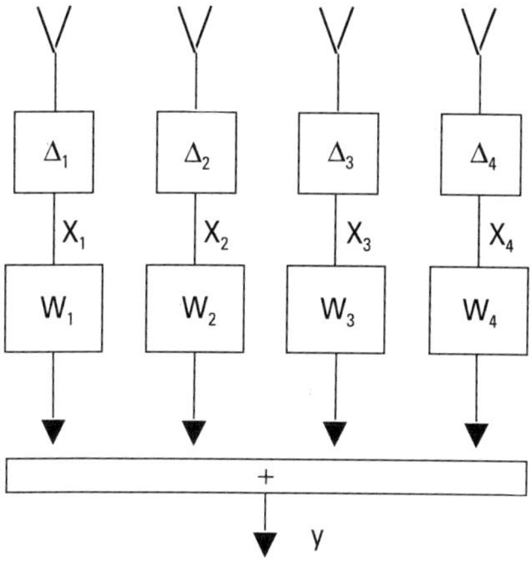

Figure 5.14 Spatial processing.

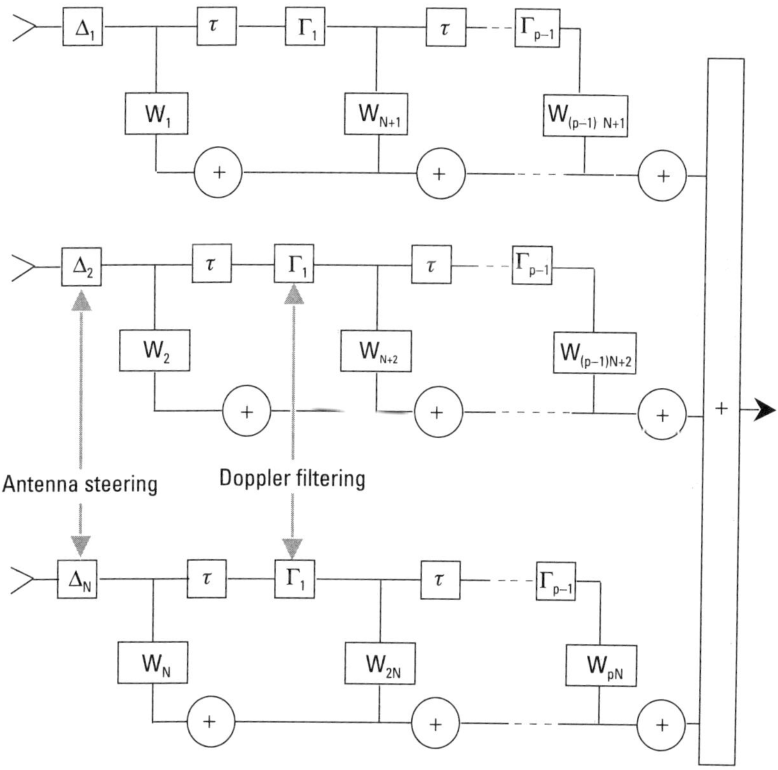

Figure 5.15 Optimum space-time processing.

$$\Delta_i = e^{-2\pi j \frac{\xi_i}{\lambda} \sin \theta_0} \quad \xi_i\text{: abscissa of sensor number } i \tag{5.105}$$

$$\Gamma_i = e^{j\omega_0 i\tau} \quad \tau\text{: sampling period} \tag{5.106}$$

Given these phase shifts, the expression for the desired signal remains:

$$\mathbf{a} = \frac{1}{\sqrt{pN}} \begin{pmatrix} 1 \\ 1 \\ \vdots \\ 1 \end{pmatrix} \quad (p \times N \text{ rows, 1 column})$$

and (5.101) again sets the rules for calculating the gains.

It is important to note that in this space-time processing, constraint $\mathbf{W}^{\mathrm{t}}\mathbf{a}(\theta) = 1$ is itself a space-time constraint, since θ then represents both the angular and Doppler parameters.

The algorithm therefore determines, for each possible angle and Doppler shift, the vector $\mathbf{W}$ of the filters, which enables the targets or spurious signals located at the other angles and other Doppler shifts to be best rejected.

5.5.8 Equivalence Between Adaptive Suppression and Constrained Minimization

Let us return to the sensor array example described in Section 5.3.4 (Figure 5.7). Downstream of each sensor i a complex gain w_1' is applied:

$$\begin{cases} w_1' = 1 - w_1 \\ w_2' = 1 + w_1 - w_2 \\ \cdots \\ w_N' = 1 + w_{N-1} \end{cases}$$

As a result, these gains verify the constraint:

$$\sum_i w_1' = N$$

which is not different from constraint (5.98), $\mathbf{W}^{\mathrm{t}}\mathbf{a} = 1$, given the fact that in this case the expected signal is, after the phase shifts Δ_i, identical on all the sensors.

Therefore it appears that, in spatial processing of sensor arrays, the adaptive noise subtraction carried out in Section 5.3 and the constrained minimization carried out in Section 5.5.6 are two versions of the same processing; in both cases, the adopted criterion is the minimization of the output power, but in one case the constraint is implicitly "hardwired" (by formation of the reference channel), while in the other case it is taken into account explicitly by the algorithm.

This equivalence, demonstrated under more general conditions in [6], simply translates the intuitive equivalence between two procedures:

1. Subtracting the best estimate of the noise from the received signals;
2. Minimizing the output energy under the constraint of letting the signal pass.

In both cases, we have implicitly assumed that "what is not signal is noise:" in the first case by subtraction of the sensors in pairs, and in the second case by constrained minimization. This is also the exact sense of (5.83), which enables the correlation matrix of the noise **K** to be estimated from the measured correlation matrix **R**.

5.5.9 Calibration [7, 8, 9]

5.5.9.1 Calibration Errors

The previously described algorithms implicitly assume that all the receiving channels are known in amplitude and phase [since the form of the replica $\mathbf{a}(\theta)$ is assumed to be known]. This assumption is not always valid: when no special care is taken, the fluctuations in amplitude from one receiving channel to another can be larger than 3 dB, and the phase difference larger than 10°. Furthermore, the angular diagrams of the radiating elements constituting the antenna array are not perfectly identical (or perfectly known), essentially because of interactions between the radiating elements and the extremities of the array. In order to reduce these imperfections, an internal calibration is usually implemented, reducing the unbalance between different channels to within 0.5 dB in amplitude and 1° in phase (orders of magnitude). However, the imperfect knowledge of the elementary diagrams still remains, with uncertainty (and fluctuations) often higher than the fluctuations from channel to channel.

In a first step, it is interesting to quantify the effect of these uncertainties on the array diagram, defined in (5.67) as the response of the antenna array to a single target in direction θ_0, in the absence of noise and diagram uncertainties [7]:

$$\chi(\theta,\ \theta_0) = \frac{1}{N}\,\mathbf{a}^t(\theta)\mathbf{a}(\theta_0)$$

For a linear array with elementary radiating elements, at abscissas d_i, and for the steering direction $\theta = 0$, this expression is written:

$$\chi(\theta,\ 0) = \sum_{i=0}^{N} a_i e^{-j\frac{2\pi}{\lambda}d_i \sin\theta} \tag{5.107}$$

where the coefficients a_i are a weighting function chosen to minimize the sidelobes of the antenna [in (5.67) the weighting is assumed to be uniform: $a_i = 1/\sqrt{N}\ \forall\ i$].

Let δa_i and $\delta\varphi_i$, respectively, be the zero-mean uncorrelated errors in amplitude and phase. The pattern actually computed by the receiver, when these errors are present, is written as:

$$\chi_{\text{err}}(\theta,\ 0) = \sum_{i=0}^{N}(a_i + \delta a_i)e^{-j\frac{2\pi}{\lambda}d_i \sin\theta}e^{j\delta\varphi_i} \tag{5.108}$$

The received power is written:

$$\mathrm{E}[|\chi_{\text{err}}(\theta,0)|^2] = \sum_{i=1}^{N}\sum_{k=1}^{N}\mathrm{E}[(a_i + \delta a_i)(a_k + \delta a_k)e^{j(\delta\varphi_i - \delta\varphi_k)}]e^{-j\frac{2\pi}{\lambda}(d_i - d_k)\sin\theta}$$

$$\mathrm{E}[|\chi_{\text{err}}(\theta,0)|^2] = \sum_{i=1}^{N}\sum_{k=1}^{N}\mathrm{E}[e^{j(\delta\varphi_i - \delta\varphi_k)}]e^{-j\frac{2\pi}{\lambda}(d_i - d_k)\sin\theta} + \sum_{i=1}^{N}\mathrm{E}[\delta a_i^2]$$

Expanding $e^{j(\delta\varphi_i - \delta\varphi_k)}$ to the second order: $e^{j(\delta\varphi_i - \delta\varphi_k)} \approx 1 + j(\delta\varphi_i - \delta\varphi_k) - \dfrac{(\delta\varphi_i - \delta\varphi_k)^2}{2}$, we obtain:

$$\begin{aligned}\mathrm{E}[|\chi_{\text{err}}(\theta,0)|^2] = {} & |\chi(\theta,0)|^2 \\ & - \sum_{i=1}^{N}\sum_{k=1}^{N} a_i a_k \mathrm{E}\left[\frac{(\delta\varphi_i - \delta\varphi_k)^2}{2}\right]e^{-j\frac{2\pi}{\lambda}(d_i - d_k)\sin\theta} \\ & + \sum_{i=1}^{N}\mathrm{E}[\delta a_i^2]\end{aligned}$$

with

$$
\begin{aligned}
|\chi(\theta, 0)|^2 &= \sum_{i=1}^{N} \sum_{k=1}^{N} a_i a_k \mathrm{E}\left[\frac{(\delta\varphi_i - \delta\varphi_k)^2}{2}\right] e^{-j\frac{2\pi}{\lambda}(d_j - d_k)\sin\theta} \\
\mathrm{E}[|\chi_{\mathrm{err}}(\theta, 0)|^2] &= |\chi(\theta, 0)|^2 + \sum_{i=1}^{N} \mathrm{E}[\delta a_i^2] + \sum_{i=1}^{N} a_i^2 \mathrm{E}[\delta\varphi_i^2] \qquad (5.109) \\
&\quad - \sum_{i=1}^{N} \sum_{k=1}^{N} a_i a_k \frac{\mathrm{E}[\delta\varphi_i^2] + \mathrm{E}[\delta\varphi_k^2]}{2} e^{-j\frac{2\pi}{\lambda}(d_i - d_k)\sin\theta}
\end{aligned}
$$

Assuming independent amplitude and phase errors affecting the different sensors, that is, $\forall i$

$$
\mathrm{E}[\delta\varphi_i^2] = \sigma_\varphi^2
$$

$$
\mathrm{E}\left[\frac{\delta a_i^2}{a_i^2}\right] = \sigma_a^2
$$

(standard deviation of relative amplitude error)
we obtain:

$$
\mathrm{E}[|\chi_{\mathrm{err}}(\theta, 0)|^2] = (1 - \sigma_\varphi^2)|\chi(\theta, 0)|^2 + \sum_{i=1}^{N} a_i^2 (\sigma_\varphi^2 + \sigma_a^2) \qquad (5.110)
$$

Two effects are induced by these calibration errors: a loss on the main lobe level, and an increase in the sidelobes level. The loss on the main lobe level is written:

$$
\frac{\mathrm{E}[|\chi_{\mathrm{err}}(0, 0)|^2]}{|\chi_{\mathrm{err}}(0, 0)|^2} = (1 - \sigma_\varphi^2) + \frac{\sum_{i=1}^{N} a_i^2}{\left(\sum_{i=1}^{N} a_i\right)^2} (\sigma_\varphi^2 + \sigma_a^2)
$$

For typical weighting functions a_i, $\dfrac{\sum_{i=1}^{N} a_i^2}{\left(\sum_{i=1}^{N} a_i\right)^2}$ is on the order of $1/N$, the inverse of the gain of the total array of N radiating elements (not taking into account the gain of the individual radiating element). The main lobe loss is thus written:

$$\frac{\mathrm{E}[|\chi_{\text{err}}(0, 0)|^2]}{|\chi(0, 0)|^2} = (1 - \sigma_\varphi^2) + \frac{1}{N}(\sigma_\varphi^2 + \sigma_a^2)$$

or, for large N:

$$\frac{\mathrm{E}[|\chi_{\text{err}}(0, 0)|^2]}{|\chi(0, 0)|^2} \approx (1 - \sigma_\varphi^2) \tag{5.111}$$

The average sidelobe level, compared to the main lobe level, is written:

$$\frac{\mathrm{E}[|\chi_{\text{err}}(\theta, 0)|^2]}{\mathrm{E}[|\chi_{\text{err}}(0, 0)|^2]} = \frac{(1 - \sigma_\varphi^2)|\chi(\theta, 0)|^2 + \sum_{i=1}^{N} a_i^2(\sigma_\varphi^2 + \sigma_a^2)}{(1 - \sigma_\varphi^2)\left(\sum_{i=1}^{N} a_i\right)^2 + \sum_{i=1}^{N} a_i^2(\sigma_\varphi^2 + \sigma_a^2)}$$

From (5.111), we can write, for large N:

$$\frac{\mathrm{E}[|\chi_{\text{err}}(0, 0)|^2]}{|\chi(0, 0)|^2} \approx (1 - \sigma_\varphi^2) \Rightarrow \mathrm{E}[|\chi_{\text{err}}(0, 0)|^2] \approx (1 - \sigma_\varphi^2)\left(\sum_{i=1}^{N} a_i\right)^2$$

from which the average sidelobe level can be written:

$$\frac{\mathrm{E}[|\chi_{\text{err}}(\theta, 0)|^2]}{\mathrm{E}[|\chi_{\text{err}}(0, 0)|^2]} = \frac{|\chi(\theta, 0)|^2}{\left(\sum_{i=1}^{N} a_i\right)^2} + \frac{\sum_{i=1}^{N} a_i^2(\sigma_\varphi^2 + \sigma_a^2)}{\left(\sum_{i=1}^{N} a_i\right)^2(1 - \sigma_\varphi^2)} \tag{5.112}$$

The first term on the left-hand side of this relation is the relative sidelobe level when no calibration error is present. The second term corresponds to the increase in the sidelobe level due to these calibration errors. For large N, it is equal to:

$$\frac{(\sigma_\varphi^2 + \sigma_a^2)}{N}$$

For example, an error of 1/10 (10%, or 0.8 dB) in relative amplitude is equivalent to 100 mrad (5°) error in phase, and these errors give an average increase in sidelobe level: $\frac{1}{N}((1/10)^2 + (1/10)^2) = \frac{1}{N}0.02$, that is, 17 dB below $1/N$. For a horizontal linear array with 30 elements, these errors give a sidelobe level of −17 − 15 = −32 dB in the horizontal plane, presumably tolerable for most applications.

The effect of phase-shift quantization noise can be analyzed in the same way, since phase-shift quantification on p bits amounts to adding phase noise with uniform probability density

Between $-\frac{\pi}{2^p}$ and $+\frac{\pi}{2^p}$, whose standard deviation is $2\frac{\pi}{2^p}\frac{1}{\sqrt{12}}$, that is, $\sigma_\varphi = \frac{\pi}{\sqrt{3}\,2^p}$. Phase noise variance added by phase quantization is therefore: $\sigma^2_{\varphi_{\text{quant.}}} = 5.1 - p.6$ (dB)

5.5.9.2 Adaptive Processing in the Presence of Calibration Errors

The influence of calibration errors is best understood with reference to Figure 5.7, keeping in mind the equivalence between hardwired and soft constraint minimization shown in Section 5.5.8. In Figure 5.7, calibration errors are seen to have two very different effects:

1. On the primary channel (z): Calibration errors will degrade the coherent summation of the signal samples coming from the desired direction.
2. On the reference channel (y): The noise reference, computed by successive differences between adjacent channels, will be polluted by the useful signal, which is no longer cancelled in this differentiation in the presence of calibration errors.

The first effect is generally negligible, as soon as errors are lower than 10° in phase or 1 dB in relative amplitude. The second effect is more serious, since it amounts to introducing a part of the useful signal in the noise reference: it was shown in Section 5.3.1 that the effect was to set a limit on the output signal to interference ratio [(5.18): $\rho_{out} = 1/\rho_{ref}$]. The effect will be similar for phased arrays; if the residual signal in the noise reference channel is X dB below the interference, then the output signal to interference ratio is limited to less than X dB.

Since the residual signal in the reference channel can be evaluated from (5.112), the output signal to interference is lower than:

$$\left(\frac{S}{I}\right)_{\text{output}} \leq \frac{N-1}{(\sigma_a^2 + \sigma_\varphi^2)} \tag{5.113}$$

Briefly, the effect of calibration errors can thus be interpreted as a small mismatch to useful signal, and a bound to the cancellation ratio given by (5.113). Furthermore, this analysis shows that the effect is a decrease in the signal output (due to the presence of signal on the reference channel); in other words, the jammers are still cancelled, but the signal itself is attenuated by this correlation between primary and reference channel.

This analysis also shows that it is necessary to know the pattern of each array element *in the direction of observation only*, since the algorithm only uses these values. More precisely, the algorithms do not require that the array elements pattern be identical, for instance, as is the case for autoregressive modelization.

This explanation of array elements errors is valid for adaptive processing without noise-only reference. Adaptive processing with noise-only reference is not affected by the same degradation if the noise reference is not polluted by useful signals; depending on the exact mode of calculation of this noise reference, the degradation of performances can be analyzed with the same rationale.

5.5.10 Coherent Targets

In the previous demonstrations, total decorrelation was assumed between the useful signal $\mathbf{a}(\theta)$ and the noise. This assumption, which is written $E[\mathbf{a}(\theta)\mathbf{b}^t] = 0$, though fundamental for weights calculation, may not be valid. For example, when two targets at the same Doppler frequency are present in different directions θ_1 and θ_2, one cannot assume, when looking in direction θ_1, that the target in direction θ_2 is part of a noise uncorrelated with the useful signal (for deterministic signals to be decorrelated, their Doppler shifts must be resolvable in the available observation time). Similarly, when the same jamming signal is received from two different directions through multipath propagation, the signals coming from those two directions cannot be assumed to be uncorrelated. It is therefore essential to analyze the effect of such correlated sources on the adaptive algorithms and to define appropriate procedures to deal with these situations.

When target and noise (or undesired targets in another direction) are totally correlated, the assumption of uncorrelated signal and noises is no longer valid. For instance, if two pointlike coherent sources are present, the correlation matrix takes the form:

$$|\alpha_1|^2\mathbf{a}(\theta_1)\mathbf{a}^t(\theta_1) + |\alpha_2|^2\mathbf{a}(\theta_2)\mathbf{a}^t(\theta_2) + 2\,\mathrm{Re}(\mathrm{E}[\alpha_1\alpha_2^*]\mathbf{a}(\theta_1)\mathbf{a}^t(\theta_2)) + N_0\mathbf{I}$$

One way to overcome this difficulty consists of estimating this matrix by means of a spatial averaging known as *spatial smoothing.* The principle of operation will be explained in the case of a linear array with N elementary identical antennas; the received signal is written:

$$\mathbf{z} = \alpha\mathbf{a}(\theta) + \mathbf{b}$$

with

$$\mathbf{a}^t(\theta) = [a_1(\theta), \ldots, a_N(\theta)] \text{ and } a_i(\theta) = \exp\left(2\pi j\frac{\xi_i}{\lambda}\sin\theta\right)$$

where ξ_i is the abscissa of the ith sensor, $\mathbf{b}$ is the noise vector, and α the complex amplitude of the received signal.

The linear array can be subdivided into overlapping subarrays of p sensors, sensors 1 to p constituting the first subarray, sensors 2 to $p + 1$ the second subarray, and so on, and sensors $N - p + 1$ to N constituting the $(N - p + 1)$th subarray. From the kth subarray, a correlation matrix $\mathbf{R}_k$ can be estimated. The *spatially smoothed correlation matrix* is defined as the sample mean of the subarrays correlation matrices:

$$\overline{\mathbf{R}} = \frac{1}{N-p+1}\sum_{k=1}^{N-p+1}\mathbf{R}_k$$

It can then be shown [10] that the spatially smoothed correlation matrix is nonsingular when the number of correlated sources is lower than $N - p + 1$, the number of subarrays. In other words, for example, when two coherent sources are to be observed, at least two subarrays have to be averaged.

An additional refinement consists of combining forward and backward smoothing along the array. The array with sensors ranked in decreasing order, from N to 1, has the same sources but in opposite directions, so that the correlation matrix $\mathbf{R}$, with elements $r_{ij} = \mathrm{E}(z_i z_j^*)$ can be averaged with $\mathbf{R}_B^*$, with elements $r^*_{ij} = \mathrm{E}(z_{N-i} z_{N-j}^*)$. Finally, the correlation matrix is computed as follows:

$$\mathbf{R} = \frac{1}{2}(\mathbf{R}_{\text{smoothed}} + \mathbf{J}\mathbf{R}^*_{\text{smoothed}}\mathbf{J})$$

with

$$\mathbf{R}_{\text{smoothed}} = [r_{ij}]$$

and

$$r_{ij} = \frac{1}{N-p+1} \sum_{k=1}^{N-p+1} \mathrm{E}(z_{i+k-1} z^*_{j+k-1})$$

where $\mathbf{J}$ is the reversal matrix:

$$\mathbf{J} = \begin{bmatrix} 0 & \dots & 0 & 1 \\ \dots & \dots & 1 & 0 \\ 0 & \dots & \dots & \dots \\ 1 & 0 & \dots & 0 \end{bmatrix}$$

It is important to note that this tolerance to coherent sources, which is essential in some active sonar or radar applications, is obtained at the expense of resolution. If q coherent signals are to be observed, then $q/2$ subarrays are needed (taking into account the forward-backward smoothing), each one consisting of at least $q + 1$ elementary antennas (in order to resolve q sources). Finally, $q + q/2 = 3q/2$ elements are needed to resolve q coherent sources—compared with $q + 1$ elements, for noncoherent sources. Furthermore, the angular resolution is degraded, since the number of sensors coherently processed is reduced to allow for spatial smoothing.

5.5.11 Sidelobe Blanking

Sidelobe blanking [11] is an alternative procedure to alleviate problems caused by strong targets or jammer signals received through the angular sidelobes. The technique uses an ancillary antenna (and associated receiver channel) with low gain, and calculates, for each detected signal in the primary channel, the ratio of the levels on the ancillary and primary channels in the same range-Doppler bin; if the signal is coming through a sidelobe of the main antenna, the ratio will be larger than a predetermined blanking threshold—and lower on the opposite hypothesis (Figure 5.16).

The logic then inhibits the detection when the ratio is larger than the blanking threshold, which has also the effect of suppressing possible targets in this bin; the inhibition must therefore be used only for low-duty cycle jammers, for which the loss in detection probability can be tolerated.

This technique is therefore very complementary with adaptive processing, since it can use the same ancillary channel, and it is efficient in situations where adaptive processing is not adequate (low-duty cycle jammers or false targets generators). Its implementation involves careful regulation of the detection loss through optimization of the ancillary antenna pattern (preferably giving a null in the main antenna steering direction) and the receiving channels calibration.

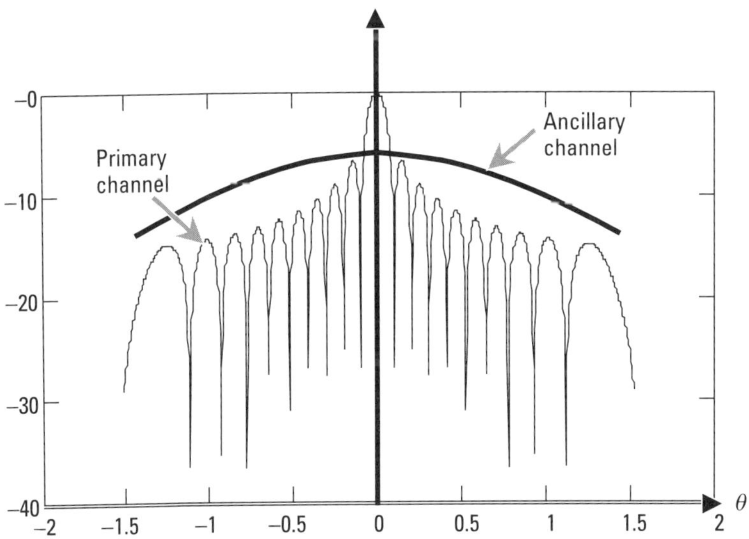

Figure 5.16 Ancillary antenna pattern.

5.6 Discussion

This chapter described the rationale leading from the optimum processing described in Chapter 4 to the adaptive processing used in practice, depending on the application and the assumptions made.

Three structures were reviewed consecutively:

1. Adaptive subtraction, by transversal filter inserted in a noise reference channel;
2. Whitening lattice filter, set according to adjacent bins;
3. Space-time processing (constrained minimization).

These three approaches lead to closely related structures, differing mainly by the algorithm used. They do not constitute, by far, an exhaustive catalogue of the adaptive techniques used in active radar and sonar [7]. However, the review of these techniques enables us to understand how the basic problem of knowledge of the colored noise characteristics (without which it is not possible to improve the processing that was optimum in white noise) can be solved.

Two approaches can be used: the use of a noise-only reference (if available), or constrained minimization, which is also the optimum procedure in the absence of a noise-only reference.

Basically, the first approach amounts to subtracting from the received noisy signal the best possible estimate of the interference, as shown in the jammer suppression example.

The second approach amounts to forcing the processing to let the useful signal pass, while minimizing the output signal power—which amounts, at least intuitively, to assuming that the noise-only reference consists of "all that is not useful signal." This is formalized by the noise correlation matrix estimation rule (5.83), which is, in terms of principles, the basis for constrained minimization processing.

In active radar or sonar, time processing by whitening lattice filter is being used in practice. As sensor arrays (which are often used in sonar, but more rarely in radar) develop, and considering the rapid progress in processing capabilities, the optimum space-time processing described in Section 5.5.5 [receiver forming the quantity (5.89), or (5.101)] could be adopted, in various versions, by future systems.

5.7 Passive Listening

5.7.1 Problem Statement

As we have seen in Chapter 1, passive monitoring includes several types of situations—in the electromagnetic domain, it involves the reception of signals transmitted by independent transmitters such as radio or radars; in sonar, it involves listening to noises emitted by ships or submarines; in aerial acoustics, it involves listening to noises emitted by vehicles or aircraft. In these diverse situations, the form of the signals received by the observation sensors is not completely known; the bandwidth occupied by the signals may be broad, so a delay does not simply represent a phase shift of the received signal; and the interference noise is generally neither white nor uncorrelated from one sensor to the next.

A rather general model of these situations represents the time signals (complex) $x_i(t)$ received by the N different sensors by:

$$x_i(t) = s_i(t) + b_i(t), \qquad i = (1, \ldots N) \tag{5.114}$$

where $b_i(t)$ represents all the interferences, that is, the signals that interfere with the considered application. Generally, these interferences are assumed to be uncorrelated with the useful signals. In what follows, we shall further assume that the observed sources are point sources, and that the deformation produced by the propagation medium is reduced to an attenuation, identical on all source-sensor paths. Although not necessary for all the arguments to come, this approximation enables the principle of the various algorithms to be described more simply. Under these conditions, if D_i is the delay due to the propagation between the source and sensor number i, the signal received by this sensor is given by:

$$x_i(t) = s_i(t - D_i) + b_i(t) \tag{5.115}$$

For convenience, the time origin will be chosen so that:

$$D_1 = 0 \tag{5.116}$$

thus, D_i represents the delay between the signal arriving on sensor i and that arriving on sensor number 1.

In the spectral domain, this signal is given by:

$$\tilde{x}_i(f) = \tilde{s}(f)e^{-2\pi jfD_i} + \tilde{b}_i(f) \tag{5.117}$$

Let us write, in vector form,

$$\mathbf{X}(t) = \mathbf{S}(t) + \mathbf{B}(t) \tag{5.118}$$

where

$$\mathbf{S}(t) = \begin{bmatrix} s(t - D_1) \\ \dots \\ s(t - D_N) \end{bmatrix}$$

$$\tilde{\mathbf{X}}(f) = \tilde{s}(f)\mathbf{a}(\theta) + \tilde{\mathbf{B}}(f) \tag{5.119}$$

with $\mathbf{a}(\theta)$, the vector characterizing the direction of the source (if the sensors are aligned, θ characterizes the angle between the direction of the source and that of the array; if the array is not linear, θ represents the two parameters, elevation and bearing, of the source relative to this array):

$$\mathbf{a}(\theta) = \begin{bmatrix} 1 \\ e^{-2\pi jfD_2(\theta)} \\ \dots \\ e^{-2\pi jfD_N(\theta)} \end{bmatrix} \tag{5.120}$$

In what follows, we shall call $\mathbf{R}(f)$ the cross-spectral density matrix:

$$\tilde{\mathbf{R}}(f) = \mathrm{E}[\tilde{\mathbf{X}}(f)\tilde{\mathbf{X}}^t(f)]$$

This is, at each frequency, the correlation matrix of the signal amplitudes and phases received on the sensors. Similarly, this is the matrix whose elements are the Fourier transforms of the elements of the correlation matrix:

$$\mathbf{R}(t) = \mathrm{E}[\mathbf{X}(t)\mathbf{X}^t(t)]$$

Given $\mathbf{X}(t)$ or $\tilde{\mathbf{X}}(f)$ received, different types of problems may arise:

- *Estimation* of the signal arrival direction, or directions if several signals arrive simultaneously;
- *Detection* of the presence of a signal, in a given direction, known or unknown a priori;
- *Extraction* of the signal from one direction, given a priori or not, excluding interference (listening or reconstruction of the signal, for later identification of the corresponding noise source, for example).

In what follows, we shall first discuss these three aspects in the maximum likelihood sense, using the tools previously developed in this chapter. Then we shall return to the examination of the first two points by a very different approach, based on the use of the properties of the correlation matrix $\mathbf{R}(f)$. Finally, we shall describe a location procedure, now commonly used, based on signals recorded on two sensors.

5.7.2 Maximum Likelihood (Signal + Noise)

5.7.2.1 Method

Let us return to the expression of the received signal vector, in the spectral domain:

$$\tilde{\mathbf{X}}(f) = \tilde{s}(f)\mathbf{a}(\theta) + \tilde{\mathbf{B}}(f) \quad \tilde{\mathbf{X}}, \mathbf{a}, \tilde{\mathbf{B}}\text{: vectors}$$

In this expression, $\tilde{s}(f)$ is an unknown scalar that will be determined if the third problem mentioned above (signal extraction) is posed, $\mathbf{a}(\theta)$ is the vector of known form that characterizes the direction of the source [see (5.120)], and $\tilde{\mathbf{B}}(f)$ is an additive Gaussian noise. As a result, at a given frequency, the problem is formally identical to the radar problem examined previously: detection or estimation of the parameters of a signal of known form $\mathbf{a}(\theta)$ in the presence of an a priori colored noise $\tilde{\mathbf{B}}$ [see (5.59)]. As no noise reference is available (since noise and signals are present simultaneously on the sensors), the optimum processing, according to Section 5.5.4, estimates θ by determining the maximum of:

$$\frac{1}{\mathbf{a}^t(\theta)\tilde{\mathbf{R}}^{-1}(f)\mathbf{a}(\theta)} \text{Max}/\theta \tag{5.121}$$

or, for detection, in setting a threshold on this same quantity. Furthermore, it was shown in the same section that the optimum estimator for $\tilde{s}(f)$ [which plays the role played by α in this section—see (5.59) and (5.119)] is (5.86):

$$\tilde{s}(f) = \frac{\mathbf{a}^t(\hat{\theta})\tilde{\mathbf{R}}^{-1}(f)\tilde{\mathbf{X}}(f)}{\mathbf{a}^t(\hat{\theta})\tilde{\mathbf{R}}^{-1}(f)\mathbf{a}(\hat{\theta})} \tag{5.122}$$

Thus, the three previously mentioned problems [detection, estimation of θ, and reconstruction of the signal $s(t)$ or $\tilde{s}(f)$] are solved.

The procedure thus defined is not, however, optimal, as the different frequencies are processed independently of one another—which implies the definition of a final detection test and a final estimation of θ combining the results obtained by (5.121) at the different frequencies. The simplest technique consists of summing the obtained quantities (5.121), and determining the maximum of this sum to estimate θ (or setting a threshold on this sum to perform detection): this technique is also optimal when the signals received at different frequencies are statistically independent, which is a fairly general situation.

This procedure can solve the three problems—detection, estimation, and signal extraction—mentioned in Section 5.7.1. It is based on a model [summarized in (5.119)] that, in a situation where the noise sources are multiple, according to the model, the sources located in the other directions are equivalent to Gaussian noise and are taken into account by $\tilde{\mathbf{B}}(f)$. In so doing, no use is made of the information provided by the multisource nature of the noise.

In contrast, the procedure described in the next paragraph makes full use of this information, while abandoning the possibility of reconstructing, at least directly, the signals received from the sources.

5.7.3 Spectral Decomposition (Multiple Sources + Noise): MUSIC

In this approach, in which the main objective is to improve the resolution capability of the array, the directions and powers of the sources are determined from the properties of the eigenvalues and eigenvectors of the cross-spectral density matrix $\tilde{\mathbf{R}}(f)$ [12, 13]. Let us assume that the received signals come from p independent point sources located in directions $\theta_1, \theta_2, \ldots, \theta_p$, with complex amplitudes (and phases) $\alpha_1, \alpha_2, \ldots, \alpha_p$ at frequency f, with the addition of noise $\tilde{\mathbf{B}}(f)$. These signals are then given by:

$$\begin{bmatrix} \tilde{X}_1 \\ \tilde{X}_N \end{bmatrix} = [a(\theta_1) \ldots a(\theta_p)] \begin{bmatrix} \alpha_1 \\ \alpha_p \end{bmatrix} + \begin{bmatrix} \tilde{B}_1 \\ \tilde{B}_N \end{bmatrix} \tag{5.123}$$

(with implicit dependence on f).

Given: $\tilde{\mathbf{X}} = \mathbf{A}\boldsymbol{\alpha} + \tilde{\mathbf{B}}$ with $\mathbf{a}(\theta_i)$ defined as previously:

$$\mathbf{a}(\theta_i) = \begin{bmatrix} 1 \\ e^{2\pi f D_2(\theta_i)} \\ \cdots \\ e^{2\pi f D_N(\theta_i)} \end{bmatrix} \tag{5.124}$$

If the sources are independent of one another and independent of the noise, the cross-spectral density matrix is written:

$$\tilde{\mathbf{R}}(f) = \mathbf{A}(f)\mathbf{P}(f)\mathbf{A}^t(f) + \tilde{\mathbf{K}}(f) \tag{5.125}$$

where $\tilde{\mathbf{K}}(f)$ is the noise spectral density matrix:

$$\tilde{\mathbf{K}}(f) = \mathrm{E}[\tilde{\mathbf{B}}(f)\tilde{\mathbf{B}}^t(f)] \tag{5.126}$$

and $\mathbf{P}(f)$ is a diagonal matrix (independent sources) whose diagonal elements are the powers of the p sources at frequency f:

$$\mathbf{P}(f) = \begin{bmatrix} P_1(f) & & 0 \\ & \cdots & \\ 0 & & P_p(f) \end{bmatrix}, \; P_i(f) = E[\alpha_i \alpha_i^*] \tag{5.127}$$

In what follows, we shall assume that the noise is white $[\tilde{\mathbf{K}}(f) = \sigma^2\mathbf{I}]$, to simplify the discussion [the results remain valid for any $\tilde{\mathbf{K}}(f)$, subject to prior whitening].

As a result, the matrix $\tilde{\mathbf{R}}(f)$ can take the form shown below (with implicit dependence on f):

$$\tilde{\mathbf{R}} = \sum_{i=1}^{p} P_i \mathbf{a}(\theta_i)\mathbf{a}^t(\theta_i) + \sigma^2\mathbf{I} \tag{5.128}$$

Then we can show that $\tilde{\mathbf{R}}(f)$ has $N - p$ eigenvalues equal to σ^2, and that this multiple eigenvalue, associated with the eigen subspace known as the "noise subspace," is the minimum eigenvalue. The eigenvalues of $\tilde{\mathbf{R}}(f)$ are of the form $\lambda_k + \sigma^2$, where λ_k is an eigenvalue of $\mathbf{APA}^t$, hence positive [$\mathbf{APA}^t$ and $\tilde{\mathbf{R}}(f)$ are hermitian definite positive matrices].

Moreover, the eigenvectors of this noise subspace are orthogonal to the steering vectors of the sources. Indeed, let **V** be the eigenvector associated to the eigenvalue σ^2. **V** verifies:

$$\mathbf{V}^t\tilde{\mathbf{R}}\mathbf{V} = \sigma^2\mathbf{V}^t\mathbf{V} \tag{5.129}$$

but, following (5.126):

$$\mathbf{V}^t\mathbf{R}\mathbf{V} = \sum_{i=1}^{p} P_i\mathbf{V}^t\mathbf{a}(\theta_i)\mathbf{a}^t(\theta_i)\mathbf{V} + \sigma^2\mathbf{V}^t\mathbf{V}$$

$$\Rightarrow \sum_{i=1}^{p} P_i\mathbf{V}^t\mathbf{a}(\theta_i)\mathbf{a}^t(\theta_i)\mathbf{V} = 0 \Rightarrow \mathbf{V} \perp \mathbf{a}(\theta_i)\ \forall\ i \tag{5.130}$$

So, the noise subspace vectors are orthogonal to the steering vectors of the sources.

The algorithm for determining the positions θ of the possible sources then consists of:

- Obtaining the eigenvalues of matrix $\tilde{\mathbf{R}}(f)$;
- Setting a threshold λ such that the eigenvalues inferior to λ are assumed to be equal to σ^2, noise power at frequency f;
- Obtaining the eigenvectors associated with these "noise" eigenvalues;
- Calculating, as a function of θ, the following function.

$$P(\theta) = \frac{1}{\mathbf{a}^t(\theta)\mathbf{R}_B\mathbf{R}_B^t\mathbf{a}(\theta)} \tag{5.131}$$

where $\mathbf{R}_B$ is the matrix whose columns are the eigenvectors of the noise subspace. This function has a maximum when $\mathbf{a}(\theta)$ is orthogonal to the noise subspace, hence when θ is the direction of a source, since its denominator is the length of the projection of $\mathbf{a}(\theta)$ in the noise subspace. Once the directions of the sources have been obtained, it is possible to calculate their power [14].

When the observation time is infinite and the sources and noise are stationary, this algorithm leads to a perfect estimation of the direction of the sources—at the cost, obviously, of non-negligible processing complexity. It is shown in [13] that it is also, for a limited observation time, optimum in terms of maximum likelihood criteria.

This algorithm can be repeated at each frequency f in the useful band, and the results combined as in the case of the procedure described in Section 5.6.2.1. It is also possible to build an algorithm of the same type, using all the frequencies in the band on a global basis [13].

A variant of MUSIC algorithm, for a linear array with regular sensor spacing d, is known as "root-MUSIC" [15], and consists of replacing the search over θ required by (5.131) by a search for the roots of the polynomial in z: $\mathbf{a}^t(\theta)\mathbf{R}_B\mathbf{R}_B^t\mathbf{a}(\theta)$, with $z = e^{2\pi jf\frac{d\sin\theta}{\lambda}}$.

If there are p eigenvalues associated with sources, and $N-p$ with noise, the polynomial to analyze is written as:

$$\sum_{k=1}^{N-p}\left(\sum_{m=1}^{N} v_{mk} z^{-(m-k)}\right)\left(\sum_{m=1}^{N} v_{mk}^{*} z^{+(m-K)}\right)$$

where v_{mk} is the mth coordinate of the kth noise eigenvector.

Different approaches have been proposed for reducing the number of operations, by replacing the systematic search over θ by a direct, explicit solution [16]. ESPRIT is such a method, appropriate for arrays consisting of doublets of identical sensors [17]. In the special configuration of linear regular arrays, GEESE [18] is another variant, with an excellent resolution even for few samples and low computation load. However, calibration requirements may prove to be a limiting factor for these procedures, since the basic assumption is the identity of sensor diagrams. Application in the time domain (for Doppler analysis) should be more promising, since the time samples are collected through the same channel.

In the same endeavor toward complexity reduction, a promising technique prefilters the spatial samples in order to restrict the search to a smaller angular domain. This approach is especially convincing for search modes in low target density situations, in which it is obviously inefficient to search with high accuracy in regions where no target is present; a preliminary Fourier digital beamforming allows high-resolution algorithms to be applied only after predetection, and possibly on fewer channels if few interfering signals are expected [16, 19]. The angular resolution of such techniques has been shown to be equivalent to that of direct high-resolution operation, at much lower computation cost [16].

5.7.4 Comparison

The maximum likelihood algorithm (5.121) and the spectral decomposition algorithm known in the literature as MUSIC have been compared in simulations under many conditions. For example, [20] presents the results of simulations of signals received by the seven-sensor array described in Figure 5.17. Figure 5.18, taken from this reference, presents the results obtained by these two algorithms in the case of observation of two incoherent sources of power 30 dB greater than that of the isotropic white noise, for sources located at an azimuth of 45° and at elevation deviations of 10.5° and 4°. The better resolution obtained by the spectral decomposition algorithm clearly appears in this figure.

An interpretation of this gain in resolution can be given by writing the maximum likelihood algorithm in the form [21]:

$$\text{Determination of max}/\theta \text{ of } \frac{1}{\sum_{k=1}^{N} (1/(\lambda_k + \sigma^2))\left|\mathbf{V}_k^t \mathbf{a}(\theta)\right|^2},$$

where V_k is the kth eigenvector of the correlation matrix.

If the eigenvalues of this matrix are ranked in increasing order, and if p is the number of sources estimated from these eigenvalues ($N-p$ minimum values equal to σ^2), the spectral decomposition algorithm is:

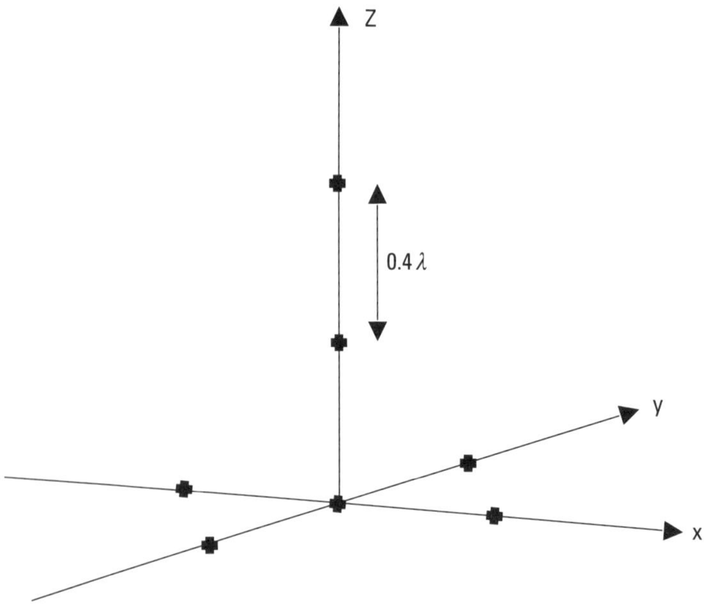

Figure 5.17 Seven-sensor array. (*From:* [21].)

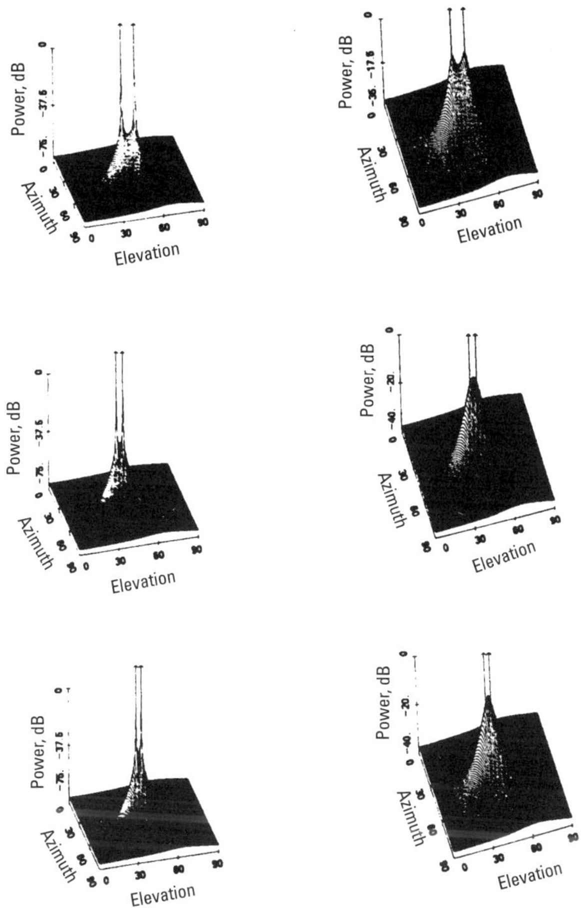

Figure 5.18 Comparison between maximum likelihood method and spectral decomposition method. SNR = 30 dB, $\delta\theta$ = 10°, 5°, 4°. MLM: maximum likelihood method; MUSIC: spectral decomposition. © 1986 IEEE [20].

$$\text{Determination of } \max/\theta \text{ of } \frac{1}{\sum_{k=1}^{N} |\mathbf{V}_k^t \mathbf{a}(\theta)|^2}$$

It appears that the structure of the two algorithms is comparable, with the spectral decomposition algorithm giving more importance to the eigenvectors of the noise subspace than the maximum likelihood algorithm, as it only takes into account the eigenvectors of the noise subspace (summation limited to $N - p$) whereas the maximum likelihood algorithm takes into account all the eigenvectors [with, however, weighting according to $1/(\lambda_k + \sigma^2)$, favoring the eigenvectors of the noise subspace].

From the user's point of view, it is important to note that the maximum likelihood algorithm is the only one that enables the received signal to be restored in amplitude and phase, according to (5.122); for some applications that require further detailed analysis of the source characteristics, this property may be decisive.

Finally, let us note that in the case where a noise-only reference is available (pulse sources, for example, in which case a noise reference exists before the pulse), a reasoning similar with that developed in Section 5.7.2.1 shows that the optimum receiver seeks θ by maximization of

$$\frac{|\mathbf{a}^t(\theta)\tilde{\mathbf{K}}^{-1}(f)\tilde{\mathbf{X}}(f)|^2}{\mathbf{a}^t(\theta)\tilde{\mathbf{K}}^{-1}(f)\mathbf{a}(\theta)} \tag{5.132}$$

and calculates $\tilde{s}(f)$ by:

$$\tilde{s}(f) = \frac{\mathbf{a}^t(\hat{\theta})\tilde{\mathbf{K}}^{-1}(f)\tilde{\mathbf{X}}(f)}{\mathbf{a}^t(\hat{\theta})\tilde{\mathbf{K}}^{-1}(f)\mathbf{a}(\hat{\theta})}$$

where $\tilde{\mathbf{K}}(f)$ is the cross spectral density matrix of the noise only.

5.7.5 Generalized Correlation [22]

Let us return to the problem posed in Section 5.7.1, in the case of two sensors. Let us note the signal received on these two sensors in the form:

$$\begin{aligned} x_1(t) &= s(t) + n_1(t) \\ x_2(t) &= s(t - D) + n_2(t) \end{aligned} \tag{5.133}$$

where $n_1(t)$ and $n_2(t)$ are colored noises, independent from one sensor to the other, and $s(t)$ is the received random useful signal.

Let $\tilde{\mathbf{X}}(f) = \begin{bmatrix} \tilde{x}_1(f) \\ \tilde{x}_2(f) \end{bmatrix}$ be the vector consisting of Fourier transforms of $x_1(t)$ and $x_2(t)$ calculated over a time period T, long relative both to D and to the correlation time of $s(t)$.

We can then demonstrate:

$$\mathrm{E}[\tilde{\mathbf{X}}(f)\ \tilde{\mathbf{X}}^t(f')] = S(f) \text{ for } f = f' \qquad (5.134)$$
$$\mathrm{E}[\tilde{\mathbf{X}}(f)\ \tilde{\mathbf{X}}^t(f')] = 0 \text{ for } f \neq f'$$

Let $N_1(f)$ and $N_2(f)$ be the noise spectra on the two sensors. When adopting the maximum likelihood criterion to estimate D, one is led to seek $\hat{D}$, which maximizes:

$$\int \tilde{\mathbf{X}}^t(f)\tilde{\mathbf{R}}^{-1}(f)\tilde{\mathbf{X}}(f) + \log \det[\tilde{\mathbf{R}}(f)]df \qquad (5.135)$$

where $\tilde{\mathbf{R}}(f)$ is the previously introduced matrix of cross spectral densities [$\tilde{\mathbf{X}}(f)$ is Gaussian as a linear form of Gaussian process $\mathbf{X}(t)$, and the integral in f is justified by the independence of frequencies (5.134)]. In this expression, the sought parameter D is only involved in $\tilde{\mathbf{R}}(f)$:

$$\tilde{\mathbf{R}}(f) = \begin{bmatrix} S(f) + N_1(f) & S(f)e^{2\pi jfD} \\ S(f)e^{-2\pi jfD} & S(f) + N_2(f) \end{bmatrix}$$

(as noise is supposedly independent from one sensor to the next).

Under these conditions, $\det\tilde{\mathbf{R}}(f)$ does not depend on the sought parameter D, and the receiver therefore attempts to maximize: $\int \tilde{\mathbf{X}}^t(f)\tilde{\mathbf{R}}^{-1}(f)\tilde{\mathbf{X}}(f)\,df$

Given $\tilde{\mathbf{R}}(f)$, we obtain:

$$\mathrm{Max}/D \int \frac{S(f)}{S(f)[N_1(f) + N_2(f)] + N_1(f)N_2(f)}$$
$$\times\ \mathrm{Re}[\tilde{x}_1^*(f)x_2(f)e^{2\pi jfD}]\,df$$

The receiver seeking the maximum relative to D of this expression is therefore optimum in the maximum likelihood sense, for a random signal with a known spectrum $S(f)$ and noise decorrelated from one sensor to the other and with known spectra $N_1(f)$ and $N_2(f)$. When these spectra are not known, it is possible to derive a usable procedure, noting:

$$S(f) = \left|\mathrm{E}[\tilde{x}_1(f)\tilde{x}_2^*(f)]\right| \qquad (5.136)$$

(owing to the assumption of independent noise on both sensors), and:

$$S^2 + S(N_1 + N_2) + N_1N_2 = \mathrm{E}|\tilde{x}_1(f)|^2\mathrm{E}|\tilde{x}_2(f)|^2 \qquad (5.137)$$

Finally, we obtain the following rule for estimating D:

$$\text{Max}/D \int \frac{\mathrm{E}[\bar{x}_1(f)x_2^*(f)]}{\mathrm{E}|\bar{x}_1(f)|^2\mathrm{E}|\bar{x}_2(f)|^2 - |\mathrm{E}\bar{x}_1(f)x_2^*(f)|^2} \times \mathrm{Re}[\bar{x}_1^*(f)x_2(f)e^{2\pi jfD}]\,df \tag{5.138}$$

In this form, the procedure is applicable in the absence of noise-only reference. This procedure is compared in [17] with a number of other procedures proposed elsewhere, which allow a trade-off between estimation accuracy and robustness to noise.

5.7.6 Discussion

In all cases, the generalized correlation procedures essentially aim at measuring D, that is, the delay between sensors, despite spectral characteristic noise more or less known but independent from one sensor to the next. Consequently, these procedures will generally have no special property (superresolution) with regard to directive interference, for which this independence between sensors is not verified.

This noise source directivity property is only taken into account by the maximum likelihood methods—through the colored noise assumption—and especially by the spectral decomposition methods, which explicitly take into account the existence of multiple pointlike sources. As the complexity of the corresponding algorithms increases from generalized correlation to maximum likelihood, then on to spectral decomposition methods, the selected procedure to be used will generally result from a trade-off between complexity and performance characteristics, which vary from one application to another.

5.8 Conclusion

Based on the optimum receivers structures in the presence of colored noise with known characteristics obtained in Chapter 4, this chapter has enabled us to build receivers usable in actual situations where the noise characteristics are both not well known and variable. This approach has emphasized several basic aspects of radar and sonar signal processing in the presence of external interference. They are as follows:

- The knowledge of noise characteristics, both in terms of receiver definition and the quality of the information output by the receiver (Section 5.3.1) is of key importance.
- There is equivalence between the two main receiver construction techniques: On the one hand, extraction of a noise-only reference from the received signals, to be used in an "optimum" receiver and on the other, minimization of the receiver output energy level, subject to the constraint that the useful signal is not affected by the receiver (Sections 5.5.4 and 5.5.6). After all, these two techniques are nothing but two different ways of extracting the noise-only reference.
- It is possible to derive an adaptive receiver from an optimum receiver, when optimality can be defined by an energy minimization criterion (which, in principle, is attributable to the Gaussian character of the noise, which implies an equivalence between the least squares and maximum likelihood criteria). Transition to an adaptive receiver is generally a two-stage process: first, selection of a receiver structure (type of filtering), then, replacement of the average values by instantaneous values.
- Finally, in the highly different situation of passive monitoring, the choice of the received signal model is important. In particular, the performance obtained by the spectral decomposition method, which is interpreted as the application of the maximum likelihood method to a multiple point source model [13], clearly shows how the angular resolution of a sensor array can be improved by a more precise modeling of the physical phenomena.

This chapter especially shows, through the diversity of the addressed procedures, the importance of a clear definition of the functions to be fulfilled by the receiver (detection/location, or extraction of an interference-free signal) and of the usable information (available duration, exact nature of the targets and interference). The next chapters, by addressing more specifically the nature of the received signals, will at least partly address this concern, which should govern, in the long run, in each special case, the definition of the transmitter/receiver system—at the cost of indispensable trade-offs between functions but also between performance characteristics and system simplicity.

Finally, we must admit that this chapter is highly incomplete, for at least three reasons.

1. The types of processing addressed are still little used in practice, owing to their complexity. This situation is obviously changing rapidly, as the capacity of the processing devices (digital, or even analog) increases. This situation, however, leads to focusing our statement on the principles, with limited examples of implementations, which vary from one manufacturer to another.
2. Within the vast area of passive monitoring, the great diversity of problems posed prevents us from reviewing all the situations, models, and usable systems, which there again led us to focusing the statement on principles—which is, after all, the main purpose of this book.
3. Each addressed processing method implies, whether explicitly or not, the estimation of a correlation function (or matrix). This estimation is all the better when the phenomena are more stationary and the available observation time is longer. In practice, the estimation accuracy of the correlation function determines, to a large extent, the final performance characteristics of processing. There again, in each particular situation, processing must be optimized, taking into account this primary influence.

References

[1] Levine, M., *Fondements Théoriques de la Radiotechnique Statistique*, Vol. 3, Mir (Translation from Russian), 1973.

[2] Widrow, B., and S. D. Stearns, *Adaptive Signal Processing*, Englewood Cliffs, NJ: Prentice-Hall, 1985.

[3] Burg, J. P., "Maximum Entropy Spectral Analysis," Ph.D. Dissertation, Dept. of Geophysics, Stanford University, Stanford CA, Nov. 1981.

[4] Gibson, C., and S. Haykin, "Radar Performance Studies of Adaptive Lattice Clutter-Suppression Filters," *IEE Proceedings*, F. 130, No. 5, Aug. 1983.

[5] Zolésio, J. L., "Filtrage Autorégressif d'Elimination de Fouillis," 10ème GRETSI, Nice, France, May 1985.

[6] Jim, C. W., "A Comparison of Two LMS Constrained Optimal Array Structures," *Proc. IEEE*, Vol. 60, No. 8, Dec. 1977.

[7] Farina, A., *Antenna-Based Signal Processing Techniques for Radar Systems*, Norwood, MA: Artech House, 1992.

[8] Steinberg, B. D., *Principles of Aperture and Arrays System Design*, New York: Wiley, 1976.

[9] "Special Calibration d'Antenne," *Traitement du Signal*, Vol. 10 No. 5–6, 1993.

[10] Shan, T. J., M. Wax, and T. Kailath, "On Spatial Smoothing for Direction-of-Arrival Estimation of Coherent Signals," *IEEE Trans. ASSP*, Vol. 33, Aug. 1985.

[11] O'Sullivan, M. R., "A Comparison of Sidelobe Blanking Systems," *Proc. IEE International Radar Conference, RADAR '87*, London, Oct. 1987.

[12] Bienvenu, G., "Propriétés Haute Résolution de la Matrice de Corrélation Spatiale," 9ème GRETSI, Nice, France, Mai 1983.

[13] Bienvenu, G., and L. Kopp, "Optimality of High-Resolution Array Processing Using the Eigensystem Approach," *IEEE Trans. ASSP*, 3 1, No. 5, Oct. 1983.

[14] "Special Issue on Adaptive Arrays," *IEEE Trans. AP*, Vol. AP34, No. 3, March 1986.

[15] Rao, B. D., and K. V. S. Hari, "Performance Analysis of Root-MUSIC," *IEEE Trans. ASSP*, Vol. 37, No. 12, Dec. 1989.

[16] Calvary, P., "FFC Explicite," *Journée d'Etude SEE sur le Traitement d'Antenne*, Paris, 1994.

[17] Roy, R., and T. Kailath, "ESPRIT Estimation of Signal Parameters via Rotational Invariance Techniques," *IEEE Trans. ASSP*, Vol. 37, No. 7, July 1989.

[18] Pillai, S. U., and B. H. Kwon, "GEESE (Generalized Eigenvalues Utilizing Signal Subspace Eigenvectors)," *Proc. IEEE ICASSP*, New York, 1988.

[19] Bienvenu, G., and L. Kopp, "Decreasing High-Resolution Method Sensitivity by Conventional Beamformer Preprocessing," *Proc. IEEE ICASSP*, San Diego, CA, April 1984.

[20] Johnson, R. L., and G. E. Miner, "Comparison of Superresolution Algorithms for Radio Direction Finding," *IEEE Trans. AES*, 22, No. 4, July 1986.

[21] Kates, J. M., "Generalized Approach to High-Resolution Array Processing," *IEEE ICASSP* 83, Boston, 1983.

[22] Knapp, C. M., and G. C. Carter, "The Generalized Correlation Method for Estimation of Time Delay," *IEEE Trans. ASSP*, No. 4, Aug. 1976.

Appendix 5A: Matrix Inversion Lemma

Let $\mathbf{A}$ and $\mathbf{B}$ be two invertible matrices $(n \times n)$, $\mathbf{C}$ a matrix $(m \times n)$, and $\mathbf{D}$ an invertible matrix $(m \times m)$.

The "inverse matrix lemma" can be stated as follows:

$$\mathbf{A}^{-1} = \mathbf{B}^{-1} + \mathbf{C}^t\mathbf{D}^{-1}\mathbf{C} \tag{5A.1}$$

$$\mathbf{A} = \mathbf{B} - \mathbf{B}\mathbf{C}^t(\mathbf{C}\mathbf{B}\mathbf{C}^t + \mathbf{D})^{-1}\mathbf{C}\mathbf{B} \tag{5A.2}$$

Let us multiply (1) to the left by $\mathbf{A}$:

$$I = \mathbf{A}\mathbf{B}^{-1} + \mathbf{A}\mathbf{C}^t\mathbf{D}^{-1}\mathbf{C}$$

Let us multiply to the right by **B**:

$$\mathbf{B} = \mathbf{A} + \mathbf{A}\mathbf{C}^t\mathbf{D}^{-1}\mathbf{C}\mathbf{B} \tag{5A.3}$$

Let us multiply to the right by $\mathbf{C}^t$:

$$\mathbf{B}\mathbf{C}^t = \mathbf{A}\mathbf{C}^{t\prime} + \mathbf{A}\mathbf{C}^t\mathbf{D}^{-1}\mathbf{C}\mathbf{B}\mathbf{C}^t = \mathbf{A}\mathbf{C}^t\mathbf{D}^{-1}[\mathbf{D} + \mathbf{C}\mathbf{B}\mathbf{C}^t]$$

Let us multiply to the right by $[\mathbf{D} + \mathbf{C}\mathbf{B}\mathbf{C}^t]^{-1}$

$$\mathbf{B}\mathbf{C}^t[\mathbf{D} + \mathbf{C}\mathbf{B}\mathbf{C}^t]^{-1} = \mathbf{A}\mathbf{C}^t\mathbf{D}^{-1}$$

Let us multiply to the right by **CB**:

$$\mathbf{A}\mathbf{C}^t\mathbf{D}^{-1}\mathbf{C}\mathbf{B} = \mathbf{B}\mathbf{C}^t[\mathbf{D} + \mathbf{C}\mathbf{B}\mathbf{C}^t]^{-1}\mathbf{C}\mathbf{B} \tag{5A.4}$$

Let us subtract (5A.4) from **B**:

$$\mathbf{B} - \mathbf{A}\mathbf{C}^t\mathbf{D}^{-1}\mathbf{C}\mathbf{B} = \mathbf{B} - \mathbf{B}\mathbf{C}^t[\mathbf{D} + \mathbf{C}\mathbf{B}\mathbf{C}^t]^{-1}\mathbf{C}\mathbf{B}$$

Per (5A.3), we therefore have:

$$\mathbf{A} = \mathbf{B} - \mathbf{B}\mathbf{C}^t[\mathbf{D} + \mathbf{C}\mathbf{B}\mathbf{C}^t]^{-1\prime}\,\mathbf{C}\mathbf{B}\text{: QED}$$

For example, for $\mathbf{K} = \mathbf{R} - |\alpha|^2\mathbf{s}\mathbf{s}^t$ we obtain:

$$\mathbf{K}^{-1} = \mathbf{R}^{-1} - \frac{\mathbf{R}^{-1}\mathbf{s}\mathbf{s}^t\mathbf{R}^{-1}}{\mathbf{s}^t\mathbf{R}^{-1}\mathbf{s} - \dfrac{1}{|\alpha|^2}}$$

6

Target and Background Signatures

6.1 Summary

Defining a model of the signal received by an active radar or sonar system, as described in Chapter 1, has made it possible to build the optimum receiver in white noise, and then in colored noise. In reality, however, the interaction between the wave transmitted and the target is more complex than the simple reflection assumed so far, and the echoes returned by the various elements of the landscape are not always correctly modeled by an additive colored noise. This is the reason why the signal processing described so far is sometimes supplemented by postprocessing—the justification for which is generally more or less intuitive—to reduce the performance loss caused by these deviations with respect to the original model. Obviously, however, the ideal would be to specify the target model in order to take the exact nature of the received signals into account from the start; moreover, a more detailed analysis of the received signal could supply valuable information to the operator, such as the type of target observed and a detailed analysis of the environment. Keeping this in mind, the scope of this chapter is to provide an overview of the existing target and background models and to examine the possibilities of application to signal processing.

The first improvement consists of taking into account, for radar applications, the polarizations of the transmitted and received waves. The polarization of a plane monochromatic wave characterizes the evolution of the direction of the electric field vector: linear polarization if this vector remains in the same plane, circular polarization if the vector extremity describes a circular helix, elliptical polarization in the general case. As any plane wave can be decomposed into the sum of two plane waves with "orthogonal" polarizations, the reflection from the target is characterized by a backscattering matrix relating the waves received with these two polarizations to the waves transmitted with the same two polarizations. An RCS can thus be defined for each transmission-reception polarization pair in a way that relates the powers in the received wave and the transmitted wave.

In sonar, although the interaction between the transmitted wave and the target is often extremely complex, the nature of the transmitted wave and of the received wave is simpler, since these are longitudinal waves. The sonar equivalent of the RCS (to within a coefficient) is the target strength.

Once these definitions have been established, the most appealing approach to modeling the received signal consists of building a physical model of the wave-target interaction. Unfortunately, this operation is usually impossible without approximation. As the dimensions of the targets observed are usually large with respect to the wavelengths used (in both radar and sonar), it is possible to try to model this interaction within the scope of a high-frequency approximation.

The first high-frequency approximation is geometrical optics—or acoustics—which, starting from an asymptotic expansion for large frequencies, describes the propagation of the wave in the form of rays. This description, very favorable to intuitive reasoning, makes it possible to analyze relatively simply the propagation of a wave through a medium of variable refractive index, as is the case in sonar (due to the variation of temperature, and consequently of the index, with depth) or in radar near the sea surface. When applied to the calculation of the RCS (or of the sonar target strength) of a convex surface, this theory results in a value proportional to the product of the two main radii of curvature of the surface at the so-called specular reflection point (point M of this surface minimizing the transmitter-M-receiver path). However, this theory has important limitations, among which are:

- The prediction of an infinite field at the crossing point of two rays (caustic surface);
- The prediction of an infinite RCS for a plane target or a cylindrical target (because one of the radii of curvature is then infinite).

The second high-frequency approximation is that of physical optics. Starting from a strict integral equation (Stratton-Chu) relating the incident field and the diffracted field and making a tangent plane approximation to evaluate—in this equation, the surface currents—the diffracted field is obtained in explicit form as a function of the incident field. This expression leads, in particular, to a good approximation of the RCS—or sonar target strength—of a plane surface of area *S* observed in specular reflection (cross section proportional to the ratio of the square of the area to the square of the wavelength). However, as in the case of geometrical optics, this expression does not make it possible to correctly take into account the diffraction due to edges, or the surface wave generated at the limit of the surface illuminated by the transmitter. More seriously, this approximation leads to a polarization of the received wave that is identical to that of the transmitted wave, which contradicts, for example, the simple results for the wire or the dihedral reflector.

A third approach, which is extremely fertile, consists of returning to the ray theory (geometrical optics) and making local modifications to it to model the edges, shadow boundaries, or cone vertices. This "geometrical theory of diffraction," gradually improved in recent years, can often predict with satisfactory accuracy the echo returned by a target of known shape. In contrast, this theory does not enable us to improve the model of the received signal directly, since this evaluation of the returned echo involves a decomposition of the target (e.g., into edges and corners), which is completely specific to each situation. It would thus lead to a signal model including a large number of parameters and consequently difficult to implement in signal processing (correlation with too large a number of replicas).

In the opposite case, in which the dimensions of the target are small with respect to the wavelength (raindrops, for example), the Rayleigh approximation can be used; the RCS is then proportional to the square of volume of the target, which can be validly considered as a single scatterer—thus satisfying the model adopted in Chapter 1.

Although it does not reveal a single model for the received signals, this survey of RCS modeling methods nevertheless leads to a classification of different types of reflection that have unequal importance:

- Specular reflection, resulting in high RCS levels and well predicted by geometrical optics or physical optics;
- Diffraction by edges or vertices, for example, which is correctly described by the geometrical theory of diffraction and which often results in significantly lower levels than specular reflection;
- Diffraction by elements with small dimensions with respect to the wavelength (Rayleigh domain), which is very often negligible.

To these elements are added the reflections by cavities (aircraft air intake, for instance) or by antennas, the calculation of which is very difficult, and whose level is often high.

This diversity of the situations that can be encountered leads to the examination of two other models, more empirical than strictly physical, but leading to representations of the received signal that can be used directly in signal processing: the multiple-scatterers model and the stochastic model.

According to the multiple-scatterers model, the target is represented by a collection of independent pointlike reflectors, each characterized by an amplitude and a phase. The response to an elementary pulse is then an impulse response for each aspect angle of the target. This impulse response is in fact the projection of the scatterers on the line of sight of the radar (seen as always through an apparatus function defined by the characteristics of the transmitted pulse).

Consequently, by summing the projections obtained for the different aspect angles, it is possible to obtain an image of this target by a process similar to the tomographic synthesis used in medical scanning, called "radar holography" in radar. This technique, which is a powerful target representation tool and which is now used by many laboratories, enables the electromagnetic response of the target to be made visible.

According to the stochastic model, the complex backscattering coefficient of the target at a given wavelength, as a function of aspect angle, is represented by a Gaussian stochastic process (a justification for this model can be found in the existence of numerous scatterers with levels varying with the aspect angle; a central limit argument then leads to a Gaussian model). This process is characterized by its spectrum, which is proved to be proportional to the modulus of the transverse response; this quantity, analogous to the impulse response, is the projection of the scatterers (amplitude and phase) on an axis perpendicular to the line of sight of the radar. This transverse response, like the impulse response, is directly accessible to measurement, since it is obtained by fine Doppler analysis of the signal received from the target assumed to be rotating (as the Doppler analysis is in fact a spectral analysis, it naturally delivers the spectrum of the signal received, that is, the spectrum of the Gaussian process modeling this signal). This stochastic model is extremely convenient, since it enables the complex signal received by the radar to be synthesized for a given transverse distribution, that is, for a given distribution of scatterers.

These two models—the multiple-scatterers model and the stochastic model—are the main tools available for the radar or sonar operator to represent the reflection from the target. Their description has shown, in passing, the importance of two quantities accessible to measurement: the impulse response and the transverse response, which are two fundamental tools for the analysis of targets—projections of its scatterers parallel and perpendicular to the line of sight, respectively.

The targets considered so far were assumed to be dimensionally stable; in reality, it is also necessary to take into account the specific signatures produced by moving parts, such as helicopter rotor blades, compressors and turbines of aircraft engines, or submarine propellers, which generate various modulations of the transmitted signal that can be used to recognize the type of target observed.

The echoes from various elements of the landscape, called clutter in radar or reverberation in active sonar, are added to the echoes from targets. The RCSs of surface clutter echoes are proportional to the illuminated area (or footprint), and can be modeled by random variables. For high-incidence angles, a Gaussian density is generally assumed. For low-incidence angles, a Weibull density or a K-density is generally preferred (to take into account the mixture of thermal noise in shadowed areas with effective clutter returns in illuminated areas), with parameters depending on frequency and on the type of terrain. These clutter returns sometimes have a wide Doppler spectrum, related to the relative movements of the radar and of the scattering elements (leaves, water surface). For ground clutter, the spectrum can be described as being made of three components: (1) a dominant coherent component resulting from fixed objects, at zero Doppler; (2) a slow diffuse component, with negative exponential Doppler spectrum; and (3) a much smaller fast diffuse component.

Moreover, in some configurations the target is coupled with the landscape by the existence of multiple paths of the transmitter-target-landscape-receiver type, which further modify the shape of the received signal. The dominant effect of multipath is to severely attenuate the returns from targets and clutter for low-grazing angles. Finally, propagation conditions can sometimes (more often in sonar) lead to nonrectilinear propagation, which further modifies the received signal in a specific way.

Thus we can see that, in the general case, the signal received can deviate considerably from the simple model defined in Chapter 1, which makes it necessary to supplement the optimum processing described previously by various procedures, or even (as will be seen in Chapter 7) to adapt the operating mode of the radar to the specific conditions determined by the environment.

This diversity of situations encountered in the active mode—which nevertheless offers the possibility of using certain simple models—is paralleled by still greater diversity in the passive mode, which in practical terms makes it impossible to cover all the problems arising, at least within the scope of this book. However, the Gaussian random model can often be used, as well as the periodic signal model that will be described in Chapter 7.

6.2 Radar or Sonar Reflection

In the previous chapters, the radar target was assumed to behave like a perfect pointlike reflector (refer to Chapter 1), since the signal received was identical to the signal transmitted, shifted in time and frequency and subject to a complex attenuation coefficient. In the real situation, the wave transmitted by the radar or the sonar interacts with the target in a more complex way, and generates a backscattered wave with properties depending on the physical characteristics of the target (shape, materials, moving parts). Moreover, in radar, the transmitted wave is a vector wave (electric and magnetic fields), and the polarization properties must therefore be taken into account at reception and transmission. Furthermore, these targets are moving through a given environment (ground, sea, clouds) that also returns echoes, with characteristics depending on the physical properties of the elementary reflectors. In some cases, the reflecting properties of the target and of its environment are even coupled (multiple transmitter-target-environment-receiver paths).

An initial method of taking into account this nonpoint character of the targets or of these echoes due to the environment consists of supplementing the optimum receivers defined previously with appropriate postprocessing, based on variable thresholds, on tracking of detected echoes and on various normalizations.

It is obvious, however, that an approach taking into account from the start the complex nature of the echoes makes it possible to come closer to the optimal solution, and thus to improve the overall performance of the system. Moreover, it then becomes possible to take advantage of the specific nature of the received signals to discriminate between the targets and the backgrounds, or even to perform an automatic classification of the targets according to their radar or sonar echoes.

It thus appears that it would be interesting to have a more accurate model of the received radar or sonar signal than the single-scatterer model used so far, while remaining sufficiently general. The first idea that comes to mind is simply to use a model with a few scatterers; the signal received will then depend on a greater number of parameters, since the relative positions and amplitudes of the different scatterers will be added to the range and Doppler shift of the target. The estimation of these parameters will enable the target to be located and identified by the positions and amplitudes of its scatterers.

In reality, however, it is of course not obvious that a multiple-scatterers model is the most correct or even the simplest to describe the received signal. It is therefore necessary to return to the physics of phenomena to try to define signal models that can be used for signal processing. In particular, the model should not be specific to each situation but should cover all (or most) situations encountered.

Thus the ideal solution would be to define a model, with a few parameters, for all the observable signals (this point of view is, in particular, different from that of system evaluation, in which the purpose is to examine the behavior of a given radar facing a given target; in this case, the target model need not be a general model, provided that a model for each type of target can be found).

In this chapter, the main properties of the targets and of their environments will be examined in order to generate target and background models, which can be used to define signal processing procedures adapted to the real "signatures." After establishing a few basic definitions (polarization, RCS, sonar target strength), we shall describe the main properties of targets and backgrounds, and we shall summarize them through random or deterministic models before defining, in Chapter 7, processing procedures adapted to the detection of targets in ground or surface echoes, nonpoint target tracking procedures, and target classification procedures.

6.2.1 Polarization [1–3]

The polarization of a plane monochromatic electromagnetic wave, in a homogeneous medium such as air, characterizes the evolution of the electric field vector during a wave period. A wave is said to be linearly polarized if the electric field—and, consequently, the magnetic field too—remains parallel to itself (Figure 6.1).

A linearly polarized wave whose vector **E** remains parallel to the axis *Oy* is said to be horizontally polarized; it is said to be vertically polarized if **E** remains parallel to *Oz* (*Ox* being the direction of propagation).

In the general case, a plane monochromatic wave can be broken down into the sum of two waves with horizontal and vertical polarizations, as follows: let **E** be the electric field at time t; its components E_y and E_z are written:

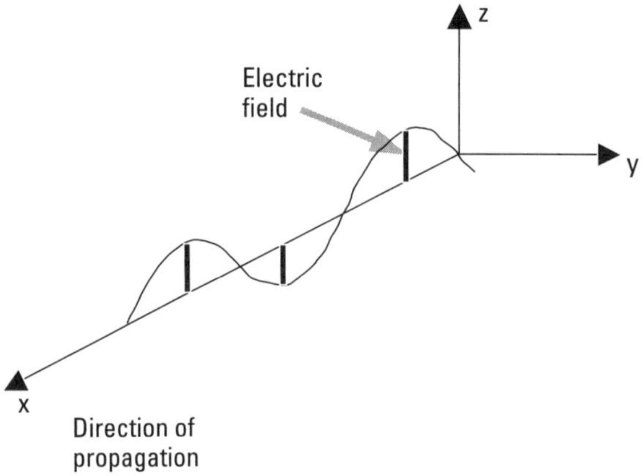

Figure 6.1 Plane wave, vertical polarization.

$$E_y = E_y^0 \exp[+j(\omega t - kx)]$$
$$E_z = E_z^0 \exp[+j(\omega t - kx + \varphi)]$$

where φ is the phase shift between the two components.

Then elementary calculations show that, in a duration equal to one wave period, the extremity of **E** describes an ellipse (Figure 6.2), which is generally characterized by two parameters (Figure 6.3): the direction angle ϕ, $0° \le \phi \le 180°$, which is the angle between the horizontal and the main axis of the ellipse, and the ellipticity angle τ, $0° \le \tau \le 45°$ (or its tangent, which is the ellipticity ratio).

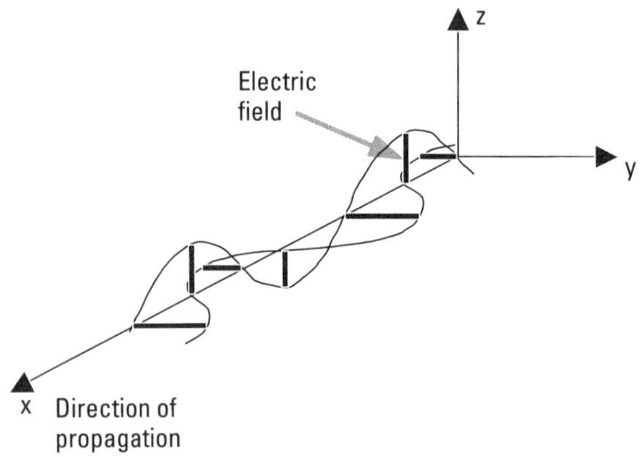

Figure 6.2 Plane wave, elliptical polarization.

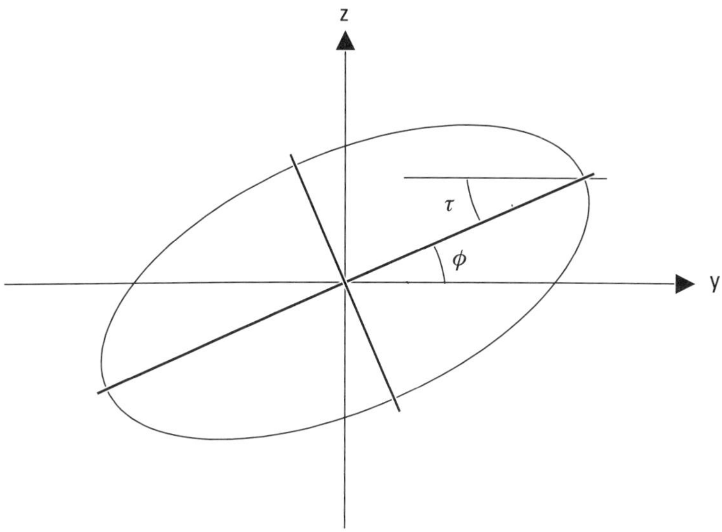

Figure 6.3 Characteristics of an elliptical polarization.

An interesting special case is obtained for $E_y^0 = E_z^0$, $\varphi = -\pi/2$, for which the vector **E** describes a circle clockwise, or for $E_y^0 = -E_z^0$ and $\varphi = +\pi/2$, for which **E** describes a circle counterclockwise: these waves are said to be circularly polarized. Generally speaking, it can be seen that a wave is characterized by its (complex) polarization vector: $\begin{bmatrix} E_y^0 \\ E_z^0 e^{j\varphi} \end{bmatrix}$ which is usually normalized to unity.

The polarization vector of the linearly polarized wave is then $\begin{bmatrix} 1 \\ 0 \end{bmatrix}$ (horizontal polarization) or $\begin{bmatrix} 0 \\ 1 \end{bmatrix}$ (vertical polarization); the polarization vector of the circular polarization wave is (according to the IEEE convention [1]) $\begin{bmatrix} 1 \\ -j \end{bmatrix} \frac{1}{\sqrt{2}}$ (right-hand circular polarization) or $\begin{bmatrix} 1 \\ +j \end{bmatrix} \frac{1}{\sqrt{2}}$ (left-hand circular polarization). With the latter two vectors also being orthogonal, they can be used as a basis on which any plane wave can be decomposed.

Different representations of the polarization state of a wave have been proposed; one of them, the Poincaré sphere, is universally adopted (Figure 6.4). It represents the polarization state by a point on a sphere, at longitude 2ϕ (orientation of the ellipse) and at latitude 2τ (ellipticity angle). The poles

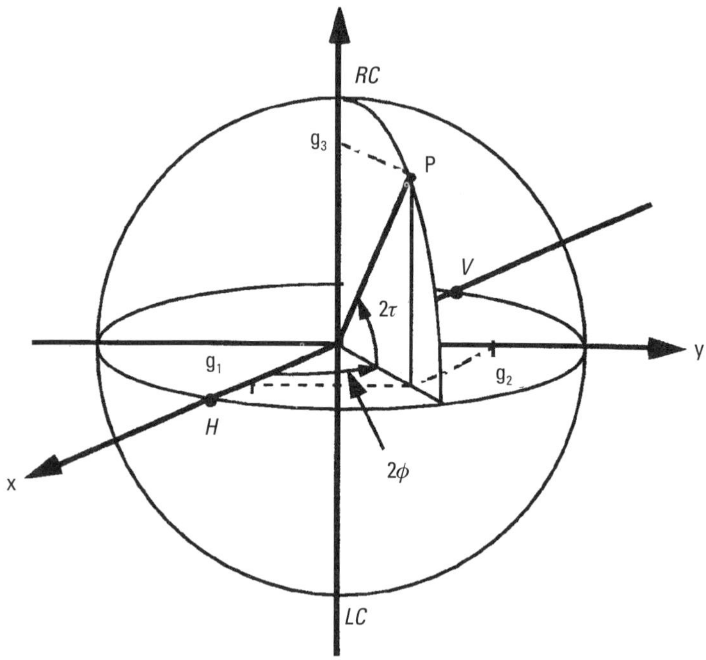

Figure 6.4 Poincaré sphere.

of the sphere then correspond to the circular polarizations, and each point on the equator to a linear polarization.

Finally, the polarization state can also be represented by a vector called the *Stokes vector,* with four components, only three of which are independent

$$\mathbf{g} = \begin{bmatrix} g_0 \\ g_1 \\ g_2 \\ g_3 \end{bmatrix} = \begin{bmatrix} |H|^2 + |V|^2 \\ 2\,\mathrm{Im}(H^*V) \\ |H|^2 - |V|^2 \\ 2\,\mathrm{Re}(H^*V) \end{bmatrix} \tag{6.1}$$

where H and V are the complex components of the field on the y and z axes ($H = E_y$ and $V = E_z e^{j\varphi}$). It can be shown by elementary geometrical calculations that g_1, g_2, and g_3 are also the coordinates of the representative point on Poincaré sphere. Therefore, for a wave with a total power of unity ($g_0 = 1$), the extremity of the Stokes vector is the representative point of the polarization state on the Poincaré sphere.

6.2.2 RCS

In radar, the power of the echo reflected by a target is characterized by an area (not necessarily related to its physical area) called the RCS (refer to Chapter 1). The RCS is the area that would have to be present, if the target scattered energy isotropically in all directions, to return an echo with the same power as the echo actually received by the receiver. Thus, if $|\mathbf{E}_T|^2$ is the power delivered by the radar, the target intercepts, at range R, the fraction $\frac{\sigma}{4\pi R^2}$ that it scatters isotropically; the power density of the field backscattered by the target, measured at the target location, is then given by:

$$|\mathbf{E}_R|^2 = |\mathbf{E}_T|^2 \frac{\sigma}{4\pi R^2}$$

and, by definition of σ (isotropic scattering), this field is assumed to undergo an attenuation $1/R$ on the return path.

Furthermore, this definition is set for far field conditions, so the definition of the RCS is:

$$\sigma = \lim_{R \to \infty} 4\pi R^2 \left| \frac{\mathbf{E}_R}{\mathbf{E}_T} \right|^2 \tag{6.2}$$

where $\mathbf{E}_T$ is the field transmitted by the radar and $\mathbf{E}_R$ the field backscattered by the target (measured at the target location).

Obviously, this definition implicitly assumes that the polarizations of the transmitter and of the receiver are defined; the same definition can be made for any transmission and reception polarization pair.

In addition, this definition only specifies the power of the echo returned, without providing information on the phase shift caused by the reflection. This is the reason why the complex backscattering matrix $\mathbf{S}$ of the target is defined more precisely, for a given polarization vector base (linear or circular), as follows:

$$\mathbf{S} = \begin{pmatrix} S_{AA} & S_{AB} \\ S_{BA} & S_{BB} \end{pmatrix}$$

where S_{AB} is the complex backscattering coefficient of the target, illuminated in polarization B and observed in polarization A, where $|S_{AB}|^2$ is the RCS for this polarization pair, and where Arg $[S_{AB}]$ is the phase shift undergone by the wave when reflected (with similar definitions for S_{AA}, S_{BB} and S_{BA}).

Thus, for example, the field backscattered by a given target—for a given aspect angle and wavelength—is written as follows, if E_H^T and E_V^T are the components, in linear polarization, of the plane wave transmitted, and if E_H^R and E_V^R are the components of the wave received in the same polarizations:

$$\begin{pmatrix} E_H^R \\ E_V^R \end{pmatrix} = \begin{pmatrix} S_{HH} & S_{HV} \\ S_{VH} & S_{VV} \end{pmatrix} \begin{pmatrix} E_H^T \\ E_V^T \end{pmatrix} \tag{6.3}$$

The matrix relations of this type, which can be written for any orthonormal polarization pair, simply express the linear character of the wave reflection on the target.

By changing the base, it is then possible to derive, from the backscattering matrix in one polarization pair, the analogous matrix in another polarization pair; for example, if $\vec{\mathbf{H}}$ and $\vec{\mathbf{V}}$ are the base vectors in linear polarization and $\vec{\mathbf{L}}$ and $\vec{\mathbf{R}}$ the base vectors in circular polarization, the components of $\vec{\mathbf{L}}$ and $\vec{\mathbf{R}}$ are (in the linear base):

$$\vec{\mathbf{L}}: \begin{pmatrix} 1 \\ j \end{pmatrix} \frac{1}{\sqrt{2}} \qquad \vec{\mathbf{R}}: \begin{pmatrix} 1 \\ -j \end{pmatrix} \frac{1}{\sqrt{2}}$$

and the change from the field components E_L, E_R in circular base to the components E_H, E_v in linear base can be written:

$$\begin{pmatrix} E_H \\ E_V \end{pmatrix} = \frac{1}{\sqrt{2}} \begin{pmatrix} 1 & 1 \\ j & -j \end{pmatrix} \begin{pmatrix} E_L \\ E_R \end{pmatrix} = \mathbf{U} \begin{pmatrix} E_L \\ E_R \end{pmatrix} \quad (\mathbf{U}\text{: unitary matrix})$$

or inversely

$$\begin{pmatrix} E_L \\ E_R \end{pmatrix} = \frac{1}{\sqrt{2}} \begin{pmatrix} 1 & -1 \\ j & +j \end{pmatrix} \begin{pmatrix} E_H \\ E_V \end{pmatrix} = \mathbf{U}^{-1} \begin{pmatrix} E_H \\ E_V \end{pmatrix}$$

At reception, the following can be written to take the reversal of the propagation direction[1] into account:

$$\begin{pmatrix} E_H^{\text{received}} \\ E_V^{\text{received}} \end{pmatrix} = \mathbf{U}^* \begin{pmatrix} E_L^{\text{received}} \\ E_R^{\text{received}} \end{pmatrix} \quad \text{*indicating the conjugate-matrix}$$

As a result, (6.3) defining the backscattering matrix in linear polarization becomes:

$$\begin{pmatrix} E_L^{\text{received}} \\ E_R^{\text{received}} \end{pmatrix} = \mathbf{U}^t \mathbf{S}^{\text{linear}} \mathbf{U} \begin{pmatrix} E_L \\ E_R \end{pmatrix}$$

when the relation (where t denotes the conjugate transposed matrix):

$$\mathbf{S}^{\text{circular}} = \mathbf{U}^t \mathbf{S}^{\text{linear}} \mathbf{U}; \qquad \mathbf{U} = \frac{1}{\sqrt{2}} \begin{pmatrix} 1 & 1 \\ j & -j \end{pmatrix}; \ \mathbf{U}^{-1} = \mathbf{U}^t \qquad (6.4)$$

This characterization of the target by an RCS or by a backscattering matrix is valid for a given transmitter-target-receiver geometrical configuration and for a given wavelength. In fact, for a nonpoint target (extended target), the coefficients of the backscattering matrix depend on the aspect angles of the target, as shown in Figure 6.5, which represents the evolution of the RCS of an aircraft as a function of the aspect angle and frequency, in a horizontal plane, with horizontal polarization at transmission and at reception.

Similarly, Figure 6.6 shows the evolution of the RCS of a conducting sphere as a function of the frequency; here again, the variations in the RCS are a consequence of the nonpoint character of the target and may be taken into account when defining the observation radar system.

6.2.3 Simple Examples

To illustrate the previous definitions, it is useful to consider the case of simple nonpoint targets, such as the elementary dipole, the flat plate or the rectangular dihedral.

---

1. This means that the polarization is defined with respect to the antenna system and not intrinsically with respect to the wave propagation direction.

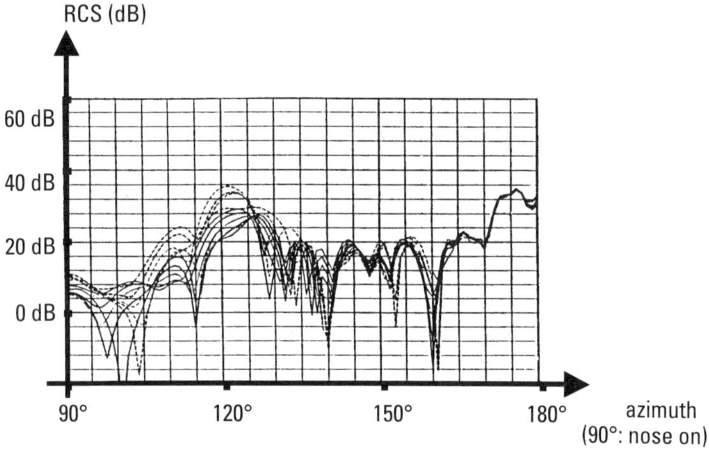

Figure 6.5 RCS of an aircraft, for 10 frequencies between 250 and 260 MHz, as a function of aspect angle (90°-nose-on aspect).

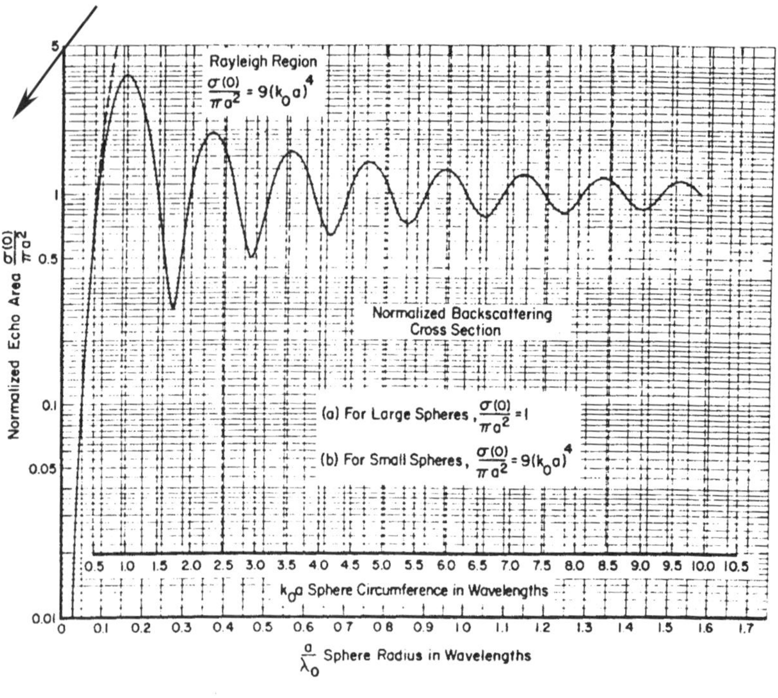

Figure 6.6 RCS of a conducting sphere. (*From:* [4].)

A dipole of length $\lambda/2$, illuminated in linear polarization, carries an induced current that is a function of the angle between the transmitted electric field and the dipole, and the radiated field is oriented parallel to the dipole. If this dipole is located in a plane perpendicular to the axis of propagation and makes an angle γ with the horizontal (as shown in Figure 6.7), and if the transmitter and the receiver are located at the same place (monostatic system, as opposed to bistatic systems in which the transmitter and the receiver are located at different places), the backscattering matrix can be written, to within a real coefficient:

$$\mathbf{S}_{\text{dipole}}^{\text{linear}} = \begin{pmatrix} \cos^2\gamma & \sin\gamma\cos\gamma \\ \sin\gamma\cos\gamma & \sin^2\gamma \end{pmatrix}$$

According to (6.4), the backscattering matrix in circular polarization can be written:

$$\mathbf{S}_{\text{dipole}}^{\text{circular}} = \begin{pmatrix} e^{2j\gamma} & 1 \\ 1 & e^{-2j\gamma} \end{pmatrix}$$

Let us consider now an infinite and conducting planar plate, perpendicular to the direction of observation: for the total field to be zero at the surface of this plate, the reflected field must be opposite to the incident field. Therefore, the backscattering matrix in linear polarization can be written:

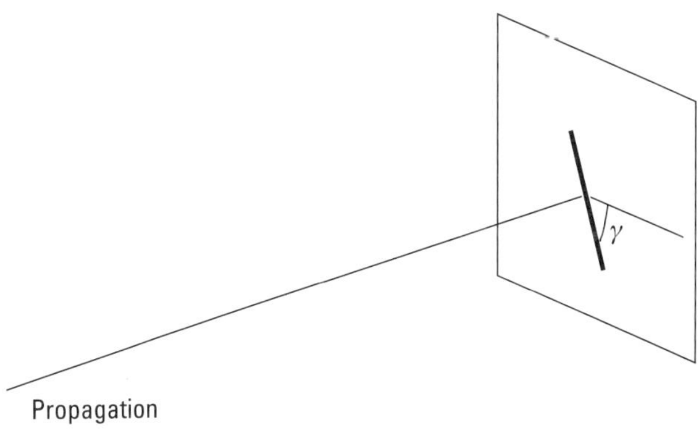

Figure 6.7 Dipole.

$$\mathbf{S}_{\text{plate}}^{\text{circular}} = \begin{pmatrix} -1 & 0 \\ 0 & -1 \end{pmatrix}$$

Applying (6.4), we can derive the backscattering matrix in circular polarization:

$$\mathbf{S}_{\text{plate}}^{\text{circular}} = \begin{pmatrix} 0 & -1 \\ -1 & 0 \end{pmatrix}$$

This means that, for a flat plate, reception in circular polarization must be performed on an antenna other than that used for transmission, since the received polarization is orthogonal to the transmitted one.

Finally, let us consider the case of a right dihedral, with dimensions assumed to be large with respect to the wavelength, and with the ridge contained in a plane parallel to the plane $y\,0\,z$ (that is, orthogonal to the direction of propagation Ox), and making an angle α with the horizontal.

Let $\vec{a}$ be a vector parallel to the ridge of the dihedral, and $\vec{b}$ a vector perpendicular to $\vec{a}$ and such that $\vec{b} \wedge \vec{a} = \vec{x}$; we obtain:

$$\begin{cases} \vec{a} = \cos\alpha\vec{y} + \sin\alpha\vec{z} \\ \vec{b} = \sin\alpha\vec{y} - \cos\alpha\vec{z} \end{cases} \tag{6.5}$$

The incident field is then given by:

$$\vec{E}_i = (H_i\cos\alpha + V_i\sin\alpha)\vec{a} + (H_i\sin\alpha - V_i\cos\alpha)\vec{b}$$

where H_i and V_i are the field components in horizontal and vertical linear polarizations.

After the two reflections on the faces of the dihedral, the component along $\vec{a}$ will have been phase-shifted twice by 180° (to meet the zero condition of the field on the surface). A similar argument shows that the component along $\vec{b}$ will have been phase-shifted by 180°. The backscattered field is thus given by:

$$\vec{E}_r = (H_r\cos\alpha + V_r\sin\alpha)\vec{a} + (-H_r\sin\alpha + V_r\cos\alpha)\vec{b}$$

that is, according to (6.4):

$$\begin{pmatrix} H_r \\ V_r \end{pmatrix} = \begin{pmatrix} \cos 2\alpha & \sin 2\alpha \\ \sin 2\alpha & -\cos 2\alpha \end{pmatrix}\begin{pmatrix} H_i \\ V_i \end{pmatrix}$$

The backscattering matrix in linear polarization is therefore:

$$\mathbf{S}_{\text{dihedral}}^{\text{linear}} = \begin{pmatrix} \cos 2\alpha & \sin 2\alpha \\ \sin 2\alpha & -\cos 2\alpha \end{pmatrix}$$

And, according to (6.4), the backscattering matrix in circular polarization is:

$$\mathbf{S}_{\text{dihedral}}^{\text{circular}} = \begin{pmatrix} e^{2j\alpha} & 0 \\ 0 & e^{-2j\alpha} \end{pmatrix}$$

Thus we can see that in circular polarization the effect of the double reflection on the dihedral is to return an echo in the polarization used at transmission. Generally, the use of circular polarization makes it possible to distinguish single reflectors (plates) or triple reflectors (trihedrals) from double reflectors (dihedrals).

6.2.4 Sonar Reflection [5]

The reflection of a monochromatic plane acoustic wave from a rigid sonar target is a phenomenon completely analogous to radar reflection, since the two types of waves obey a similar propagation equation:

$$\nabla^2 f + k^2 f = 0$$

with, in electromagnetism:

- f vectorial (electric field $\vec{E}$, or magnetic field $\vec{H}$);
- $k = \dfrac{\omega}{c}$, ω: wave pulsation (carrier);
- $c = \dfrac{1}{\sqrt{\epsilon\mu}}$: wave propagation velocity; ϵ: permittivity; μ: permeability.

And in acoustics:

- f scalar (pressure p or particles velocity u);

- $k = \frac{\omega}{c}$, ω: wave pulsation (carrier);
- $c = \sqrt{\frac{B}{\rho}}$: wave propagation velocity; B: modulus of elasticity, or Young modulus; ρ: density.

In each case, an impedance of the medium is defined.

In electromagnetism: $\vec{u} \wedge \vec{E} = Z\vec{H}$, $Z = \sqrt{\frac{\mu}{\epsilon}}$, $\vec{u}$ unitary vector directed the direction of propagation.

In acoustics: $p = Zu$, $Z = \sqrt{\rho B}$, p: pressure and u: velocity.

Considering these analogies, it would be desirable of course to characterize the reflection on the target in a similar way. Unfortunately, this is not what is done in practice, since the quantity analogous to the RCS is the target strength in acoustics, defined as the ratio of the intensity reflected in the direction of the receiver, measured at 1m from the center of the target, to the incident intensity on the target:

$$\sigma = \frac{I_{r,1\mathrm{m}}}{I_i} = \frac{P_{r,1\mathrm{m}}^2}{P_i^2} = \frac{u_{r,1\mathrm{m}}^2}{u_i^2}$$

This is thus a dimensionless quantity, numerically equal to $t = \sigma_{\text{section}}/4\pi$, where σ_{section} is the sonar cross section defined like the RCS in Section 6.2.2. (In this relation, the coefficient 4π is indeed $4\pi \cdot 1\mathrm{m}^2$, since the definition is given at 1m distance from the target.)

6.3 Modeling of Nondeformable Targets

6.3.1 Analytical Models of RCS or Sonar Target Strength

The best way to take into account precisely the exact nature of radar or sonar targets would doubtless be to have a general analytical model of the received signal that could replace the simple reflection model used in the preceding chapters. The survey of the available analytical models carried out in this chapter will show that such a general model has yet to be invented and that consequently only models that are less accurate but more easily used in signal processing—multiple-scatterers model, stochastic model—can be used in practice. This overview of modeling techniques will also provide a more precise idea of the nature of the wave-target interaction, by describing the approximations (high or low frequency) that can be adopted. In contrast, it will not mention the numerical methods for calculating RCSs, such as the method of moments, based on the numerical solution of the Stratton-Chu equations (6.25) and (6.26). Although these methods cannot be used directly to define radar processing, they do enable the RCS or sonar target strength of targets of any shape to be calculated, at least for large wavelengths.

This survey of models starts with an examination of the geometrical optics—or acoustics—approximation (ray calculation), which enables description of the propagation in a medium of variable refractive index, frequently encountered in sonar, then calculation of the RCS—or sonar target strength—of a convex object with large radii of curvature (compared to the wavelength). Examination of the physical optics approximation and of the geometrical theory of diffraction will then provide a deeper analysis of the wave-target interaction for this short wavelength situation.

6.3.1.1 Geometrical Optics and Acoustics [6]

The propagation of an electromagnetic wave or an acoustic wave in a medium characterized by the wave number $k = \dfrac{\omega}{c}$ obeys the Helmholtz equation:

$$\nabla^2 \vec{E} + k^2 \vec{E} = 0 \text{ (electric field)} \tag{6.6}$$

or

$$\nabla^2 p + k^2 p = 0 \text{ (pressure)} \tag{6.7}$$

When k is not identical at all points in the medium, this equation generally cannot be solved. An often justified approximation consists of accepting that, as the changes in the nature of the medium are on a large scale compared with the wavelength, the equation remains valid locally. Moreover, as the wave number of the medium varies slowly around $k = k_0$, the electric field or the pressure is expressed for large k in the form of an expansion in $1/k_0^m$ (whose validity is justified only a posteriori by comparison with the experimental results):

$$\begin{cases} \vec{E}(R) = e^{-jk_0\psi(R)}\left[\vec{E}_0(R) + \dfrac{\vec{E}_1(R)}{k_0} + \ldots\right] \\ \text{with: } \nabla^2\vec{E} + k_0^2 n^2 \vec{E} = 0,\ n = \dfrac{k}{k_0}: \text{refractive index of the medium} \end{cases} \tag{6.8}$$

or

$$\begin{cases} p^{(R)} = e^{-jk_0\psi(R)}\left[p_0(R) + \dfrac{p_1(R)}{k_0} + \ldots\right] \\ \text{with: } \nabla^2 p + k_0^2 n^2 p = 0,\ n = \dfrac{k}{k_0}: \text{refractive index of the medium} \end{cases} \tag{6.9}$$

(The dependence on $e^{+j\omega t}$ is implicit, as in all that follow.)

Let us consider the case of the electromagnetic wave (a parallel argument can be applied in the case of the acoustic wave).

Substituting (6.8) for the field in the Helmholtz equation (6.6) and annulating the coefficients of each power of k_0, we obtain for E_0:

$$|\vec{\nabla}\psi|^2 = \frac{k^2}{k_0^2} = n^2 \tag{6.10}$$

$$(\vec{\nabla}\psi \cdot \vec{\nabla})\vec{E}_0 + \frac{1}{2}\nabla^2\psi\vec{E}_0 = 0 \tag{6.11}$$

In addition, the direction of the field is given by $\vec{\nabla}\vec{E} = 0$ (Maxwell's equation), which implies $\vec{\nabla}\psi \cdot \vec{E}_0 = 0$.

$\vec{\nabla}\psi$ is perpendicular to the constant-phase surfaces (wave fronts) $\psi(R)$ = Const. Therefore:

$$\vec{\nabla}\psi \cdot \vec{\nabla} = \frac{\partial}{\partial\tau} \tag{6.11a}$$

where τ specifies the position on a curve perpendicular to the surface (ray). Equation (6.11) becomes:

$$\frac{\partial}{\partial \tau}\vec{E}_0 + \frac{1}{2}\nabla^2\psi\vec{E}_0 = 0$$

and is integrated as:

$$\begin{cases} \vec{E}_0(\tau) = \vec{E}_0(\tau_0)\exp\left[-\frac{1}{2}\int_{\tau_0}^{\tau}\nabla^2\psi du\right] \\ \text{with: } \vec{E} = e^{-jk_0\psi(\tau)}\vec{E}_0(\tau) \\ \text{and: } \vec{\nabla}\psi \cdot \vec{E}_0 = 0 \end{cases} \tag{6.12}$$

These relations thus define the propagation of the wave along rays perpendicular to the wave fronts $\psi(R)$ = Const. Before considering the case of an isotropic homogeneous medium, let us examine more closely the appearance of these rays in the case of a medium of variable refractive index—a frequent situation in submarine or aerial acoustics but also encountered in electromagnetism, for example in grazing propagation above the surface.

The rays, perpendicular to the wave fronts $\psi(R)$ = Const, are therefore tangent to the vector $\vec{\nabla}\psi$, whose modulus is the index n according to (6.10). The tangent unit vector is therefore:

$$\vec{s} = \frac{1}{n}\vec{\nabla}\psi \tag{6.13}$$

Let us calculate the curvature r at a point of the ray [6]. If $\vec{N}$ is the unit vector in the direction of the radius of curvature, we have:

$$r\vec{N} = \frac{d\vec{s}}{ds}$$

where s is the curvilinear abscissa on the ray.

However

$$\frac{d\vec{s}}{ds} = (\vec{s} \cdot \vec{\nabla})\vec{s} = -\vec{s} \wedge \text{rot}\vec{s}$$

$$r = -\vec{N} \cdot \vec{s} \wedge \text{rot}\vec{s}$$

Since $\vec{N}$ is perpendicular to $\vec{\nabla}$, and given the relation:

$$\vec{\nabla}\left(\frac{1}{n}\right) = -\frac{1}{n}\vec{\nabla}\log n$$

we obtain

$$r = \vec{N} \cdot \vec{\nabla}\log(n) \tag{6.14}$$

Consequently, the variation in the refractive index n is positive toward the center of curvature, and therefore the ray turns toward the regions of higher index (or of lower velocity). If the refractive index of the medium is constant, $r = 0$ and the ray is a straight line.

The rays can then be calculated as follows: the tangent vector $\vec{s}$ has components $\frac{dx}{ds}$, $\frac{dy}{ds}$, and $\frac{dz}{ds}$, where:

$$\left(\frac{dx}{ds}\right)^2 + \left(\frac{dy}{ds}\right)^2 + \left(\frac{dz}{ds}\right)^2 = 1 \tag{6.15}$$

According to (6.1.) we have:

$$n\frac{dx}{ds} = \frac{\partial\psi}{\partial x};\ n\frac{dy}{ds} = \frac{\partial\psi}{\partial y};\ n\frac{dz}{ds} = \frac{\partial\psi}{\partial z} \tag{6.16}$$

ψ can be eliminated by writing:

$$\frac{d}{ds}\left(n\frac{dx}{ds}\right) = \frac{d}{ds}\left(\frac{\partial\psi}{\partial x}\right)$$
$$= \frac{\partial^2\psi}{\partial x^2}\frac{dx}{ds} + \frac{\partial^2\psi}{\partial x\partial y}\frac{dy}{ds} + \frac{\partial^2\psi}{\partial x\partial z}\frac{dz}{ds}$$

given (6.16):

$$\frac{d}{ds}\left(n\frac{dx}{ds}\right) = \frac{\partial^2\psi}{\partial x^2}\frac{1}{n}\frac{\partial\psi}{\partial x} + \frac{\partial^2\psi}{\partial x\partial y}\frac{1}{n}\frac{\partial\psi}{\partial y} + \frac{\partial^2\psi}{\partial x\partial z}\frac{1}{n}\frac{\partial\psi}{\partial z}$$
$$= \frac{1}{2n}\frac{\partial}{\partial x}[|\nabla\psi|^2]$$

given (6.10):

$$\begin{cases} \dfrac{d}{ds}\left(n\dfrac{dx}{ds}\right) = \dfrac{\partial n}{\partial x} \\ \text{and similarly:} \\ \dfrac{d}{ds}\left(n\dfrac{dy}{ds}\right) = \dfrac{\partial n}{\partial y} \\ \dfrac{d}{ds}\left(n\dfrac{dz}{ds}\right) = \dfrac{\partial n}{\partial z} \end{cases} \tag{6.17}$$

This system of differential equations defines the rays when the law of variation of the refractive index n of the medium is known.

An example of variable-index medium frequently encountered in practice is the case of a layered medium. This type of medium is a good approximation of the real situation, both in submarine acoustics, where the index is essentially a function of depth, and in radar, where the refractive index near the surface of the sea varies essentially with height. Let us calculate these rays in the case in which the index is a function of height or depth z. The calculation is performed in the plane xOz since, as the medium is layered, the calculation in another plane would be deduced from it by rotation about the Oz axis. Using (6.17), defining the rays, we have:

$$\frac{d}{ds}\left(n\frac{dx}{ds}\right) = \frac{\partial n}{\partial x} = 0 \tag{6.18}$$

$$\frac{d}{ds}\left(n\frac{dz}{ds}\right) = \frac{dn}{dz} \tag{6.19}$$

Therefore

$$\frac{dx}{ds} = \frac{C}{n(z)} \qquad C\text{: constant}$$

As

$$\left(\frac{dx}{ds}\right)^2 + \left(\frac{dz}{ds}\right)^2 = 1$$

$$\frac{dz}{ds} = \pm\sqrt{1 - \frac{C^2}{n^2(z)}}$$

If the Oz axis is directed upward, the ray will therefore be upward for $dz/ds > 0$ and downward in the converse case.

For an upward ray:

$$\frac{dz}{dx} = \sqrt{\frac{n^2(z)}{C^2} - 1}$$

and for a downward ray:

$$\frac{dz}{dx} = -\sqrt{\frac{n^2(z)}{C^2} - 1}$$

Hence the equation for an upward ray:

$$\mathbf{x} = \int_{z_0}^{z} \left(\frac{n^2(z)}{C^2} - 1\right)^{-1/2} dz \tag{6.20}$$

the sign being inverted for a downward ray. This equation can be used for numerical calculation of the points of the different rays, subject to the basic approximations of geometrical optics or acoustics. An example of the results thus obtained [7] is shown in Figure 6.8. It can be seen that for this type

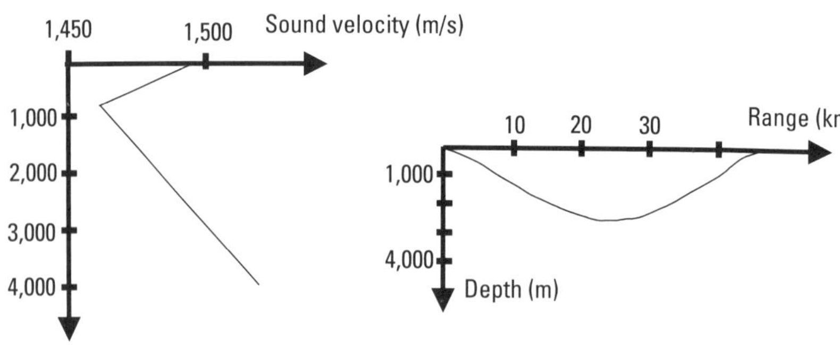

Figure 6.8 Submarine acoustic propagation: (a) velocity profile; (b) path of the corresponding ray.

of index profile, often encountered in submarine acoustics, the propagation follows quite markedly curved rays. The acoustic illumination of the submarine space by a given source is consequently not very homogeneous; the sound intensity accumulates in some regions whereas others remain almost silent. For completeness, the evolution of the intensity along a ray should be calculated; this calculation is performed assuming that the energy flux is constant in a tube of rays, since the energy propagates along the rays (locally-planar wave).

We shall not carry out this calculation in the general case but rather in the special case of an isotropic homogeneous medium; from it, we shall derive the expression for the RCS—or the target strength—of a convex surface whose radius of curvature is large compared with the wavelength.

In the case of an isotropic homogeneous medium, the rays are thus straight lines perpendicular to the wave fronts; we can therefore write:

$$d\psi = |\nabla \psi| ds = n ds \tag{6.21}$$

$$\psi = \psi_0 + ns \tag{6.22}$$

Taking into account (6.11a), we obtain: $ds = n d\tau$.

Let ρ_1 and ρ_2 be the main radii of curvature of the wave front ψ = Const; it can then be shown that, if $G = 1/\rho_1 \rho_2$, we have:

$$\frac{dG}{ds} = -G\frac{\nabla^2 \psi}{n} \Rightarrow \frac{dG}{d\tau} = -G\nabla^2 \psi$$

which can be integrated to give:

$$\frac{G(\tau)}{G(\tau_0)} = \exp\left[-\int_{\tau_0}^{\tau} \nabla^2 \psi dt\right]$$

Substituting in (6.12), we finally obtain:

$$\vec{E}(\tau) = \vec{E}(\tau_0) e^{-jk_0 \psi_0} \sqrt{\frac{G(\tau)}{G(\tau_0)}} e^{-jks}$$

It appears that the field, orthogonal to the direction of propagation, propagates along the rays (abscissa s) with a phase shift proportional to the abscissa on the ray (e^{-jks}) and an attenuation that is a function of the wave front curvature variation.

This representation of the electric field is conventionally known as the geometrical optics field, although strictly speaking the geometrical optics field resulting from Fermat's principle (optical path of extreme length $\int n\, ds$) and from the conservation of the energy flux along a tube of rays has neither a phase term nor a direction, which are generally added to it a posteriori.

The term G/G_0 can be expressed as (Figure 6.9):

$$G(\tau)/G(\tau_0) = \frac{\rho_1 \rho_2}{(\rho_1 + s)(\rho_2 + s)}$$

where the expression of the field:

$$\vec{E}(\tau) = \vec{E}(\tau_0) e^{-jk_0 \psi_0} e^{-jks} \sqrt{\frac{\rho_1 \rho_2}{(\rho_1 + s)(\rho_2 + s)}} \tag{6.23}$$

If the field is computed on a caustic (defined as a point of intersection between rays), ρ_1 or $\rho_2 = -s$, an infinite field is obtained; therefore the geometrical optics approximation is not valid in the vicinity of caustics.

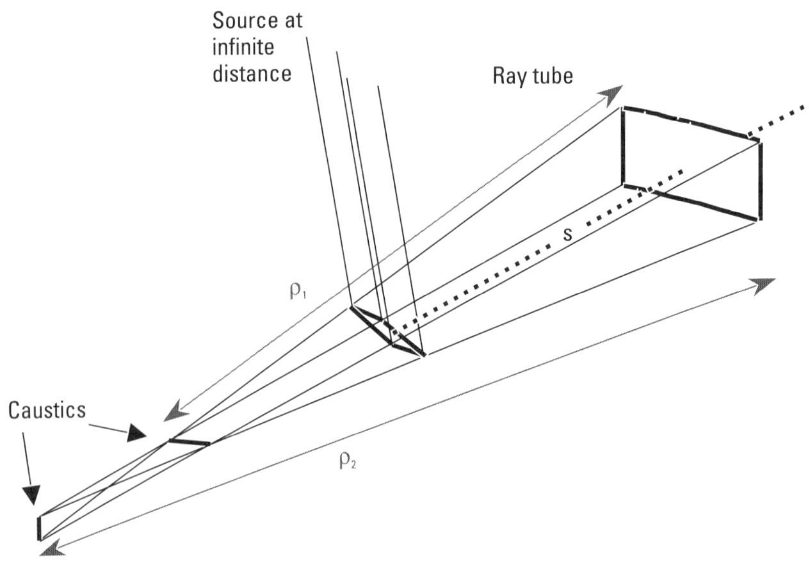

Figure 6.9 Reflected ray tube.

Finally, let us calculate the reflected field of a planar wave incident on a convex surface of large radii of curvature compared with the wavelength. Fermat's principle indicates that the incident rays are reflected in the specular reflection direction, that is, at a point such that the normal to the surface bisects the angle formed between the incident and reflected directions. If R_1 and R_2 are the main radii of curvature of the surface at this point, it can be shown that $\sqrt{\rho_1 \rho_2} = 1/2\sqrt{R_1 R_2}$; hence, the expression of the field at long range ($s >> \rho_1$, $s >> \rho_2$):

$$\vec{E}(\tau_0)e^{-jk_0\psi_0} \frac{e^{-jks}}{s} \frac{\sqrt{R_1 R_2}}{2}$$

Substituting in the RCS definition of (6.2), we obtain:

$$\sigma = \pi R_1 R_2 \tag{6.24}$$

This relation gives the RCS of a convex surface with large radii of curvature compared with the wavelength. This relation generally applies well to the ellipsoidal parts of radar targets (aircraft nose or fuselage) at the usual wavelengths (1 cm < λ < 10 cm).

Given Section 6.2, a similar argument applied to the sonar case leads to a target strength t for such a convex surface:

$$t = \frac{R_1 R_2}{4}$$

This expression is generally valid for submarine hulls, for example (at least to within an order of magnitude).

In summary, although the geometrical optics approach thus defined enables some radar and sonar reflection problems to be dealt with, for monostatic or bistatic (transmitter and receiver at different locations) configurations, and for surfaces with radii of curvature greater than a few (three to four) wavelengths, it cannot be used to deal with surfaces with a single radius of curvature (that is, the cases in which one of the radii of curvature becomes infinite: cylinder, cone, plane), since it then predicts an infinite RCS, nor diffracting surfaces (radius of curvature less than a few wavelengths). Furthermore, it has the disadvantage of predicting an infinite field on a caustic.

Applied to electromagnetic or acoustic propagation, this geometrical optics approximation results in a very intuitive and often valid representation (using rays) of the phenomena observed in variable-index media. The energy is propagated at constant flux in a "tube of rays" in which each ray is defined by the system of (6.17) in the general case or by (6.20) for layered media.

6.3.1.2 Physical Optics

The second approximation conventionally used for calculating high frequency radar or sonar echoes is the physical optics approximation [4, 8, 9].

To avoid unnecessary complication, the following account will be limited to the case of radar; a fully parallel argument could be developed in the case of sonar.

This approximation is derived in a relatively straightforward manner from the integral equations of Stratton and Chu [10], which are an exact formulation of Maxwell's equations in a source-free region. Let S be the closed surface of a reflector. The diffracted field at a point P outside this surface is given by Figure 6.10:

$$\vec{E}^d(\vec{r}) = \frac{1}{4\pi}\int_S \left[(\vec{n}'\Lambda\vec{E}^T)\Lambda\vec{\nabla}'\psi + j\omega\mu(\vec{n}'\Lambda\vec{H}^T)\psi + (\vec{n}'\cdot\vec{E}^T)\vec{\nabla}'\psi \right] dS' \tag{6.25}$$

$$\vec{H}^d(\vec{r}) = \frac{1}{4\pi}\int_S \left[(\vec{n}'\Lambda\vec{H}^T)\Lambda\vec{\nabla}'\psi + j\omega\epsilon(\vec{n}'\Lambda\vec{E}^T)\psi + (\vec{n}'\cdot\vec{H}^T)\vec{\nabla}'\psi \right] dS' \tag{6.26}$$

where $\vec{E}^t$ and $\vec{H}^t$ are the total fields (incident + diffracted) at $\vec{r}$.

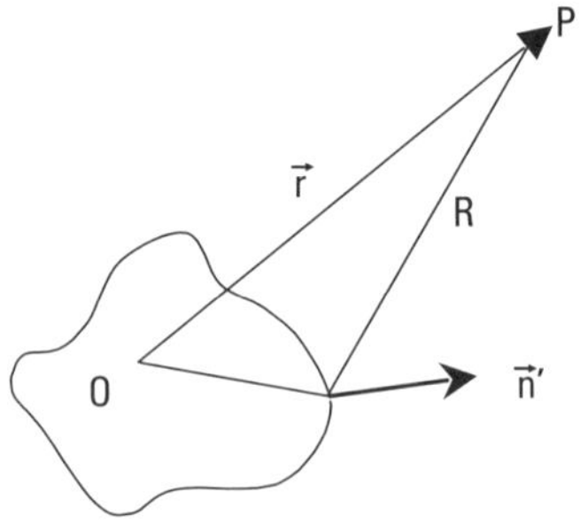

Figure 6.10 Coordinate system.

$$\psi = \frac{e^{jk|\vec{r} - \vec{r}'|}}{|\vec{r} - \vec{r}'|} \text{ is Green's function in free space}$$

$$\vec{\nabla}' \text{ indicates the gradient with respect to } \vec{r}' \tag{6.27}$$

$$\vec{E}^d \text{ and } \vec{H}^d \text{ are the fields diffracted by the target}$$

In these equations, the diffracted field $\vec{H}^d$ (or $\vec{E}^d$), which is unknown, is present in both members (since $\vec{H}^T = \vec{H}^d + \vec{H}^i$). The physical optics approximation reduces these equations to an explicit expression of the backscattered field. To apply the physical optics approximation, (6.26) must first be transformed to reduce it to a surface integral on the illuminated part of the object, limited by the "shadow boundary" Γ; we show that the expression of $\vec{H}^d$ then introduces a correction term in the form of an integral over the outline Γ of the boundary (S_1 being the illuminated surface of the object):

$$\vec{H}^d(\vec{r}) = \frac{1}{4\pi} \int_{s_1} (\vec{n}' \Lambda \vec{H}^T) \Lambda \vec{\nabla}' \psi - j\omega\epsilon(\vec{n}' \Lambda \vec{E}^T)\psi + (\vec{n}' \cdot \vec{H}^T)\vec{\nabla}'\psi \, dS' - \frac{1}{4\pi j\omega\mu} \int_{\Gamma} \vec{\nabla}'\psi \vec{E}^T \cdot \vec{dl} \tag{6.28}$$

The far field approximation is then used, so we can write:

$$\vec{\nabla}\psi = -jk\psi\vec{k}^d$$

$\vec{k}^d$: unit vector along the observation direction.

Thus converting the curvilinear integral into a surface integral, we obtain:

$$\vec{H}^d(\vec{r}) = \frac{-j\omega\epsilon e^{jkR_0}}{4\pi R_0} \tag{6.29}$$

$$\int_{s_1} \left[(\vec{n}' \Lambda \vec{H}^T) \Lambda \vec{k}^d \sqrt{\frac{\mu}{\epsilon}} + \vec{n}' \Lambda \vec{E}^T - \vec{k}^d \cdot (\vec{n}' \Lambda \vec{E}^T)\vec{k}^d \right] e^{-jk(\vec{k}^d \cdot \vec{r}')} dS'$$

The currents induced on the surface are then estimated by approximation of a tangent plane, that is, in the case of a conducting body:

$$\begin{cases} \vec{n}' \Lambda \vec{E}^T = \vec{0} \\ \vec{n}' \Lambda \vec{H}^T = 2\vec{n}' \Lambda \vec{H}^i \end{cases}$$

The expression of the diffracted field then becomes:

$$\vec{H}^d(\vec{r}) = \frac{-jke^{jkR_0}}{2\pi R_0} \int_{s_1} ((\vec{n}' \Lambda \vec{H}^i) \Lambda \vec{k}^d) e^{-jk(\vec{k}^d \cdot \vec{r}')} dS' \tag{6.30}$$

from which the electric field is derived (planar wave):

$$\vec{E}^d(\vec{r}) = \sqrt{\frac{\mu}{\epsilon}} \vec{k}^d \Lambda \vec{H}^d(\vec{r})$$

Equation (6.30) thus gives, according to the physical optics approximation, the explicit expression of the field $\vec{H}^d(\vec{r})$ diffracted by the object at a point $\vec{r}$, as a function of a surface integral of the incident field $\vec{H}^i$ over the illuminated surface S_1 of the object.

This expression is simplified in the case of a monostatic system, in the far field of the target ($\vec{k}^d = -\vec{k}^i$, $\vec{k}^d \cdot \vec{H}^d = \vec{k}^i \cdot \vec{H}^i = 0$)

$$\vec{H}^d(\vec{r}) = \frac{-jke^{jkR_0}}{2\pi R_0} \int_{s_1} e^{-j(\vec{k}^d \cdot \vec{r}')k} (\vec{k}^d \cdot \vec{n}') \vec{H}^i dS' \tag{6.31}$$

In particular, this expression shows that, according to physical optics, the polarization of the diffracted field is the same as that of the incident field, in the case of a perfectly conducting target observed in a monostatic configuration. We have seen in the simple examples in Section 6.2.3 that this rule was not always valid.

Let $\vec{H}_0^i$ be the incident field at R_0; we can write:

$$\vec{H}^i = \vec{H}_0^i e^{+j\vec{k}^i \cdot \vec{r}'}$$

$$\Rightarrow \vec{H}^d(\vec{r}) = \frac{-jke^{jkR_0}}{2\pi R_0} \vec{H}_0^i \int_{s_1} e^{-2j(\vec{k}^d \cdot \vec{r}')k} \vec{k}^d \cdot \vec{n}' \, dS'$$

Using the diffraction direction $\vec{k}^d$ as axis Oz', this expression can be written:

$$\vec{H}^d(\vec{r}) = \frac{-jke^{jkR_0}}{2\pi R_0} \vec{H}_0^i \int_0^L e^{+2jk_0 z'} \frac{dA(z')}{dz'} dz' \quad (6.32)$$

where

$$\vec{z} \cdot \vec{n}' = \vec{k}^d \cdot \vec{n}'$$

and

$$(\vec{z} \cdot \vec{n}')\, dS' = \vec{n}'_z dS' = \frac{dA(z')}{dz'} dz'$$

if $A(z')$ is the projection, parallel to Oz', of the surface of the object for z larger than z'. This expression, which shows that the amplitude of the diffracted field, as a function of the wavenumber, is proportional to the Fourier transform of $dA(z')/dz'$, is particularly useful for calculation of RCS of objects with known shape.

In conclusion, the physical optics approximation has the important advantage of generating explicit expressions of the diffracted field, and thus of the RCS by (6.32). However, it is based on the approximation of the tangent plane for the surface currents ($\vec{n}' \Lambda \vec{H}^T = 2\vec{n}' \Lambda \vec{H}^i$), which is obviously inadequate for edges or for surface waves that propagate beyond the illuminated part of the object (creeping waves). It can be shown that if the integral giving the diffracted field of the physical optics is approximated by the stationary phase method, the field of the geometrical optics approximation is obtained. This explains that, for "smooth" surfaces (high radii of curvature compared with the wavelength), physical optics leads to satisfactory results along directions close to the specular reflection direction.

Moreover, for situations in which geometrical optics is not a valid approximation, as surfaces with a single radius of curvature, physical optics may offer an appropriate solution. In particular, for a flat plate with cross section A observed at normal incidence, the RCS

$$\sigma = \frac{4\pi A^2}{\lambda^2}$$

is obtained, which is a correct result when the dimensions of the plate are greater than a few wavelengths.

This physical optics theory can be improved if an additional field generated by current distributions on the edges is added to the field it describes: this is the *physical theory of diffraction,* in which these additional fields are calculated using currents on the edges obtained by solution of canonical problems (equivalent currents).

Apart from the problems of edges and creeping waves, physical optics does not satisfy the reciprocity theorem, which states that swapping the transmitter and receiver around must not change the RCS. Moreover, we have seen that, in the monostatic configuration when observing a conducting object, it leads to a polarization of the received wave that is identical to that of the transmitted wave. This does not correspond to reality, as shown for example by the RCS calculations for the dipole and the dihedral performed in Section 6.1.3. In an attempt to remedy these faults, J. B. Keller developed the "geometrical theory of diffraction," the principles of which we shall now describe.

6.3.1.3 Geometrical Theory of Diffraction [11]

This essentially empirical theory, continuing the historic evolution of optics (gradual incorporation of the experimentally observed wave nature of the phenomenon), proposes the addition of diffracted rays, obeying laws analogous to those of reflection and refraction, to geometrical optics. Far from diffracting surfaces, these rays behave like other rays; the proposed theory thus introduces only local modifications (on the surface of the body) into the principles of geometrical optics (Fermat's principle and conservation of energy in a tube of rays).

The field diffracted by an edge (straight or curved), a corner, or a surface wave beyond the shadow boundary is obtained by applying a coefficient to the incident field. In radar situations this coefficient is generally matricial in radar situations to take into account the effects of polarization. These coefficients have the following properties:

- They tend to 0 with the wavelength (with $\lambda^{1/2}$ for edge diffraction, with λ for corner diffraction, exponentially for the surface wave).
- They are determined locally, and can therefore be obtained by solution of a canonical problem (such as diffraction by a half-plane, by the vertex of a cone, by an infinite cylinder).

The diffracted rays thus defined also obey a modified Fermat's principle: a ray diffracted from a transmission point P to a reception point Q by a diffracting element is a curve of minimum optical length, among all the curves from P to Q having a point on the diffracting element.

For example, in the case in which the diffracting element is a straight edge (and where all the rays are in the same homogeneous medium), the diffracted rays belong to the same cone (see Figure 6.11) whose axis is the edge and whose vertex half-angle is equal to the angle between the incident ray and the diffracting edge (this principle is also equivalent to generalizing the notion of specular reflection to the case of diffracting edges).

This theory, which is a very useful addition to geometrical optics or physical optics by taking into account diffraction, nevertheless remains limited by the local character of the modifications that are introduced, and above all by the constraint that the diffracted rays belong to the Keller cone.

6.3.1.4 Extensions

Various approaches are used to overcome the limitations of the geometrical theory of diffraction [8]. In particular, to avoid limiting the diffracted rays to the Keller cone, it is possible to introduce electric and magnetic "equivalent currents" along the contour of the singularity, induced by the incident field, whose value is determined by comparison with the geometrical theory of diffraction. This improvement enables cases in which the geometrical theory of diffraction is inadequate to be dealt with, at the price of introducing currents that have no physical existence, since their value also depends on the direction of observation. However, it appears that these theories are fundamentally limited by their local character, which is a direct consequence of the high frequency approximation.

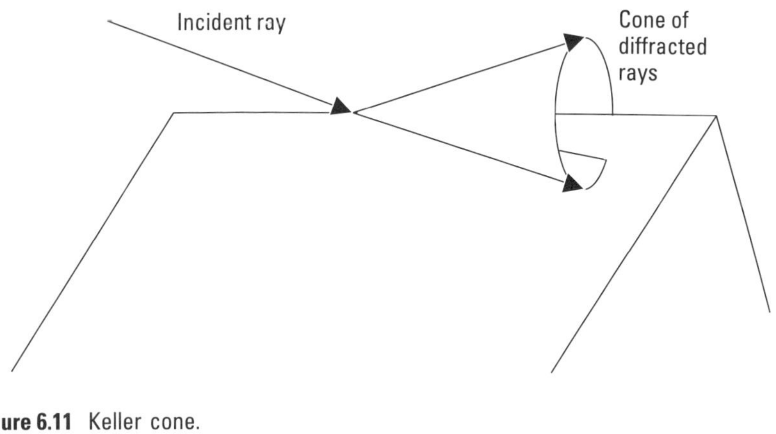

Figure 6.11 Keller cone.

6.3.1.5 Rayleigh Domain

When the wavelength is large compared with the dimensions of the target, this target causes only a local singularity of the electromagnetic field. For this frequency domain, known as the Rayleigh domain, it can be shown that the RCS is given, to a good approximation, by:

$$\sigma = \frac{4}{\pi} k^4 F^2 V^2 \tag{6.33}$$

and the sonar target strength by $t = \dfrac{k^4 F^2 V^2}{\pi^2}$, where $k = 2\pi/\lambda$, V is the volume of the target and F is a form factor depending on the slenderness ratio of the body ($F = 9/8$ for the sphere, or 0.5 for a missile-shaped target). For instance, for a dielectric sphere, the RCS is $\sigma = 4\pi a^2 \left(\dfrac{\epsilon_c - 1}{\epsilon_c + 2}\right)^2 (ka)^4$, where ϵ_c is the dielectric permittivity; this expression is appropriate for raindrops RCS evaluation. The RCS in the Rayleigh region is thus relatively independent of the precise characteristics of the body and decreases in proportion to the fourth power of the frequency. This implies in particular that a target of these dimensions can legitimately be considered as a single scatterer, and thus satisfies the assumptions adopted in Chapters 1 through 5.

6.3.1.6 Discussion

This overview of the different procedures for analytical evaluation of RCSs or sonar target strengths shows clearly that the complexity of the diffraction phenomenon prevents the replacement of the single-scatterer model that we have used in the previous chapters by a sufficiently realistic single model. Only physical optics provides an explicit analytical expression of the RCS, at the price of a tangent plane approximation that is often unacceptable in practice. Nevertheless, this examination has shown that there are various reflection diffraction phenomena of different levels of importance:

- Specular reflection is based on geometrical optics but is better processed by physical optics; this reflection generates generally strong echoes from the smooth surfaces of the targets (such as aircraft fuselages, ship superstructures, and submarine hulls).
- Diffraction by edges or corners is taken into account by the geometrical theory of diffraction, then more precisely by the equivalent current technique; these diffractions are generally an order of magni-

tude less than those of specular reflection—as far as such a general statement has any meaning, given the diversity of real situations—but can be observed on wider aspect angles.

- Surface waves generate currents beyond the shadow boundary, possibly diffracted by the discontinuities they encounter on the surface, and generally produce echoes still weaker than the previous ones; more precisely, these surface wave phenomena generally "modulate" the effects of diffraction by modifying the surface currents in the neighborhood of the diffracting element.
- Diffraction by elements that are small compared with the wavelength, for which the Rayleigh approximation is applicable; these phenomena must usually be taken into account only when the three preceding ones are not present, and are thus significant mainly when the dimensions of the targets are small compared with the wavelength.

To these three types of phenomenon are added reflections in cavities (such as air intakes and cockpits), which often generate strong echoes, at least for wavelengths shorter than the aperture of the cavity. This type of phenomenon is generally difficult to evaluate analytically, except at short wavelengths, for which a calculation based on geometrical optics with multiple reflections can give an appropriate order of magnitude.

As an illustration, Figure 6.12 shows the example of an aircraft and a missile whose RCSs have been calculated at 270 MHz (using finite elements method) and 3 GHz (using physical theory of diffraction, see Section 6.3.1.2).

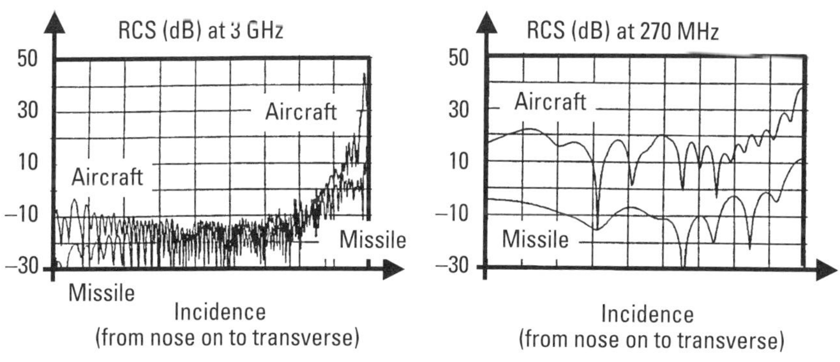

Figure 6.12 RCS of an aircraft and a missile, as a function of aspect angle (azimuth): (a) 3 GHz; (b) 270 MHz.

The current inability to construct a realistic analytical model of radar or sonar reflection means that in what follows we shall examine less ambitious but more directly usable models: the multiple-scatterers model and the stochastic model.

6.3.2 Multiple Scatterers Target Model: Radar Holography [12]

A conventional empirical approach to the reflection phenomenon is to model the target as a set of independent and isotropic scatterers. When the target is observed in a small angular sector and in a narrow frequency band this approximation is partially equivalent to the geometrical theory of diffraction if the multiple interactions between diffracting elements are neglected.

According to this hypothesis, for a pair of given transmission-reception polarizations and for a given angular sector and frequency band, the complex backscattering coefficient of the target (element of the backscattering matrix S defined in paragraph 6.2.2) is given by (Figure 6.13):

$$Q(\vec{k}) = \int I(\vec{x})\, e^{2j\vec{k}\cdot\vec{x}}\, d\vec{x} \tag{6.34}$$

where $\vec{k}$ is the observation wave vector of length $2\pi/\lambda$ and oriented in the transmission/reception direction, and where $I(\vec{x})$ is the distribution of the complex backscattering coefficients of the scatterers making up the target. (For simplicity, the discussion is limited to the case of monostatic observation where incident direction and observation direction are the same and where the target is located in far field.)

C. Pouit [12] has shown that by adopting this multiple-scatterers model one can represent the target as an image of scatterers. This representation proved to be very convenient in analyzing radar signatures and modeling

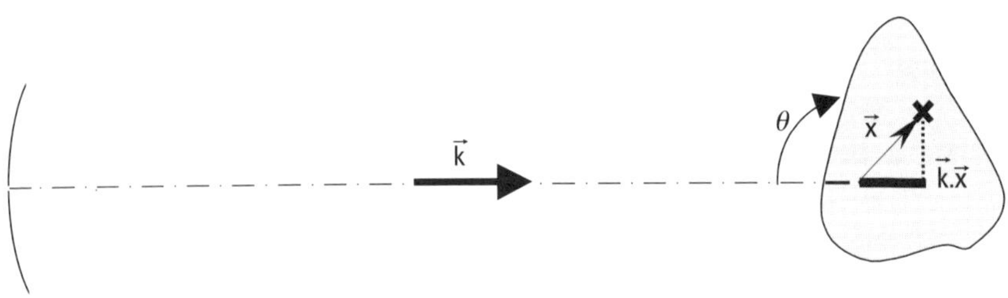

Figure 6.13 Radar target measurement.

radar systems. Equation (6.34) can be inverted directly by Fourier transform to give:

$$I(\vec{x}) = \int q(\vec{k})\, e^{-2j\vec{k}\cdot\vec{x}}\, d\vec{k} \tag{6.35}$$

(to within a coefficient, usually obtained in practice by calibration on a reference body).

This relation simply expresses the fact that, in order to evaluate the contribution $I(\vec{x})$ of the scatterer located at $\vec{x}$, it is necessary to:

- Consider the contribution of this scatterer $\vec{x}$ to the global coefficient $q(\vec{k})$, and compensate for the phase shift due to the position of this scatterer (for this observation vector $\vec{k}$), which is equal to $e^{2j\vec{k}\cdot\vec{x}}$;
- Sum up all these contributions in phase, for every available frequency and observation angle (i.e., for every vector $\vec{k}$). In this summation, only the contributions of scatterer $\vec{x}$ will add in phase (since the phase shift compensation is correct only for this position $\vec{x}$).

In practice, this image is obtained by limited scanning of the vector $\vec{k}$ in a given frequency band and angular sector. The image obtained is thus the convolution of the "real" image and an equipment function, $A(\vec{x})$, Fourier transform of the "observation aperture" $P(\vec{k})$.

To obtain a better understanding of the following examples, it is useful to specify the relations between these two domains $A(\vec{x})$ and $P(\vec{k})$ in the case where the target is observed in a plane (assumed horizontal), over an angular sector $\Delta\theta$ and in a frequency band Δf about the center frequency f_0 (Figure 6.14).

Under these conditions, the aperture $P(\vec{k})$ is approximately a rectangle of zero thickness (observation in a plane), width $f_0\Delta\theta$, and depth Δf. Given (6.35) (Fourier transform with conjugate variables $2\vec{k}$ and $\vec{r}$), the resolutions obtained in the radial direction, δx, and in the transverse direction, δy, are known to be the inverse of the extents of the observation field:

$$\begin{cases} \delta x = \dfrac{c}{2\Delta f} \\ \delta y = \dfrac{c}{2f_0\Delta\theta} = \dfrac{\lambda_0}{2\Delta\theta}, \quad \delta z\text{: } \infty \end{cases} \tag{6.36}$$

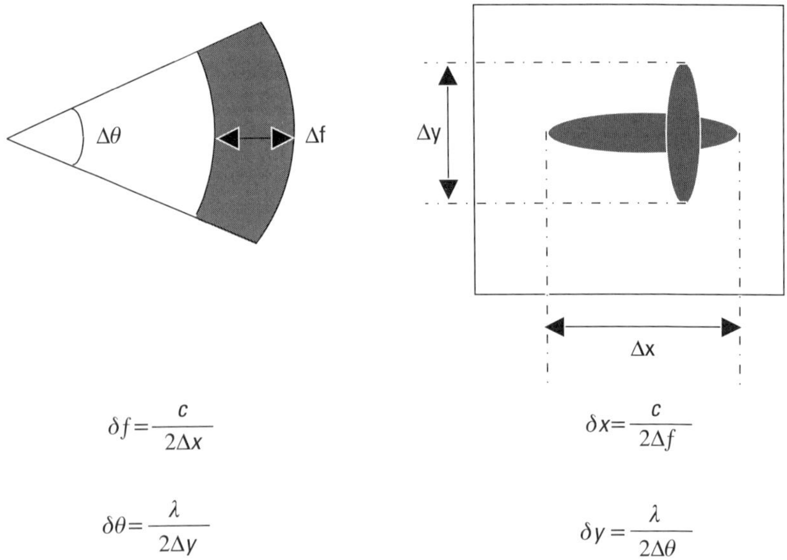

Figure 6.14 Relationship between sampling and target dimensions.

A similar (but converse) argument is used to calculate the sampling interval in angle, $\delta\theta$, and in frequency, δf, assuming that the total extent of the target is limited by Δx and Δy:

$$\begin{cases} \delta f = \dfrac{c}{2\Delta x} \\ \delta\theta = \dfrac{\lambda_0}{2\Delta y} \end{cases} \tag{6.37}$$

(In this case of observation in a single plane, the obtained image is not resolved with respect to z, that is, we obtain the projection of the real, three-dimensional image on the observation plane.)

An example of an image obtained using this procedure is given in Figure 6.15. The observed object is an actual CT 20 unmanned aerial vehicle, of overall length approximately 6m, measured at ONERA in the 8- to 10-GHz frequency band. Figure 6.15(a) shows the *measurement hologram* (in the 8- to 12-GHz band), the plot of $|q(\vec{k})|^2$ (target RCS) in gray levels for all the aspect angles in the horizontal plane; high levels are obtained for aspect angles corresponding to specular reflections (90°, 270°), and for nose-on and rear aspect angles. Figure 6.15(b) is the image $|I(\vec{r})|^2$, plotted in levels

(a)

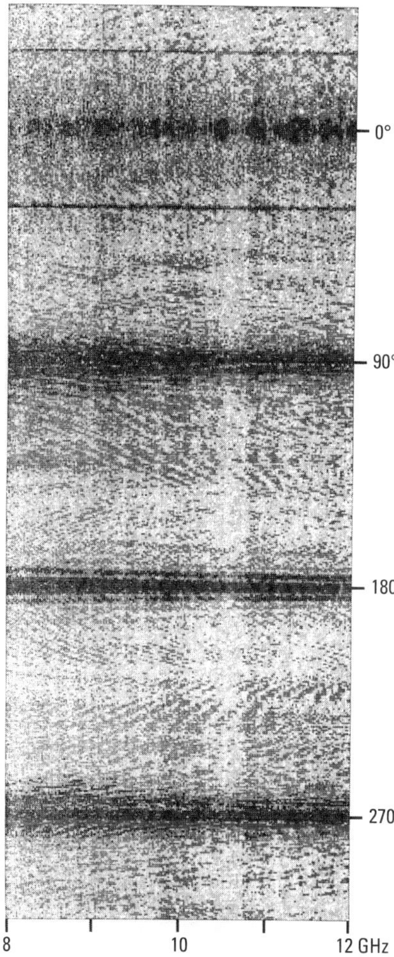

Figure 6.15 (a) Radar imaging of a CT 20 measurement hologram (ONERA measurements); (b) radar imaging of a CT20 image, 30° presentation (ONERA measurements); and (c) radar imaging of a CT 20 image around the 90° aspect angle (ONERA measurements).

of gray, obtained for $\Delta\theta = 20°$ about the aspect angle 40°, with the characteristic high return on the leading edge of the wing; the image resolution, given (6.36), is 7.5 cm in depth and 4.8 cm transversally. In particular, apart from multiple scatterers on the target outline (nose, wing dihedral, air intake lips), there is a cluster of scatterers apparently placed outside of the target, due to waves that were reflected after propagation in the air duct. This example is a good illustration of the complex character of radar reflections.

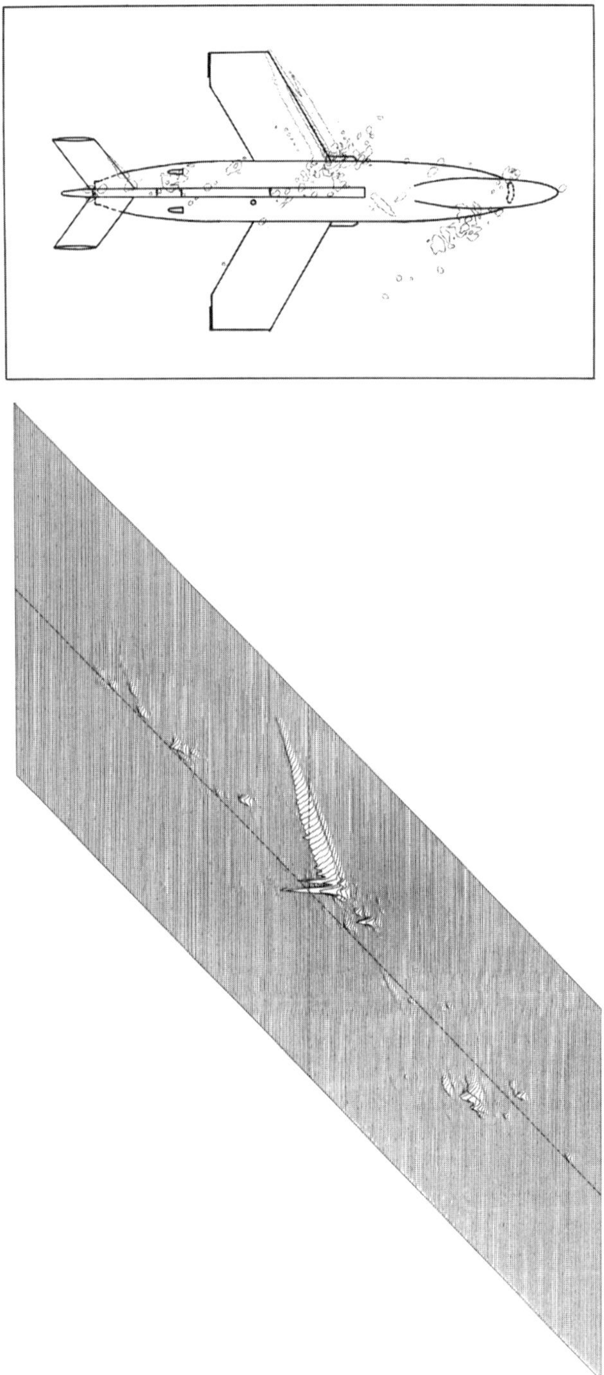

(b)

Figure 6.15 (continued).

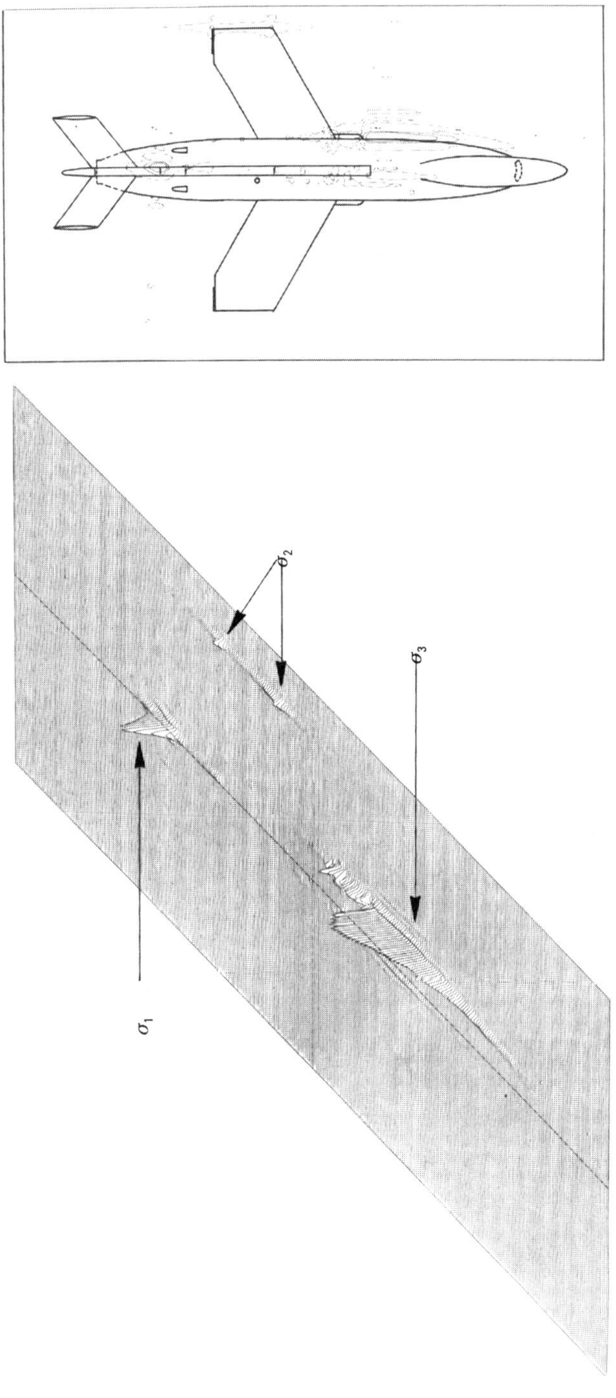

Figure 6.15 (continued).

The reflections of the geometrical optics type appear on the measurement hologram as the high levels observed for the transverse aspect angles (90° and 270°), the diffraction (which could be modeled by equivalent currents) on the wing dihedrals or on the air intake lips appears on the image, and the waves guided in the air duct generate echoes on the same image.

To obtain a rough idea of the actual nature of radar reflections, it is important to note that the same target observed in a different frequency band or at a different aspect angle might have a very different image. For example, Figure 6.15(c) shows the image of the same target, in the same frequency band, but for an observation sector shifted by 60° with respect to Figure 6.15(b) (20° around the aspect angle 90°): the major "scatterer" here is the specular effect on fuselage and fin.

This real example clearly demonstrates the difficulty of the analytical approach summarized in Section 6.2.1. Given this complexity of the wave-target interaction, Section 6.3.3 will enable a sufficiently realistic, yet analytically tractable, target model to be defined.

The argument that has been developed in this section also applies to the case of sonar reflection, using the same scatterer hypothesis, although the corresponding measurements are obviously more difficult to carry out.

6.3.3 Stochastic Models [13]

Another model that is often used for both radar and sonar applications consists of assuming that, as a function of the observation angle and for a pair of transmission-reception polarizations and a given transmitted frequency, the backscattering coefficient is a complex Gaussian random process, locally stationary, and with zero mean. The Gaussian characteristic is usually justified—with no great concern for rigor—by an argument of the "central limit" type (large number of scatterers on the target). Such a process is fully characterized by its (local) power spectrum in the neighborhood of a given aspect angle θ.

The spectrum of this process can be calculated by associating it with the distribution of scatterers, $I(\vec{x})$, of the model defined in Section 6.3.2—which also enables the locally stationary character of the process to be specified.

According to the multiple-scatterers model, (6.34), the backscattering coefficient $q(\vec{k})$ of the target is given by:

$$q(\vec{k}) = \int I(\vec{x}) \exp(2j\vec{k} \cdot \vec{x})\, d\vec{x}$$

$\vec{k}$: wave vector of length $2\pi f/c$.

Let us consider the case where $q(\vec{k})$ is observed in a plane of polar coordinates (f, θ) (generalization to a three-dimensional space, where the complex backscattering coefficient is a Gaussian process depending on the two variables θ and φ, does not raise any particular problem); if (r, α) is the location in polar coordinates on the target: $x = r\cos\alpha$, $y = r\sin\alpha$, (6.34) is written as:

$$q(f, \theta) = \int I(r, \alpha)\exp((4\pi j/\lambda)r\cos(\theta - \alpha))\,r dr d\alpha$$

that is, for small observation angles θ (assuming $\cos\theta = 1$ and $\sin\theta \approx \theta$):

$$q(f, \theta) = \int I(r, \alpha)\exp\left(4\pi j\frac{r}{\lambda}\cos\alpha\right)\exp\left(4\pi j\frac{r}{\lambda}\theta\sin\alpha\right) r dr d\alpha \tag{6.38}$$

It is important to quantify the notion of a "small" observation angular sector; this condition is given by:

$$4\pi(r/\lambda)(\cos\alpha)\frac{\theta^2}{2} << 2\pi$$

For a target of dimension L, since $r\cos\alpha < L$, this condition can be written:

$$\theta^2 << \frac{\lambda}{2L} \Rightarrow (\Delta\theta)^2 << \frac{\lambda}{2L} \tag{6.39}$$

$\Delta\theta$: width of the observation angular sector.

Returning to Cartesian coordinates on the image:

$$q(f, \theta) = \int I(x, y)\exp\left(4\pi j\frac{x}{c}f\right)\exp\left(4\pi j\theta\frac{y}{c}f\right)dx\,dy \tag{6.40}$$

where c is the velocity of light.

Thus, for small observation angles θ, the image $I(x, y)$ is obtained by a double Fourier transform of the backscattering coefficient $q(f, \theta)$, one on the frequency f and the other on the angle θ (the small observation angle approximation has simply enabled the arc of the circle at frequency f in Figure 6.14 to be replaced by its chord, so the measurement domain over which the Fourier transform is performed can be assumed rectangular).

As indicated in the previous paragraph, the resolutions obtained on the image, in depth δx and transversally δy, are the inverse of the extent of the observation field:

$$\delta x = \frac{c}{2\Delta f},\ \delta y = \frac{\lambda}{2\Delta\theta}$$

Taking into account (6.39) limiting the angular extent of the observation field, it can be seen that the transverse resolution of the image is limited by:

$$\delta y >> \sqrt{\lambda L/2} \tag{6.41}$$

a condition similar to that governing the operation of "unfocused" side-looking radar. This similarity is obviously not a coincidence; it can be shown that the imaging produced by radar holography is the same as that generated by a focused side-looking radar moving on a circle of large radius centered on the target.

This relation, (6.40), obtained under conditions (6.39) or (6.41), defines the spectrum of the Gaussian process modeling, in the stochastic framework, the backscattering coefficient of the target. The Fourier transform $\tilde{q}(f, y)$ of $q(f, \theta)$ is given by:

$$\tilde{q}(f, y) = \int q(f, \theta)\exp(-2\pi j\theta y)\,d\theta$$

$$\tilde{q}(f, y) = \int I(x, y')\exp\left(4\pi j\frac{x}{\lambda}\right)\exp\left[4\pi j\left(y' - \frac{\lambda}{2}y\right)\theta/\lambda\right]dx\,dy'\,d\theta$$

That is, defining $\Phi = \dfrac{2\theta}{\lambda}$:

$$\tilde{q}\left(f, \frac{2y}{\lambda}\right) = \frac{\lambda}{2}\int I(x, y')\exp\left(4\pi j\frac{x}{\lambda}\right)\exp(2\pi j\Phi(y' - y))\,dx\,dy'\,d\Phi$$

Defining: $\Delta\Phi = \dfrac{2}{\lambda}\Delta\theta$ ($\Delta\theta$: extent of the angular sector of observation)

and: $\operatorname{sinc} x = \dfrac{\sin \pi x}{\pi x}$

$$\Rightarrow \tilde{q}\left(f, \frac{2y}{\lambda}\right) = \frac{\lambda}{2} \int I(x, y') \exp\left(4\pi j \frac{x}{\lambda}\right) \Delta\Phi \operatorname{sinc}[\Delta\Phi(y' - y)]\, dx\, dy' \tag{6.42}$$

In this expression, the integral with respect to y' corresponds to a convolution and simply represents the "degradation" of the image $I(x, y)$ by the equipment function in terms of sinc of width $\lambda/2\Delta\theta$ due to the limited angular sector $\Delta\theta$ over which the spectrum is calculated; let the degraded image be $I_{\Delta\theta}(x, y)$:

$$I_{\Delta\theta}(x, y) = \int I(x, y')\Delta\Phi \operatorname{sinc}[\Delta\Phi(y' - y)]\, dy' \tag{6.43}$$

$$\tilde{q}\left(f, \frac{2y}{\lambda}\right) = \frac{\lambda}{2} \int I_{\Delta\theta}(x, y) \exp\left(4\pi j \frac{x}{\lambda}\right) dx \tag{6.44}$$

where the power spectrum:

$$\mathrm{E}\left[\left|\tilde{q}\left(f, \frac{2y}{\lambda}\right)\right|^2\right] = \left(\frac{\lambda}{2}\right)^2 \mathrm{E}\left[\int I_{\Delta\theta}(x, y) I^*_{\Delta\theta}(x', y) \exp\left(4\pi j \frac{x - x'}{\lambda}\right) dx\, dx'\right]$$

If the pixels are independent, which we have assumed, the equation becomes:

$$\mathrm{E}\left[\left|\tilde{q}\left(f, \frac{2y}{\lambda}\right)\right|^2\right] = \left(\frac{\lambda}{2}\right)^2 \int \mathrm{E}\left|I_{\Delta\theta}(x, y)\right|^2 dx \tag{6.45}$$

The power spectrum of the process $q(\theta)$ is thus—to within a factor of $(\lambda/2)^2$ for the level and a factor of $2/\lambda$ on the y scale—the transverse response (in power) of the target, the projection of the mean intensities of the scatterers onto an axis perpendicular to the line of sight, as these scatterers appear when observed over an angular sector of width $\Delta\theta$.

This proof also clearly reveals the problem presented by the nonstationary nature of $q(\theta)$, which is modeled by a locally stationary process (over an extent $\Delta\theta$). This imposes a certain resolution $\lambda/(2\Delta\theta)$ on the transverse response $T(y) = \int \mathrm{E}\left|I_{\Delta\theta}(x, y)\right|^2 dx$. In fact, strictly speaking, the spectrum varies, depending on the extent over which the process $q(\theta)$ is assumed to be stationary.

The main property of this stochastic model is that the target can be represented by a continuous RCS density, in contrast to the multiple-scatterers model, which concentrates the RCS at perfect scatterers. It is this property in particular, more than its random character, that makes it of interest for modeling the behavior of radar location or identification systems.

This model is also suitable for implementation as a numerical simulation and can be used to generate a complex backscattering coefficient as follows:

- Generation of consecutive independent samples of Gaussian white noise, of variance equal to the mean RCS of the target in the angular sector under consideration [of extent $\Delta\theta$ limited by (6.39)];
- Calculation of the Fourier transform of this series of samples;
- Multiplication of the moduli of the samples thus obtained by the square roots of the values of the transverse distribution (measured elsewhere, or assumed a priori); for this operation, the choice of a sampling interval in angle $\delta\theta$ for the backscattering coefficient leads to a length $\lambda/2\Delta\theta$;
- Inverse Fourier transform of the series of samples obtained.

The result of these operations is a series of samples that constitutes the series of values of the target backscattering coefficient $q(\theta)$, sampled at intervals of $\delta\theta$ over the chosen angular sector $\Delta\theta$.

As an illustration, Figure 6.16 compares the RCS of a real, unmanned air vehicle, measured at a wavelength of 10 cm, with the modeled RCS (random model according to the measured transverse response), as a function of the aspect angle.

6.3.4 Conclusion

The following main conclusions can be drawn from this overview of the electromagnetic properties of dimensionally stable targets.

- The theory of electromagnetism—or that of acoustics—does not enable the single-scatterer model adopted in the preceding chapters to be replaced by a realistic model that can be used in signal theory. The realistic models generated by the theory, generally based on rays and on local diffraction or reflection, are not expressed in a simple analytical form.
- The multiple-scatterers model can be used to take into account a large part of the interactions between the electromagnetic wave

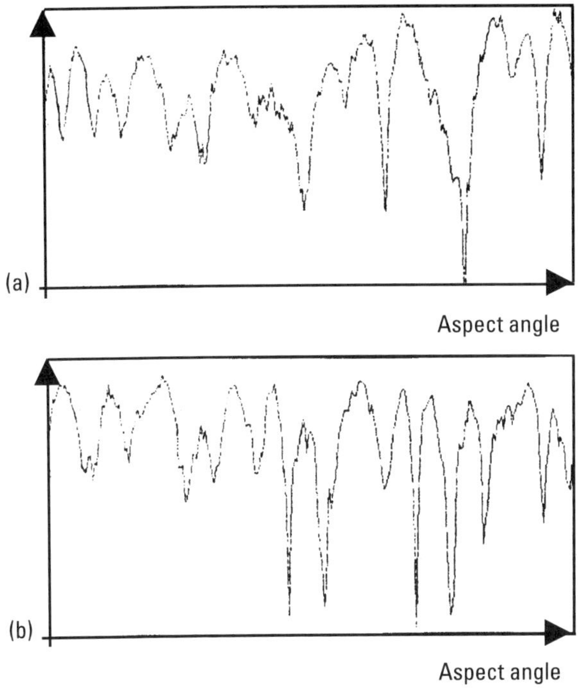

Figure 6.16 Stochastic modeling of the backscattering coefficient; (a) measured RCS, (b) RCS calculated by the model.

and the target; it also enables the development of relatively direct extensions of the previously described procedures when moving from the single-target situation to multiple targets.

- Finally, the stochastic model provides a generally quite reliable representation of the target properties and will also provide the basis for analysis of the behavior of tracking systems in the presence of non-point targets, in Section 6.6.

6.3.5 Examples: Impulse Response and Transverse Response

It is useful to supplement the examples of signatures of dimensionally stable targets shown in Figures 6.5 (RCS diagram), and 6.14 (holographic images) by two types of signature that current radars and sonars can be expected to process: the impulse response and the transverse response or differential Doppler analysis. These are in fact the two components of the double

Fourier transform (6.40) relating the target image $I(x, y)$ to the complex backscattering coefficient $q(f, \theta)$ for a small observation angular sector $\Delta\theta$.

The impulse response is, as its name suggests, the response of the target to a pulse, necessarily limited to the frequency band Δf, narrow with respect to f_0, for a given aspect angle θ; it is thus also the Fourier transform, over this bandwidth, of the frequency response $q(f, \theta_0)$ of the target (considered as a linear system):

$$R(t) = \int q(f, \theta_0) \exp(-2\pi jft)\, df$$

Adopting the multiple-scatterers mode described in Section 6.3.2, and choosing (with no loss of generality) $\theta_0 = 0$ as the angle origin, we obtain [according to (6.40)]:

$$R(t) = \int I(x, y) \exp\left[4\pi j \frac{f}{c}\left(x - \frac{ct}{2}\right)\right] dx\, dy\, df$$

where the integral with respect to frequency is limited to the domain of width Δf centered on f_0. Integrating with respect to frequency.

$$R(t) = \exp(2\pi jf_0 t) \int I(x, y) \Delta f \operatorname{sinc}\left[\frac{2\Delta f}{c}\left(x - \frac{ct}{2}\right)\right] dx\, dy$$

The integral with respect to x expresses (as in Section 6.3.3) the degradation of the image $I(x, y)$ resulting from its convolution with the equipment function in sinc of width (with respect to x) $c/2\Delta f$, due to the frequency limitation of the observation field.

Calling this degraded image $I_{\Delta f}$:

$$I_{\Delta f}(x, y) = \int I(x', y) \frac{2\Delta f}{c} \operatorname{sinc}\left[\frac{2\Delta f}{c}(x - x')\right] dx' \qquad (6.46)$$

we obtain, with the change of variable $d = ct/2$ (depth):

$$R(d) = \frac{c}{2} e^{2\pi jf_0 t} \int I_{\Delta f}(d, y)\, dy \qquad (6.47)$$

The impulse response of the target is thus, within a factor $c/2$, the depth response of the target, that is, the projection of the scatterers onto the line of sight, measured with a range resolution $c/2\Delta f$ limited by the frequency bandwidth (a similar limitation to that encountered in Chapter 3 in the case of a single scatterer, although here obtained in a deterministic framework). It is a complex quantity that integrates, at each range d, the amplitude and phase contributions of the different scatterers of the target (phase shift inherent to the nature of the scatterer, plus the phase shift corresponding to its position).

In a completely similar way, it is possible to define the transverse response of the target, as the Fourier transform with respect to θ of the backscattering coefficient $q(f, \theta)$ of the target. This quantity has already been calculated in Section 6.2.4, and led to the quantity (with $r = 2y/\lambda$):

$$T(r) = \frac{\lambda}{2} \int I_{\Delta\theta}(x, r) \exp\left(4\pi j \frac{x}{\lambda}\right) dx \tag{6.48}$$

$$I_{\Delta\theta}(x, y) = \int I(x, y') \frac{2}{\lambda} \Delta\theta \operatorname{sinc}\left[\frac{2}{\lambda} \Delta\theta (y' - y)\right] dy' \tag{6.49}$$

The transverse response, like the impulse response, is a complex quantity, taking into account the relative phase shifts of the scatterers of the target.

The target impulse response and transverse response are thus two similar quantities, Fourier transforms of the backscattering coefficient with respect to frequency and angle, respectively. They enable analysis of the target in the radial and transverse dimensions with resolutions $c/2\Delta f$ and $\lambda/2\Delta\theta$.

These two quantities can be obtained by range and Doppler analysis, respectively, of the signal received by a high-resolution radar or sonar. Just as the impulse response is the range response of a radar using a bandwidth Δf, the transverse response is the Doppler response (Fourier transform of time signal $q(f, \theta)$ received in a range bin containing the target) of the target when the latter is rotating uniformly about a fixed point. This Doppler response is obviously frequency-shifted when the target has a translation movement as well as its apparent rotation.

This identity between Doppler analysis and transverse response appears on examination of Figure 6.17. The Doppler effect of point M of the target, located at distance ρ from the center of rotation of the target, which is assumed to be rotating uniformly at angular velocity Ω, is $\frac{2\Omega}{\lambda} \rho \sin \Omega t$,

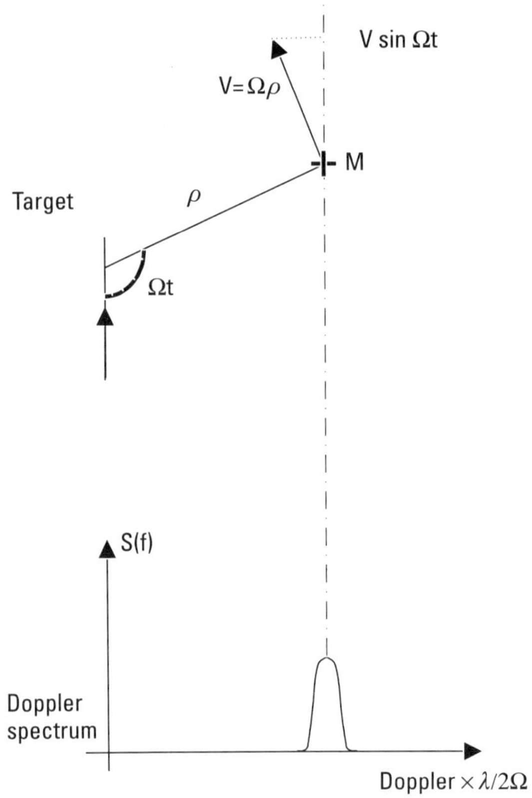

Figure 6.17 Transverse response and Doppler analysis.

which is also, to within the constant factor $\frac{2\Omega}{\lambda}$, its transverse distance (distance to the center of rotation projected onto an axis perpendicular to the line of sight). Doppler analysis can thus be used to separate the points of the target located at transverse distances differing by $\frac{\lambda}{2\Delta\theta}$, if $\Delta\theta$ is the angular extent of rotation of the target during the analysis period. The limitation (6.39) on the extent of the observation angular sector $\Delta\theta$ can be interpreted in terms of Doppler effect of point M—which varies with $\sin \Omega t$—which has not varied significantly during the analysis period T:

$$\frac{2\Omega}{\lambda}\rho \sin \Omega T << \frac{1}{T}$$

that is, for the extreme point (ρ = L and ΩT small): $(\Delta\theta)^2 = (\Omega T)^2 << \frac{\lambda}{2L}$, which is identical to (6.39).

This identity between Doppler analysis and transverse response can be simply explained if we analyze the "Doppler broadening" of a rotating target as the translation in the spectral domain of the phase modulation of the received time signal $q(f, \theta)$ during the rotation with respect to θ.

Figures 6.18 and 6.19 show, respectively, the modulus of the impulse response of a ship observed at 10 GHz with a resolution of 2.4m (ONERA measurements) and the modulus of the Doppler response of an unpiloted aircraft.

In what follows, high-resolution radar or sonar is defined as the type of system that can determine this kind of signature, given the natural movements of targets, in contrast to medium-resolution systems for which the target can reasonably be considered as a single scatterer (so long as its aspect angle does not vary more than $\delta\theta = \frac{\lambda}{2L}$ or the radar frequency does not vary more than $\delta f = \frac{c}{2L}$, if L is the target dimension).

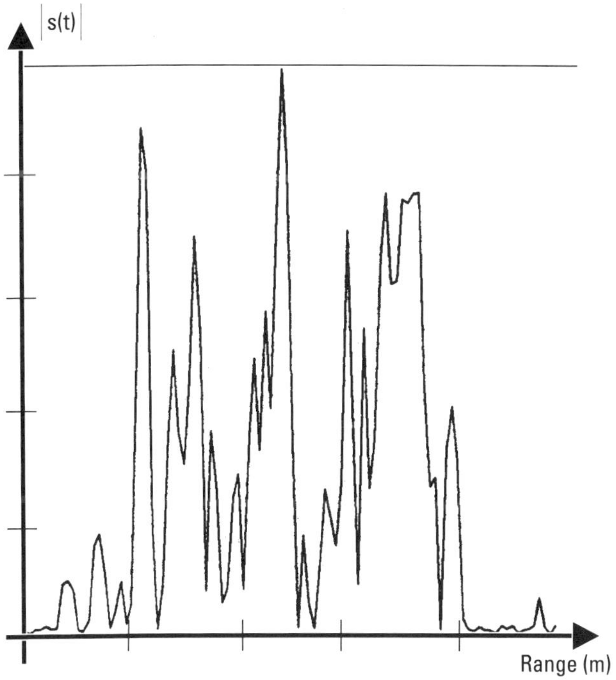

Figure 6.18 Impulse response of a ship (ONERA measurements).

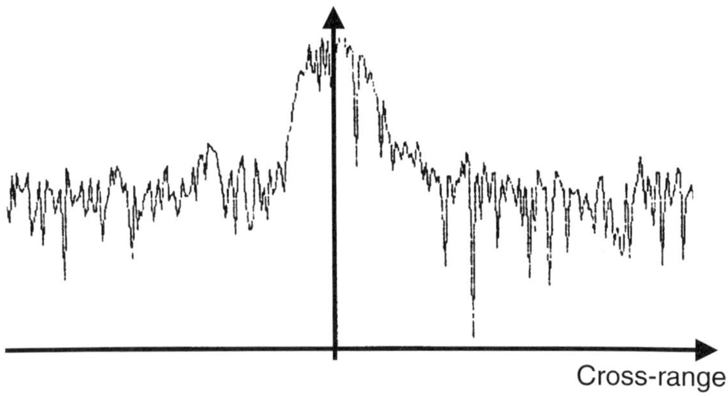

Figure 6.19 Transverse response of an aircraft (ONERA measurements).

6.4 Deformable Targets

The targets studied so far have consisted of elements of complex shape with rigid links between them. In certain cases the target also has parts that move with respect to the rest of the structure, such as helicopter rotor blades, aircraft engine compressor and turbine blades, or the propeller of a submarine.

These elements generate characteristic effects on the signals output by the radar or the sonar; these effects modify the performance of the system and can potentially be used to recognize the targets. We shall examine the radar signatures of helicopters, then those of aircraft engines.

6.4.1 Helicopter Signatures

The blades of the main rotor or the tail rotor of a helicopter, when they are conducting (metal or carbon fiber blades), can be considered to satisfy the validity conditions of the geometrical theory of diffraction; the echo that they return is mainly due to specular reflection (leading edge) or wedge diffraction (trailing edge), and thus is significant essentially when the blade is perpendicular to the observation axis (Keller cone of Section 6.3.1.3). The signal received from a helicopter thus has the appearance illustrated in Figure 6.20(a), featuring the flashes returned by the main rotor blades—seen from the leading edge side or the trailing edge side.

Figure 6.20(b) shows the appearance of the corresponding Doppler spectrum when a flash is produced during the Doppler analysis period. Considerable Doppler broadening is observed, since the blade tip approaches the observer at high speed, generally close to the velocity of sound. The time

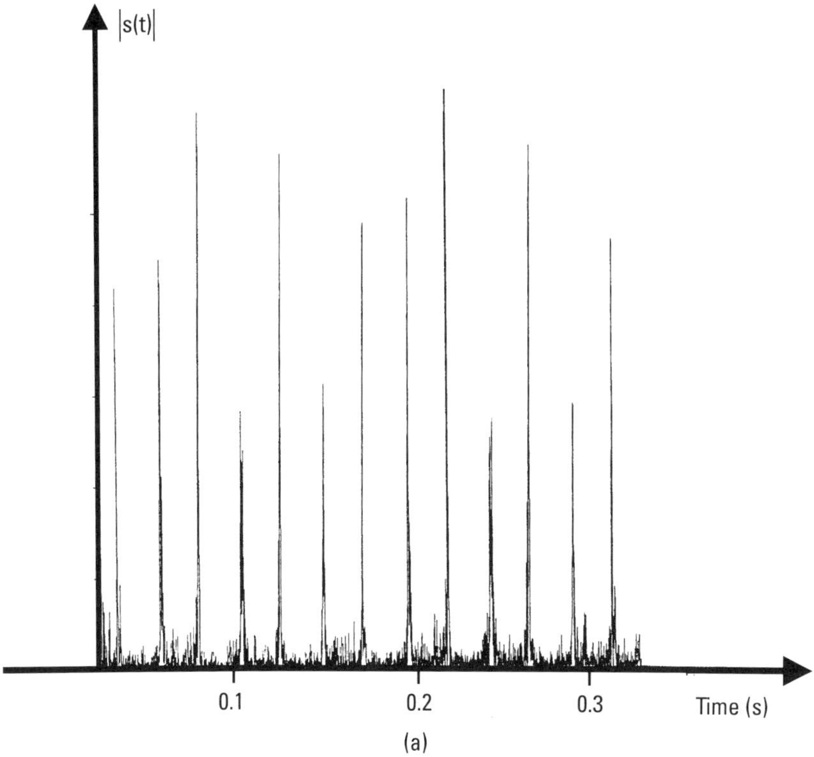

Figure 6.20 Helicopter signal (ONERA measurements): (a) time signal, (b) spectrum.

spacing between flashes is characteristic of an helicopter, together with the fact that the leading edge and trailing edge flashes are simultaneous (even number of blades) or not. However, these flashes are not always observable; the two conditions of observability are:

1. That the illumination time (pulse train duration) be long enough, that is, longer than 50 to 100 ms (the flashes repetition frequency varies from 10 to 100 Hz, depending on the helicopter);
2. That the repetition period T_r be lower than the duration of the flash, whose order of magnitude is $\lambda/(2*300)$, since the highest Doppler frequency of any scatterer on the blade corresponds to a radial velocity of 300 m/s.

Closer analysis of such spectra shows the existence of a "plateau" region, for Doppler smaller than roughly 50 m/s, where higher RCS levels can be

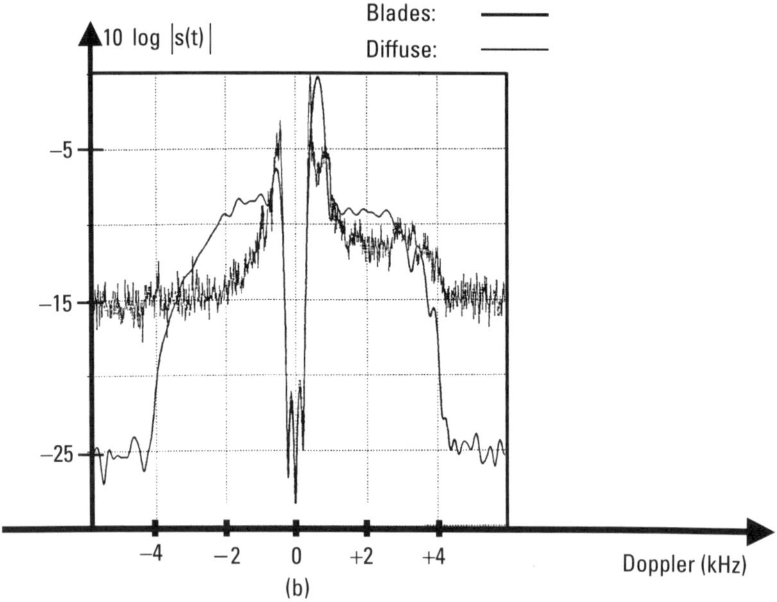

Figure 6.20 (continued).

obtained, due to the reflections from different elements of the rotor hub. These spectral components can be used for helicopter detection and classification, since they are present continuously, for nearly every position of the rotor (and not only at specific instants, like the blade flashes).

6.4.2 Jet Engine Signatures

Viewed from the nose (or from the rear), a jet engine generates a characteristic radar echo when the dimensions of the air intake (or of the engine exhaust) are at least equal to the wavelength, which is generally the case.

The waves penetrate into the duct of the engine and propagate until they reach the compressor (or turbine) blades, which reflect them to the outlet. As these blades have a periodic rotation motion, the received signal is itself periodic, of frequency $f \cdot n$, where n is the number of blades and f the rotation frequency. The spectrum of the received signal therefore consists of lines separated by this frequency: this phenomenon is also called jet engine modulation (JEM). The envelope of this line spectrum is usually asymmetrical about the transmitted frequency f_0, because of the complex interaction between the waves propagated in the duct and the blades. Given the rotation speeds and the number of blades, the interval between lines is generally relatively wide (a few kilohertz), so much so that the observed target appears as several targets at different velocities if the frequency spacing is interpreted as a Doppler effect. This effect, which hampers radars with velocity ambiguities, can obviously also enable target identification by the appearance of the line spectrum.

Frequency analysis of the received signal, in the case of jet aircraft, thus provides access to information about the type of engine. This information is in addition to that which can be obtained, as indicated in Section 6.4.1, by high-resolution Doppler analysis to determine the transverse response of the aircraft. Moreover, these two phenomena are usually totally uncoupled on the received signal, because the Doppler broadening obtained by the transverse response is much less than the interval between the engine lines.

To give some orders of magnitude, the line spacing lies between 1 and 10 kHz, and the total width of the line spectrum is generally less than 40 kHz. For an aircraft moving with an apparent rotation speed $\dot{\theta}$ of 10° per second, the Doppler broadening—independent of the echoes from moving parts such as the engine lines—is:

$$\delta f = \frac{2\dot{\theta}L}{\lambda}$$

At a wavelength of 3 cm, for an aircraft of wingspan 15m, the Doppler broadening is 166 Hz, at least an order of magnitude less than the line spacing.

As an illustration, Figure 6.21(a) shows an example of an aircraft signature where the engine lines clearly appear. Figure 6.21(b) shows the series of Doppler responses of an aircraft in flight; here, note in particular the fluctuations in the relative levels of the Doppler lines due to the scatterers of the airframe.

6.4.3 Vehicle Signatures

Moving parts can also be observed on vehicles, for which the contribution from the wheels can have nonnegligible RCS, spread in Doppler between zero Doppler and twice the vehicle speed. More specifically, the contribution from the higher part of the wheel, coupled with the vehicle through dihedral effects, frequently gives rise to a particular line in the Doppler spectrum (especially for trucks), located at twice the Doppler corresponding to the vehicle speed. This Doppler line has to be taken into account in air-surveillance radars, when defining filters for vehicles rejection. The possible Doppler

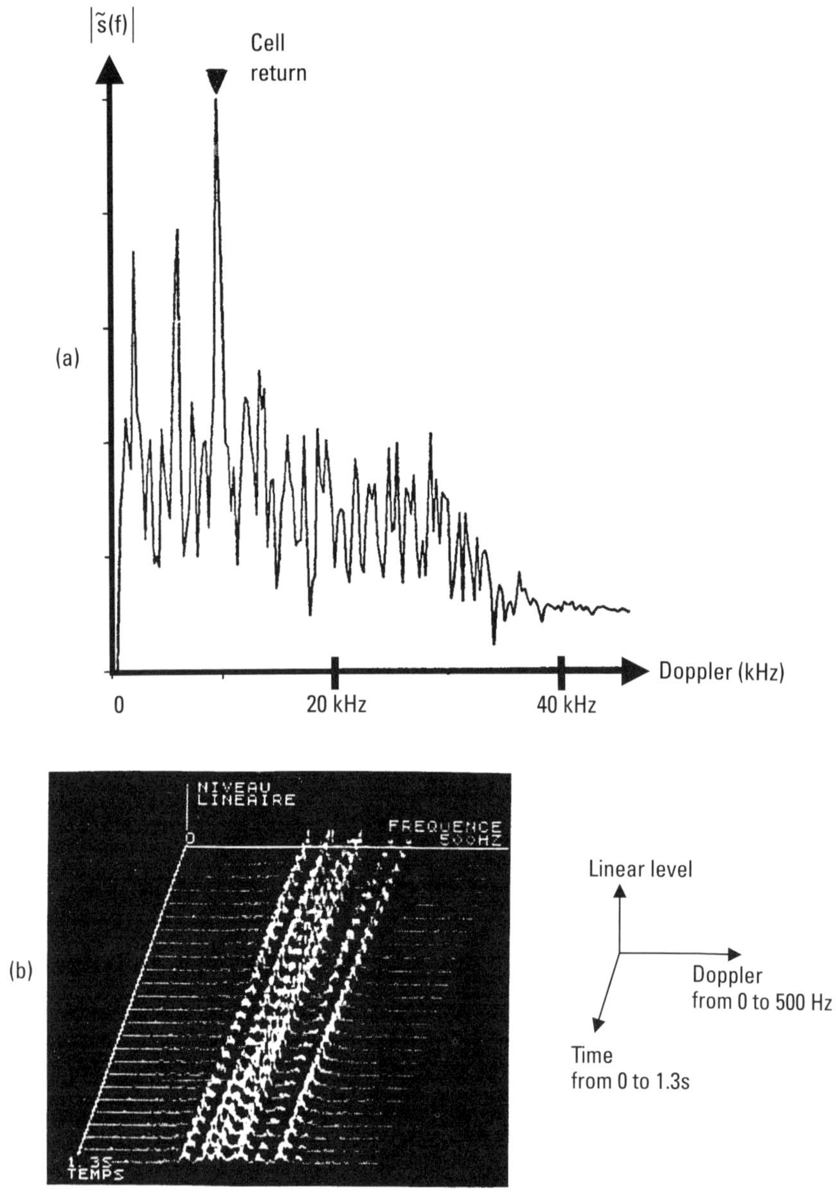

Figure 6.21 (a) Doppler signature of a jet aircraft (ONERA measurement); (b) transverse responses (Doppler) of an aircraft in flight (ONERA measurements).

for ground targets has to be assumed lower than twice the actual speed of these targets.

The Doppler spectrum of ground targets can be used for classification, as can be seen on the signature shown in Figure 6.22, resulting from successive Doppler analysis of radar returns (horizontal scale: reduced Doppler frequency; vertical scale: time; the gray levels represent the RCS, for each Doppler bin) from a vehicle observed at 10 GHz by a ground surveillance radar. The central Doppler line of clutter is clearly visible, and the returns from the vehicle spread from zero Doppler to twice the Doppler of the vehicle.

6.5 Background Signatures

The target environment obviously makes its own contribution to the signals delivered by the radar or the sonar—indeed, for mapping radars or surveillance radars, this is the main contribution. In certain cases it also produces distortions of the echoes returned by the targets, whether because of atmo-

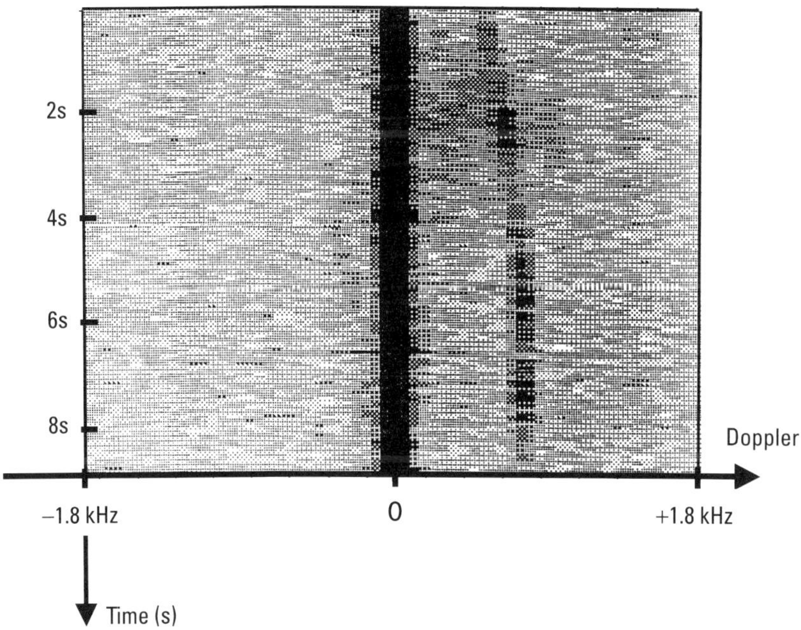

Figure 6.22 Doppler signatures from a moving vehicle.

spheric propagation phenomena (or reverberation phenomena in sonar) or because of consecutive multiple reflections from the ground or natural objects. These effects must therefore be taken into account in signal processing, in order to extract the useful part from the received signal.

In this section, we shall summarize the main properties of spurious reflections on backgrounds in order to define the principles of processing for the extraction of useful signals. This interference related to the environment will be divided into three categories: direct echoes from backgrounds (clutter or reverberation), multiple paths, and propagation anomalies.

6.5.1 Direct Echoes from Backgrounds: Clutter

In radar, the echo returned by the backgrounds—ground, water surface, natural obstacles (such as hills and trees), and artificial obstacles (buildings)—is obviously highly variable, and thus extremely difficult to predict, both in terms of average level and in terms of fluctuations (standard deviation spectrum).

Without attempting to cover the entire subject—a summary can be found in [14, 15, or 16]—it is necessary to outline the main properties of these echoes which must be taken into account in the processing of the received signals: RCS and Doppler spectra.

6.5.1.1 Clutter RCS

Like radar targets, the background echo is first characterized by its RCS. A distinction is usually made between point echoes—such as an isolated tree, house, pylon—which have characteristics similar to those of the usual radar targets, and the diffuse echo distributed over the surface.[2] To a good approximation, the RCS of the latter can generally be considered as distributed homogeneously (over a small area of terrain); it is then characterized by its reflectivity σ_0, expressed in square meters per square meters: RCS/area contained in the resolution bin (angle, range, Doppler) of the radar. The RCS of the clutter in a resolution element is therefore:

$$\sigma = \sigma_0 \cdot A$$

2. Reference [5] explains that this distinction is not really meaningful at near grazing incidence, since for usual spatial resolution (intercepted ground clutter area, or "footprint," between 10^3 and 10^6 m^2), "a more realistic construct is to imagine clutter as arising from a sea of discretes."

where A is the area of terrain contained in the resolution element, shown in Figure 6.23. If δR is the range resolution, θ_a the angular resolution in azimuth, and R the range, the RCS of the clutter is:

$$\sigma = \sigma_0 R \theta_0 \delta R \tag{6.50}$$

The reflectivity generally obeys a law of the type:

$$\sigma_0 = \gamma \sin \psi \text{ (valid for } \psi \text{ larger than } 5°) \tag{6.51}$$

where ψ is the angle of incidence of the radar beam on the ground. For ground clutter, σ_0 is on the order of 10^{-2} to 10^{-1} m^2/m^2, depending on the nature of the surface, such as its roughness. For sea clutter, a useful rule of thumb given by [14] is, for a Beaufort wind scale K_B:

$$10 \log \gamma = 6K_B - 10 \log \lambda - 64 \tag{6.52}$$

As shown in Section 6.5.2, these echoes are rapidly attenuated at very low grazing angles, because of multipath (interference between direct and reflected path), beyond a "transition distance" R_t (or below a transition grazing angle ψ_t depending on the radar height h_r and the terrain roughness, characterized by the rms height h_{rough}):

$$R_T = 4\pi h_r \frac{h_{\text{rough}}}{\lambda}, \quad \psi_t = \frac{\lambda}{4\pi h_{\text{rough}}} \tag{6.53}$$

Beyond that distance, the global power budget involves an attenuation proportional to:

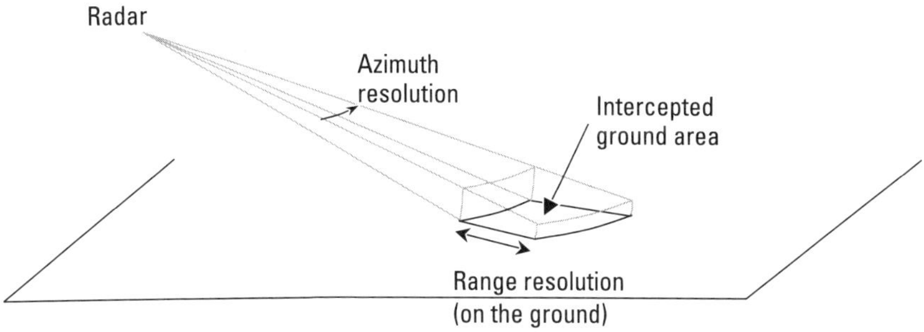

Figure 6.23 Ground clutter.

$$\frac{1}{R^4\left(1+\left(\frac{R}{R_t}\right)^4\right)}$$

Atmospheric clutter echoes, returned by rain or snow, are added to the echoes returned by the various elements of the landscape.[3] For wavelengths longer than a few centimeters, the raindrops or snowflakes are in the Rayleigh area (see Section 6.3.1.5), and therefore their RCS varies with a^6/λ^4, where a is their diameter. These echoes are therefore generally negligible at frequencies below 3 GHz (λ = 10 cm), except for long-range radars, but may become a hindrance at higher frequencies ($f \geq 10$ GHz), particularly when the radar beam has a long path in clouds. The maximum rain altitude, usually limited to 3 to 5 km, can grow up to 10 km in thunderstorms. The reflectivity of atmospheric clutter is usually described by a coefficient η, expressed in square and cubic meters (RCS per volume). A useful rule of thumb for $\lambda > 0.02$m is given in [14], for a rainfall rate r expressed in millimeters per hour:

$$\eta = 5.7 \cdot 10^{-14} \cdot r^{1.6}/\lambda^4 \tag{6.54}$$

and for dry snow:

$$\eta = 1.2 \cdot 10^{-13} \cdot r^2/\lambda^4 \tag{6.55}$$

The resulting RCS, for a range resolution δR, an azimuth resolution θ_{az}, and an elevation resolution θ_{el} is written (Figure 6.24) as:

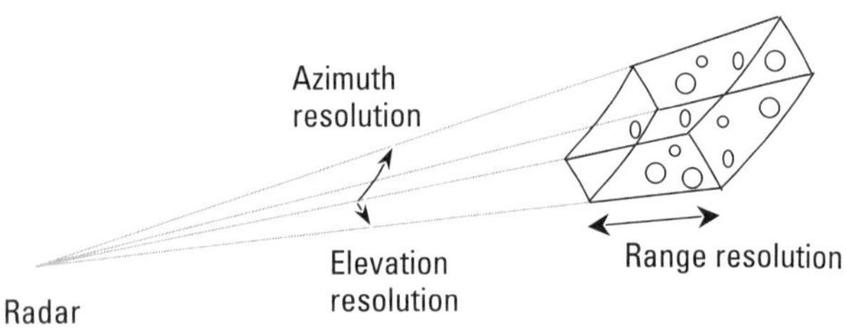

Figure 6.24 Atmospheric clutter.

3. These echoes are of special interest for meteorologists, and a detailed analysis of their properties can be found in [17].

$$\sigma = \eta R^2 \theta_{az} \theta_{el} \delta R \tag{6.56}$$

In sonar, the echoes returned by the submarine environment are known as reverberation and are divided into two categories:

- Surface reverberation and bottom reverberation, due to echoes returned by the rough surface of the sea and by the ocean floor;
- Volume reverberation, due to echoes returned by the multiple heterogeneities distributed in the ocean: fish or plankton.

Surface reverberation is closely comparable with radar clutter; the intensity of the echo is approximately proportional to the area irradiated by the sonar. For low grazing angles, reverberation is mainly due to bubbles near the surface, and as such is highly dependent on frequency (law in λ^{-4} for bubbles in the Rayleigh domain). For high grazing angles (larger than 60°), reverberation is mainly due to reflection on swell and waves, and strongly increases with grazing angle.

Volume reverberation is mainly due to returns from the *deep scattering layer,* which is a layer where the density of living organisms is high, and whose depth and thickness vary according to the hour and the season. For low frequencies, volume reverberation is generally negligible when compared to surface reverberation.

The statistics of radar clutter RCS and sonar reverberation are frequently assumed to follow an exponential model, resulting from a central limit argument (many independent scatterers giving rise to a Gaussian complex signal, with an amplitude following a Rayleigh law—Rayleigh clutter—or a squared amplitude following an exponential law). It is shown in [18], which is summed up in Appendix 6A, that for low grazing angles, the statistics become different because the received signals are the result of a combination of echoes from illuminated areas and of thermal noise due to shadowed areas. A Weibull law is more appropriate for radar ground clutter at such low incidences, as the corresponding parameters have been experimentally measured for different terrain types at different wavelengths, ranging from VHF (165 MHz) to X-band (9.2 GHz) (see [18] and Appendix 6A). K distribution has also been examined, and shown to give a best fit in many situations [19, 15].

6.5.1.2 Clutter Doppler Spectra

The inherent movements of the surface (sea, tree leaves) result in a certain Doppler broadening to these echoes, which itself depends to a large extent on the nature of the surface and on the radar configurations.

For sea clutter, a Doppler spectrum of Gaussian shape is often adopted, with a mean and a standard deviation depending on the nature of the surface, the meteorological conditions, and the seasons. The standard deviation can be assumed to be proportional to wind speed V_{wind}:

$$\sigma_{\text{wind}} = 0.1 V_{\text{wind}} \tag{6.57}$$

The mean Doppler velocity depends on the relative orientation (in azimuth) of swell θ_{swell} and of the radar line of sight θ_{radar} according to the following empirical relations [20]:

$$\begin{cases} V_{\text{mean}} = 0.13 V_{\text{wind}} \cos(\theta_{\text{swell}} - \theta_{\text{radar}})\text{: Vertical polarization} \\ V_{\text{mean}} = 0.8 \sqrt{V_{\text{wind}} \cos(\theta_{\text{swell}} - \theta_{\text{radar}})}\text{: Horizontal polarization} \end{cases} \tag{6.58}$$

For ground clutter, the mean velocity is assumed to be zero, and the standard deviation can be assumed to follow the rule [16]:

$$\sigma_{\text{ground}} = 0.015 V_{\text{wind}} \tag{6.59}$$

A more precise ground clutter spectrum model (with exponential rather than Gaussian decrease with increasing Doppler frequency), described in [18], is summarized in Appendix 6A.

Atmospheric clutter spectrum is highly dependent on altitude (wider at low altitudes), azimuth beamwidth, and rainfall rate. The mean clutter velocity is equal to the wind radial velocity at the center of the radar beam. Reference [16] gives a rule of thumb for the standard deviation of velocity spectrum for ground radars, for a wind shear k_{ws} in m/s per km altitude (3 being a common value), an elevation resolution θ_{el}, and an azimuth angle α between the radar line of sight and the wind direction:

$$\sigma_v = 0.3 k_{ws} R \theta_{\text{el}} \cos\alpha \tag{6.60}$$

Sonar surface reverberation is closely comparable with radar clutter. Its Doppler signature depends on surface waves movement, on the motion of the platform bearing the sonar, on its position with respect to the bottom and the surface, on sea surface agitation, and on the roughness of the bottom.

Volume reverberation is directly comparable with radar echoes from meteorological elements—rain, snow, and hail.

6.5.2 Multiple Paths

When a radar target is observed at an incidence close to horizontal (incidence less than a few degrees), the phenomenon of multiple paths may appear: the target is illuminated by a direct ray and one or more rays reflected from the surface, which interfere constructively or destructively depending on their relative path difference. Similarly, the echo returned to the radar is propagated simultaneously along these two paths, with the same interference effect on arrival (Figure 6.25).

If the surface is considered as planar, this phenomenon can be interpreted as an image effect; the radar receives simultaneously the target echo and that of its image on the ground, with a relative phase determined by the path difference between the rays. The complex signal received (for a given pair of transmission-reception polarizations) is then modified by a "propagation factor" F_c:

$$F_c = [1 + \rho \exp[-j(\varphi + 2\pi\delta/\lambda)]]^2 \tag{6.61}$$

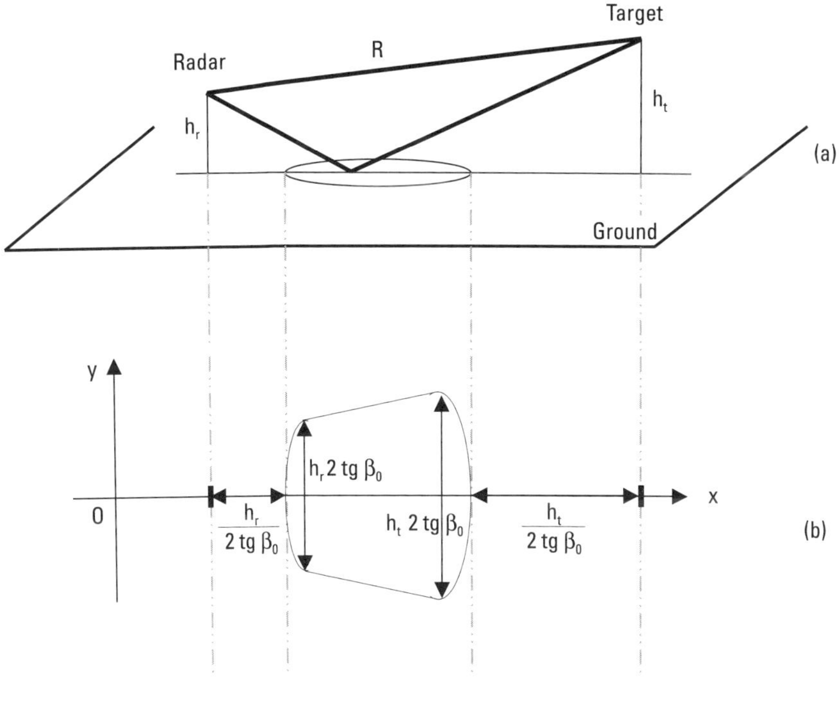

Figure 6.25 Multiple paths, glistening surface: (a) perspective view; (b) overhead view.

where φ is the phase shift due to reflection on the surface, ρ is the amplitude of the reflection coefficient on the surface, and δ is the difference in length between the direct path and the reflected path.

An elementary calculation shows that, for low incident angles and for a radar height h_r, a target height h_t, and a target-radar range R, δ becomes:

$$\delta \sim \frac{2h_r h_t}{R}$$

For grazing incidences, the phase shift is generally close to π, so that at a given range R the interference between the target and its image is destructive at zero altitude and does not become constructive until the target altitude is $h_t = \dfrac{R\lambda}{4h_r}$ (corresponding to $\delta = \lambda/2$). It then becomes destructive again at twice the altitude, and so on—this is called the lobing phenomenon, named because it produces on the received echo an effect similar to that produced by an antenna elevation pattern in the form of consecutive lobes, of 3-dB width:

$$\theta = \frac{\lambda}{4h_r}$$

Since interference is destructive at altitude zero, radar returns from very low-altitude targets will be severely attenuated. In this zone, the propagation factor is approximately equal to:

$$F_c = 2\left|\sin\left(2\pi\frac{h_t h_r}{R\lambda}\right)\right| \simeq 4\pi\frac{h_t h_r}{R\lambda} \tag{6.62}$$

which gives an attenuation, expressed in power for a two-way path, proportional to λ^{-4} for a given altitude, and to R^{-4}, which in turn means an R^{-8} attenuation in the global power budget; this is a strong incentive to the use of high-frequency carrier for low-altitude radar detection and tracking.

Therefore, for a given target altitude h_t, there exists a *transition distance* R_t beyond which $F_c < 1$, so that the target returns are attenuated in $(R/R_t)^{-4}$ (R^{-8} attenuation in the global power budget). This phenomenon is, of course, also observable for clutter returns: for a given terrain roughness characterized by an rms altitude h_{rough} such that:

$$F_c = \Leftrightarrow 4\pi \frac{h_r}{R_t} \frac{h_{\text{rough}}}{\lambda} = 1 \Leftrightarrow R_t = 4\pi h_r \frac{h_{\text{rough}}}{\lambda} \tag{6.63}$$

the clutter returns are attenuated by $(R/R_t)^{-4}$ for $R > R_t$.

This relation can be expressed in terms of ψ, the grazing angle $\left(\psi \sim \frac{h_r}{R}\right)$: for grazing angles lower than ψ_t, the *transition grazing angle,* clutter returns RCS decrease by $(R/R_t)^{-4}$ (R^{-8} attenuation in the global power budget):

$$4\pi\psi_t \frac{h_{\text{rough}}}{\lambda} = 1 \Leftrightarrow \psi_t = \frac{\lambda}{4\pi h_{\text{rough}}} \tag{6.64}$$

For grazing incidence (on the order of one degree) and relatively flat surfaces, multipath therefore consists mainly of an image effect. For rough terrain, a portion of the surface, close to the specular reflection point producing the image effect, behaves as a "glistening surface," analogous to the elongated glistening surface that appears on the sea at sunset. This phenomenon, analyzed in detail in the physical optics framework in [21], thus represents the existence of multiple paths reflected by all the points of the area shown in Figure 6.25 (where β_0 is the standard deviation of the slopes of the rough surface, assumed to be Gaussian). Generally, only a part of the contribution of this area is accepted by the radar resolution cell, because of the antenna directivity and the range resolution, which limit the difference in length between the direct path and the reflected path.

It is important to note that, in contrast to direct echoes from the backgrounds, these multiple paths often cannot be eliminated by Doppler analysis because, for the incidences at which they are produced, the Doppler effects on the direct path and on the reflected path cannot be totally separated by the radar.

The Doppler spectrum of a target observed in a multiple-path configuration typically has the form shown in Figure 6.26; the target can no longer be considered as a point, either in terms of angle, or Doppler shift, and sometimes not even in terms of range. From the point of view of signal processing, this extended target effect (added to the previously described lobing effect) is the main consequence of the multiple-path phenomenon.

The same phenomenon is observed in sonar, complicated near the surface by variations of propagation index leading, as shown in Section 6.3.1.1, to curved propagation—a phenomenon that is not unknown in

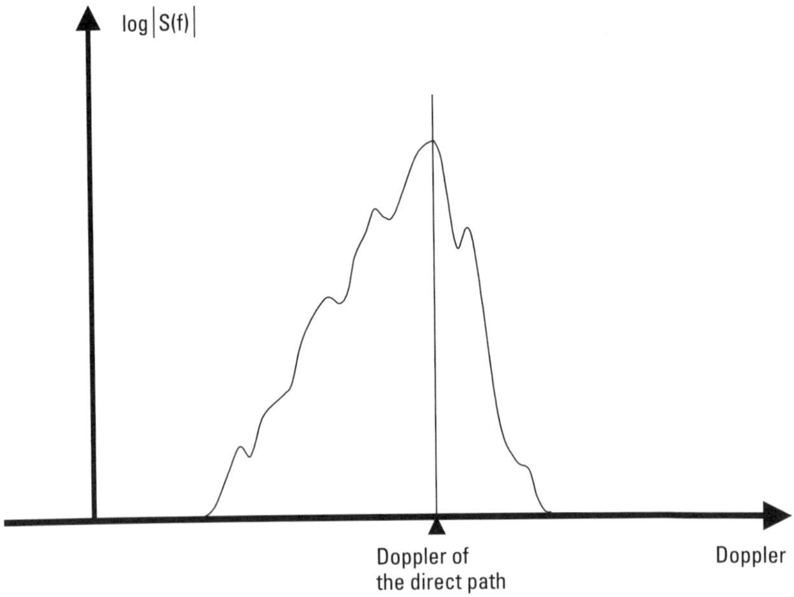

Figure 6.26 Doppler spectrum of a target in the presence of multiple paths.

radar. The signature of the sonar targets is therefore modified, in terms of range and Doppler effect, by these reflections from the surface or the bottom, reflections that are themselves modified by the nonlinear propagation of the waves. This signature modification can, in a first analysis, be described by an increase in the range and Doppler extent of the target.

6.5.3 Propagation

In radar, the transmitted wave is subjected to attenuation along the outward and return paths, measured in decibels per kilometers (exponential absorption, a consequence of the fact that the power lost in a given volume is a constant fraction of the incident power).

Figure 6.27 (which was first published by the AGARD/NATO, in Conf. Proc. No. 249, 1978 [22]) shows the (one-way) attenuation as a function of frequency for a standard atmosphere and a horizontal path, together with the attenuation due to rain (light, heavy, and tropical) and fog. Note in particular that this attenuation is generally negligible below 10 GHz and leaves propagation windows at 35 GHz, 94 GHz, 140 GHz, and so on, known as millimeter windows.

Different phenomena, such as diffraction on the terrain irregularities and evaporation ducts, affect propagation in usual conditions. To a large

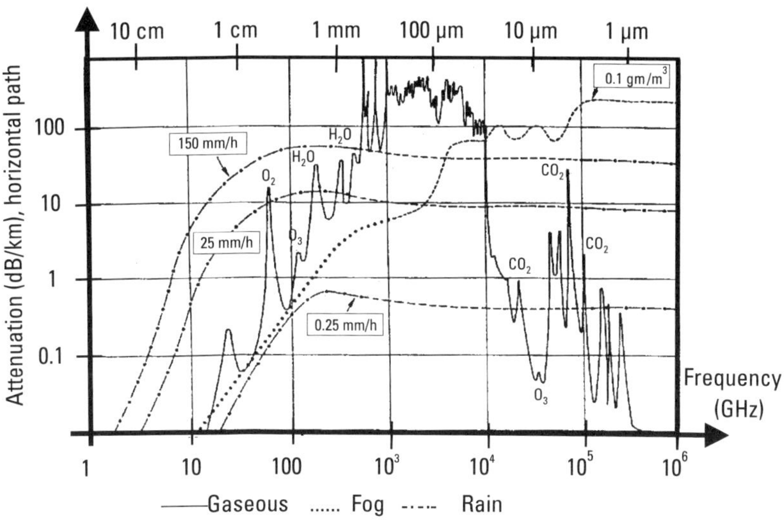

Figure 6.27 Electromagnetic attenuation temperature: 20°C; atmospheric pressure: 1 atm; water vapor: 7.5 g/m^3 [22].

extent, these nonstandard propagation situations can be predicted by available numerical techniques: parabolic equation, modes theory, or asymptotic (rays) modeling [23–25].

In sonar, the attenuation as a function of the frequency obeys a law that is shown in Figure 6.28, taken from [26]. In addition, it is important in sonar to take into account the variation in the velocity of sound in the sea (or the index) with depth [7]. Figure 6.29, shows the typical profile of the temperature and the velocity with depth. According to the analysis performed in Section 6.3.1.1, it appears in particular that the rays transmitted from a depth of approximately 1,000m will remain trapped at this depth, as the rays curve toward areas of lower velocity. The rays transmitted at other depths will follow a different path, possibly with reflection from the surface or the bottom and accumulation of acoustic energy in places.

Thus the main effect of propagation in radar is the attenuation of the received signal (attenuation varying slowly with frequency in a given band). In sonar, the effect of nonlinear propagation is often predominant; this effect disturbs the wave front illuminating the target, and thus modifies its signature in a way that is sometimes difficult to predict.

6.6 Passive Listening

The exact nature of the signals received in radar or sonar passive listening is extremely variable from one system to another. In radar, depending on

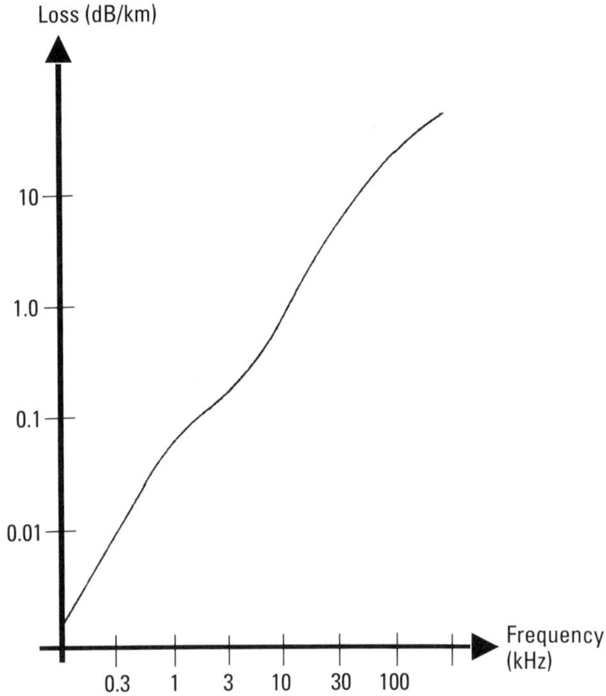

Figure 6.28 Losses by absorption in the ocean [26].

the frequency band of the analyzed signals—HF, VHF, UHF, radar bands—the function and thus the form of the signals are different (telecommunication signals, radar signals, or various spurious signals), as are their propagation conditions. Similarly, the configuration of the receiver antenna (or array) varies from one application to another. In astronomy the problems are of yet another nature, in terms of both antenna dimensions and the usable integration time. In sonar, the sources of "useful" noise are ships and submarines, whose noise is produced by the engines (in which the various rotating parts generate noises at pure frequencies and at their harmonics), by the propellers, which generate a wideband cavitation noise (creation of bubbles in the low-pressure regions around the propeller blades), and by hydrodynamic phenomena (vibrations induced in the structure of the ship or submarine by flow turbulence).

An analysis of the signatures in each particular case is obviously outside the scope of this book. However, remaining at the level of principles, it is interesting to note that in many cases the most characteristic aspect of the received signals—or that which most distinguishes them from the background

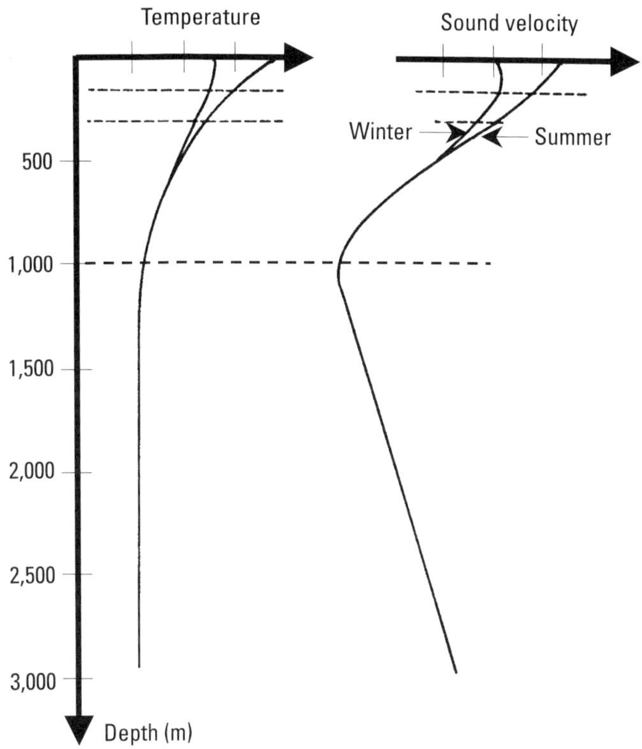

Figure 6.29 Typical profile of the velocity of sound in water [7].

noise—is their harmonic nature (periodic signal, or line spectrum). That is why the extraction of such signals will be examined in Chapter 7.

References

[1] Boerner, W. M., Ed., *Direct and Inverse Methods in Radar Polarimetry,* NATO ASI Series, Norwell, MA: Kluwer Academic Publishers, 1992.

[2] Saillard, J., editor, *Proceedings of the International Workshop on Radar Polarimetry,* IRESTE, Nantes, France, March 1990, Sept. 1992, March 1995.

[3] Krogager, E., "Aspects of Polarimetric Imaging," (doctoral thesis, Technical University of Denmark, Copenhagen, May 1993).

[4] Ruck, G. T., et al., *Radar Cross Section Handbook,* New York: Plenum, 1970.

[5] Jones, D. S., *Acoustic and Electromagnetic Waves,* Oxford: Clarendon Press, 1986.

[6] Kouyoumjian, R. G., "Asymptotic High-Frequency Methods," *Proc. IEEE,* Vol. 53, No. 8, Aug. 1985, pp. 864–876.

[7] Burdic, W. S., *Underwater Acoustic System Analysis,* Englewood Cliffs, NJ: Prentice-Hall, 1985.

[8] Knott, E. F., "A Progression of High-Frequency RCS Prediction Technique," *Proc. IEEE,* Vol. 73, No. 2, Feb. 1985, pp. 252–264.

[9] Lo, Y. T., and S. W. Lee, Ed., *Antenna Handbook,* New York: Van Nostrand Reinhold, 1988.

[10] Stratton, J. A., *Electromagnetic Theory,* New York: McGraw-Hill, 1941.

[11] Keller, J. B., "Geometrical Theory of Diffraction" *J.A.S.A.,* Vol. 52, No. 2, Feb. 1962, pp. 116–130.

[12] Pouit, C., "Imagerie Radar à Grande Bande Passante," *S.E.E. International Colloquium on Radar,* Paris, Dec. 1978.

[13] Ostrovityanov, R. V., and F. A. Basalov, *Statistical Theory of Extended Radar Targets,* Dedham, MA: Artech House, 1985.

[14] Barton, D. K., *Modern Radar System Analysis,* Norwood, MA: Artech House, 1988.

[15] Bouvier, Ch., "Modélisation du Fouillis Radar," (doctoral thesis, Université de Nice Sophia Antipolis, Sept. 1996).

[16] Nathanson, F. E., *Radar Design Principles,* New York: McGraw-Hill, 1969.

[17] Sauvageot, H., *Radar Meteorology,* Norwood, MA: Artech House, 1992.

[18] Billingsley, J. B., and H. C. Chan, "Radar Ground Clutter Measurements and Models; Part I: Spatial Amplitude Statistics; Part II: Spectral Characteristics and Temporal Statistics," *AGARD Conference Proceedings 501,* "Target and Clutter Scattering and their Effect on Military Performance," Ottawa, May 1991.

[19] Watts, S., C. S. Baker, and K. Ward, "Maritime Surveillance Radar: Detection Performance in the Sea Clutter, *IEE Proc., F, Radar Signal Processing,* Vol. 137, No. 2, 1990, pp. 63–72.

[20] Pham, K. K., "Range Ambiguous Signature Against Sea Clutter in an X-Band Multifunction Radar," *IEEE International Radar Conference,* 1990.

[21] Beckmann, P., and A. Spizzichino, *The Scattering of Electromagnetic Waves from Rough Surfaces,* Pergamon, 1963, Norwood, MA: Artech House, 1987.

[22] *AGARD Conf. Proc. No. 249,* NATO, 1978.

[23] Fournier, M., "Prévision de Propagation: Effets des Conduits à la Surface de la Mer," *AGARD Conference on Multiple Mechanisms Propagation Paths,* Rotterdam, Oct. 1993.

[24] Christophe, F., and L. Bertel, "Physical Aspects in Low-Frequency Radars," *Colloque Science et Défense,* Dunod, France, 1995.

[25] Boithias, L., *Radio Wave Propagation,* New York: McGraw-Hill, 1987.

[26] Thorp, W. H., "Deep Ocean Sound Attenuation in the Sub- and Low-Kilocycle per Second Region," *J.A.S.A.,* Vol. 33, 1961, p. 334.

Appendix 6A: A Multifrequency Ground Clutter Model

Reference [18] presents an analysis of a large multifrequency database of radar ground clutter measurement from many sites at grazing incidence, summarized in a Weibull model for RCS statistics and a three-component spectral model, convenient for radar design and performance prediction. The main characteristics of these models are described hereafter.

6A.1 Weibull Ground Clutter

Weibull distribution is a convenient model for statistical representation of many radar clutter situations because of the following features:

- Its formulation is simple enough to allow for easy false alarm calculations.
- Its variability is sufficient to take into account the different situations occurring at low grazing angles.
- It admits as a limiting case the exponential distribution, which is appropriate for clutter RCS at higher grazing angles where no shadowing occurs[4] (following a central limit argument for multiple scatterers returns).

Weibull law depends on two parameters: a Weibull spread parameter a_w, and a scale parameter s. Weibull cumulative distribution F_w and probability density f_w are written:

$$F_w(x) = 1 - \exp\left(-\frac{x^{1/a_w}}{s}\right) \tag{6A.1}$$

$$f_w(x) = \frac{x^{\left(\frac{1}{a_w}-1\right)}}{sa_w}\exp\left(-\frac{x^{1/a_w}}{s}\right)$$

The very simple form of the cumulative distribution F_w is of course particularly convenient for computing false alarm rates on clutter. The statistical moments are given by:

4. Briefly, the non-Gaussian characteristic of low grazing angle clutter can be attributed mainly to the fact that for these incidences the backscatter is due to a mixture of strong returns from discrete scatterers and low returns from shadowed areas in the same resolution cell.

$$m_i = s^{ia_w}\Gamma(ia_w + 1) \Rightarrow m_1 = s^{a_w}\Gamma(a_w + 1) \tag{6A.2}$$

where Γ is the Gamma function $[\Gamma(n) = (n - 1)!$ for integer $n]$.

For $a_w = 1$, the exponential distribution (Gaussian signal) is obtained.

The statistics of clutter for a given situation are then described by the average RCS density m_1 (in square meters per square meters) and the spread parameter a_w. The statistics depend on the terrain type, the depression angle, the frequency band, and the resolution on the ground (area resolution in m^2). From the analysis of existing databases of radar ground clutter measurements, the results obtained are given in Table 6A.1 (more detailed results and analysis are given in [18]).

The two columns under a_w in Table 6A.1 correspond to different resolutions on the ground (area resolution, or "footprint," in square meters): $10^3 m^2$ and $10^6 m^2$, respectively: it can be seen that larger footprints give smaller a_w, that is, more Gaussian clutter distribution. The exact frequencies are respectively: 165 MHz, 435 MHz, 1,230 MHz, 3,240 MHz, and 9,200 MHz.

Figure 6A.1 shows the "false alarm probability on clutter" (probability of clutter exceeding the level in abscissa) for three types of terrain at 10 GHz; the abscissa is the clutter density in square meters per square meters, and in vertical scale is plotted the probability of exceeding this level:

$$P_{fa}(x) = \exp\left(-\frac{x^{1/a_w}}{s}\right), \text{ with } s = \left(\frac{m_1}{\Gamma(a_w + 1)}\right)^{1/a_w}$$

For a Doppler radar, this value must be corrected in order to take into account the MTI clutter cancellation ratio, defined in Appendix 3C. For instance, in urban clutter, the level, $0.01 m^2$ per m^2, with a footprint of $1{,}000 m^2$ (giving an average RCS of $10 m^2$ in the resolution cell) is exceeded with probability 0.16—which means that, for a radar with a clutter cancellation ratio of −40 dB, RCS of −30 dB/m^2 may be observed with that probability of occurrence on that type of clutter. This result obviously sets a limit to the capability of detecting stealth targets, with that kind of RCS, when a reasonable false alarm rate (much lower than 0.16, and even much smaller than 0.02, which is obtained with a similar reasoning for general rural clutter with low-relief) is required.

6A.2 Three-Component Ground Clutter Spectrum

In this model, the clutter spectrum is described as being made of three components, as indicated on Figure 6A.2:

Table 6A.1
Weibull Parameters m_1 and a_W for Different Terrains and Different Frequencies

Terrain Type	Depression Angle	m_1 (dB)					a_W	
		VHF	UHF	L	S	X	10^3	10^6
General rural (low relief)	1	–30	–30	–30	–30	–30	3.0	1.8
Continuous forest (low relief)	1	–15	–19	–22	–24	–26	2.0	1.3
Open farmland (low relief)	1	–30	–30	–30	–30	–30	3.3	2.4
Desert, marsh, grassland	>0.75	–38	–40	–40	–38	–26	2.0	1.3
General rural (high relief)	0 to 2	–27	–27	–27	–27	–27	2.2	1.4
Continuous forest (high relief)	Any	–15	–19	–22	–22	–22	1.8	1.3
General urban	0.1	–20	–20	–20	–20	–20	4.3	2.8
General urban	>0.75	–20	–20	–20	–20	–20	3.0	2.0

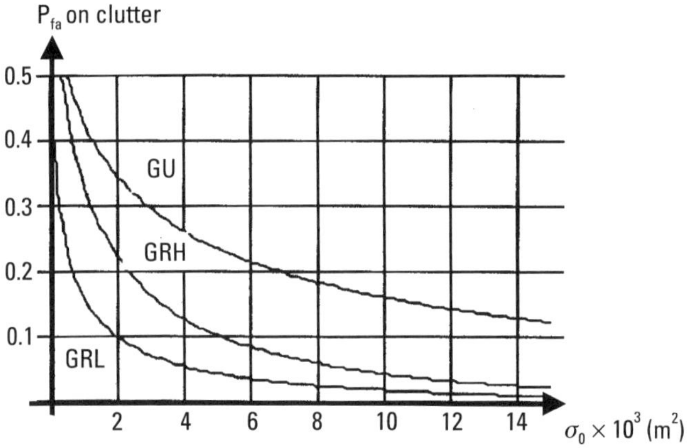

Figure 6A.1 False alarm probabilities on clutter (footprint = $10^3 m^2$). GU: General urban, GRH: General rural (high relief), and GRL: General rural (low relief).

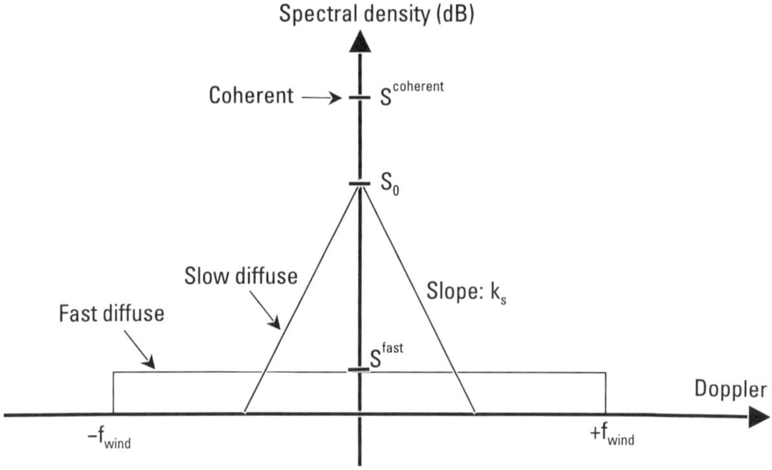

Figure 6A.2 Ground clutter spectrum.

- A dominant coherent component S^{coherent} at zero Doppler, resulting from fixed objects, and relatively higher at lower frequencies;
- A slow diffuse component, due to the movements of large objects, such as tree branches, and occupying a relative narrow region around zero Doppler, with negative exponential Doppler spectrum: $S^{\text{slow}}(f) = S_0 \exp(-k_s|f|)$ (spectral density decreasing linearly, in dB scale, with absolute value of the Doppler shift);

- A much smaller fast diffuse component, observable only at low wind speeds, and modeled as a band-limited white noise, with cutoff frequency slightly lower than the wind Doppler frequency: $S^{\text{fast}}(f) = S^{\text{fast}}$ for $|f| \leq f_{\text{wind}}$.

7

Signature Adapted Processing

7.1 Summary

The great variety of signatures of targets and backgrounds, as revealed in Chapter 6, obviously makes it impossible to imagine an optimum processing adapted to all situations. In contrast, in a given situation, it is often possible to modify the conventional processing, adapted to a point target, in order to take into consideration the specific nature of the received signals. Thus, in Chapter 5, taking into account the colored nature of the external noise led to the definition of optimum receivers by subtraction of the colored part, by whitening, or by minimization under constraint. When the representation of the clutter or of the targets by a colored noise is no longer accurate enough with respect to the radar resolution capability, it is then necessary to use processing techniques whose justification is obviously more complicated but whose performance is better. Although this processing sometimes makes use of techniques (e.g., filtering and peak clipping) used in other areas of signal processing or of pattern recognition, there are cases in which problems specific to radar or sonar (time and frequency measurement, removal of ambiguities, for instance) require a special formalization of the received signal. Five radar and active sonar problems are typical of this approach:

1. Target recognition according to their radar or sonar echoes;
2. Tracking of a nonpoint target (extended target);
3. Detection or location of targets in the presence of clutter (in radar) or of reverberation (in sonar);

4. Low frequency radar operation;
5. Wideband radar processing.

These points will be considered in turn, and a procedure will be described in each case. These procedures, even though they are not the only ones available, can nevertheless be considered and adapted to a wide range of situations. This approach is coherent with our initial objective: to remain at the level of basics, using illustrations representing a relatively general class of situations rather than a detailed series of individual examples.

7.1.1 Radar Target Recognition

To identify a target by means of radar or sonar, the first potential approach consists of making an image of this target and then presenting this image to an operator (or to a computer programmed for image recognition). Chapter 6 showed that it was possible, in the laboratory at least, to generate such images by tomographic synthesis based on projections that are impulse responses (range responses) or Doppler responses (transverse, or cross-range responses). To generate such images on a freely moving target, it is necessary to calculate the sum of these projections (for the different aspect angles) in amplitude and phase, which requires that they have the same range reference, to within $\lambda/4$ (λ: wavelength); this implies either a compensation of the target motion better than $\lambda/4$ or an "alignment," with this accuracy, of each projection in relation to the one before, by recognizing a characteristic scatterer for instance.

Although this alignment can be carried out under certain conditions, it is nevertheless an extremely delicate operation that, moreover, only provides an image that is difficult for an untrained operator to interpret (at least for air targets in radar). It is more effective to entrust a dedicated processor with the automatic recognition of this image. The generation of an image is then no longer necessary, since the processor can just as well process the signals generated by the radar directly, with no necessity for an image. These three reasons—complexity of alignment, "unnatural" character of the images, and uselessness of the image for a dedicated processor—often lead to the use of automatic classification procedures, working directly on the signals delivered by the radar.

Then, when examining examples of various radar situations in which an identification is required and seems feasible (given the target signatures and the characteristics of the radars used), it is observed that the radar generally delivers—before the quadratic envelope detection—a complex signal (amplitude and phase) depending on the range or time, in the case of pulse responses, or on frequency, in the case of Doppler responses. The signal to identify is thus a complex signal in the domain of time or of frequency, having the following characteristics:

- It fluctuates, from one measurement to the next, which makes it necessary to average the consecutive signals before attempting to identify them.
- It is altered, from one measurement to the next, by time (delay) and frequency (Doppler) shifts, which prohibit direct averaging (without alignment) of the consecutive signals received.

A possible approach then consists of extracting from the received signal a feature that is not affected by these time and frequency shifts and in averaging the consecutive features thus obtained—the identification being then performed on this average feature. Now, a simple way to eliminate a time shift is to take the autocorrelation of the time signal, and a simple way to eliminate a frequency shift is to take the modulus of the time signal. To extract an invariant feature from a signal shifted in time *and* frequency (range and Doppler), it is necessary to calculate the modulus of the autocorrelation of the signal or the autocorrelation of the modulus of the signal, and to keep the real functions thus obtained as characteristic values, which can then be averaged before attempting to identify them.

A completely similar approach can be applied to the frequency signals (Doppler) received. The autocorrelation then eliminates the frequency shift, and the modulus the time shift. This results in two features that are none other than the Fourier transforms of the two temporal features defined previously (a more detailed analysis shows that these four features are in fact simply related to the ambiguity function of the *received* signal, which is the basic discriminant invariant).

This approach enables, for a given radar and a given type of target observed, the extraction of a new signal from the received signal, referred to as the feature, which contains the essential information about the target and can therefore be used by an automatic classification procedure, which compares the feature thus obtained with similar values recorded previously. Generally this automatic classification procedure is of a conventional type, that is, not specific to the radar or sonar field (e.g., nearest neighbor or linear discrimination).

In practice, this method can be improved by various processes often used in signal processing, consisting of supplementing this feature with a few ad hoc features extracted directly from the received signal, such as mean power and duration.

7.1.2 Tracking of Extended Targets

Another situation in which it is necessary to take the exact target signature into account is that of the tracking radar or sonar, when an extreme accuracy of tracking is required. Under these conditions, the nonpoint nature of the target causes its RCS to vary with the aspect angle and consequently causes the amplitude and phase of the echo received by each antenna element to be no longer identical (because the aspect angle of the target varies from one element to another). The captured signal is then no longer maximum when the antenna is pointed in the direction of the target, and the servo control, attempting to maximize this captured signal, will direct the antenna to a wrong direction, varying with the target aspect angle. These tracking fluctuations, referred to as glint, are thus caused directly by the fact that an optimum processing for a point target is applied to an extended target.

Using the stochastic target model defined in Chapter 6, it is then possible to model the statistical characteristics of this glint. It is in particular established that, on average, the antenna remains pointed at the target center of gravity (in the sense of the distribution of scatterers on the target) and that the amplitude of the deviations from this average direction is proportional to the transverse extent of this target, which appears intuitively reasonable. Taking into account the specificities of target fluctuation, appropriate anti-glint procedures are then derived by combining temporal and angular information.

7.1.3 Detection in Clutter

Taking target signatures into account can help perform a new function (identification) or interpret errors observed in practice. Similarly, taking the background signature into account can considerably improve the operation of the radar. The conventional method for eliminating the clutter consists of taking no account of the targets detected at Doppler shifts that are too low. This method is inadequate when the detection of stationary (fixed) targets is required and also when the radar is airborne, in which case the clutter echoes observed by the antenna sidelobes have variable Doppler shifts, depending on the sidelobe involved.

To detect stationary targets, clutter echoes are a hindrance because they vary from one ground location to another (without this "spatial" fluctuation, a comparison between each resolution cell and the neighboring cells would enable the threshold to be determined in a near-optimum way) and from one instant to another. A way to reduce these fluctuations consists of consecutively transmitting signals at different carrier frequencies and averaging the results obtained after quadrature detection. This use of the "frequency agility" effectively reduces the clutter fluctuations; a detailed analysis shows, however, that this reduction is limited and not proportional to the number of frequencies used, as might have been hoped. This limitation is related to the inherent nature of the spatial fluctuations, which can really be reduced only by averaging different spatial measurements.

Finally, the use of the various characteristics of the targets and analysis of the polarization can also improve the detection of stationary targets—at the cost of a duplication of the reception channel, in order to collect signals on two orthogonal polarizations.

In the case of airborne radars, a given relative speed—and thus Doppler shift—of the clutter corresponds to a specific platform speed and antenna pointing direction if the antenna is pointed to the ground, as far as main-lobe clutter is concerned. Therefore it should be possible to filter this Doppler to keep only the targets with different Doppler. However, in practice, when searching for the presence of targets of low RCS, the echoes received by the sidelobes of the antenna pattern—hence from different directions, and with different radial velocities—are not negligible.

For a given antenna pointing direction, the radar thus delivers, besides the echo from the potential target, a range-Doppler clutter map as it appears via the antenna pattern (see Figure 7.19), with in particular strong echoes in the neighborhood of zero Doppler shift and at a range equal to the carrier altitude, originating from the ground observed (by a sidelobe) vertically below the carrier, that is, for an incidence at which this echo is very strong (altitude return).

The radar ambiguity phenomenon (described in Chapter 3) is added to the presence of these unwanted echoes; as the radar transmits equally spaced pulses, its unambiguous domain in range and velocity is limited, and the range-Doppler map of the clutter is thus "folded" according to these range and velocity ambiguities. The problem then consists of detecting the presence of targets in the clutter folded in this way and measuring the ranges and velocities of these targets.

Different operating modes of the radar enable the appropriate repetition frequency to be selected, depending on the geometrical configuration (flight altitude, antenna pointing direction) and on the required detection conditions (range, closing and receding targets). Conventionally, the following modes are distinguished: *low pulse repetition frequency* (LPRF), unambiguous in range but highly ambiguous in Doppler shift; *high pulse repetition frequency* (HPRF), unambiguous in Doppler shift but highly ambiguous in range; and *medium pulse repetition frequency* (MPRF), ambiguous in both range and Doppler shift. For this latter mode in particular, it is necessary to remove ambiguities in order to determine the target ranges and velocities. This operation is generally carried out by successively transmitting pulse trains at slightly different repetition frequencies and wavelengths; the ambiguities are then different, and can be removed using a simple algorithm.

Specific processing can be implemented for slow targets detection with an airborne side-looking radar using two or more (phased array) antennas. In this situation, it is possible to process in parallel the signals received by the two antennas and to subtract the second, appropriately delayed, from the first in order to eliminate clutter. It can be shown that this is equivalent to analyze the distribution of clutter in Doppler and range, and to detect targets in this Doppler-angle map by two-dimensional filtering (Doppler-angle, or time-space filtering). Since this procedure, known as *space-time adaptive processing* (STAP), assumes identical antennas, it has to be complemented by a calibration procedure, for example using clutter returns on the different antennas.

7.1.4 Low-Frequency Radars

At low frequencies (below 1 GHz), targets and clutter behavior have some unique properties that are interesting to analyze in order to define the appropriate signal processing techniques. We will review the specificities of these low-frequency radars, first from the standpoint of the physical phenomena involved (retrodiffusion of targets, passive and active stealth, propagation, possibility of target recognition) then from the standpoint of the radar design in itself (available technologies, angular scanning, system integration).

7.1.5 Wideband Radars

Radar processing was described in the preceding chapters with the assumption that the instantaneous bandwidth was sufficiently small so that the displacement of the target during the observation time was small compared to the range resolution cell (or, equivalently, so that the Doppler effect could be considered as a mere translation in frequency). When this assumption is not valid, the processing has to be adapted to compensate for this "range migration effect." Careful analysis of this situation shows that the ambiguity function in this situation is different from the narrowband ambiguity function, since the range migration actually removes the ambiguities in velocity of low-repetition frequency radars.

7.1.6 Passive Systems

This diversity of processing types used in active radar and sonar is paralleled by an at least equal diversity in passive systems, a field in which the procedures used are generally—the antenna processing apart—not so specific to radar or to sonar. An example of a problem frequently encountered is the detection of the presence of a periodic signal of unknown period and the measurement of its period. Knowing that a periodic signal has a spectrum formed of regularly spaced lines, one way of proceeding consists of carrying out a Fourier transform of the received signal and calculating, for every possible period of the signal, a quantity equal to the corresponding sum of the harmonic levels; the estimated period of the signal is the one for which this quantity is maximum. The theoretical analysis shows that this procedure is optimum in the sense of the maximum likelihood criterion, when the duration of the signal observed is long compared with its period.

This example of processing that is usable in the analysis of signals received by a passive system is a good illustration of the nonspecific nature of passive radar or sonar. Here we move into the more general field of processing and recognition of signals of various shapes, where the methods used are not fundamentally different according to the physical source of the signal, for example, speech and medical signals.

7.1.7 Objectives

Taking into account the exact nature of the signals received from the targets and backgrounds remains in the domain of case-by-case or ad hoc processing, even though the concern has existed for a long time. The reasons for this state of affairs can be found in the tremendous processing requirement, in terms of computation speed, that it implies, but also perhaps in the limited progress in the formalization of the problems posed, for want of target and background models that are valid in a wide enough range of situations but are still tractable in terms of signal processing complexity. In this context, the purpose of this chapter is therefore simply to describe the basis of typical problems in order to illustrate the state of the art in radar and sonar signature processing by describing some potential resolution paths.

7.2 Radar Target Recognition

7.2.1 Approaches

Confronted with the problem of target identification (ships, aircraft, vehicles) the radar operator generally first expresses the wish to obtain an image of the target. Section 6.2 showed the appearance of such images, which are generally difficult to interpret and highly dependent on the aspect angle, even though they are obtained under the best laboratory conditions. The second approach then consists of trying to form an image of the target and then, because of this rather unnatural character making it difficult for the operator to interpret, entrusting the recognition to a machine using a previously generated database. The question that then arises is whether it is necessary to use the image, which obviously does not have the same utility for the machine as for the operator, hence the definition of the automatic classification procedure making direct use of the signals delivered by the radar.

We shall successively examine the approach by imaging, whose principles were already described in Sections 3.4 (radar cartography) and 6.3 (radar holography), and the approach by direct automatic classification.

7.2.2 Radar Imaging of Moving Targets: Inverse Synthetic Aperture Radar [1–7]

The principles of imaging by side-looking radar were discussed in Section 3.4 and those of imaging by radar holography in Section 6.3. It has already been established that they are in fact equivalent apart from the instrument function (ambiguity function in the case of side-looking radar and Fourier transform of the observation aperture in the case of holography). In both cases, the target is assumed to be composed of independent scatterers, which the system resolves in range (depth) by its range resolution or its frequency analysis capability, and resolves in azimuth (transverse distance) by angular (or Doppler) analysis of the received signal—the identity between angular analysis and Doppler analysis being explained in Section 6.3.5.

The imaging of moving radar targets is based on the same principle as the side-looking radar, except that the radar is fixed and that the motion of the target is used to generate the image. Theoretically, it is possible to obtain images of the type presented in Section 6.3.2, with a resolution set by the extent of the domain of observation in angle and frequency.

In the case of a target moving freely, the problem becomes more complicated, however, due to the lack of knowledge of the exact motion of the target. If it can reasonably be assumed that the target does not move appreciably (in relation to the wavelength) during the time necessary for an impulse response to be generated, this approximation is generally not legitimate during the time required for the formation of an image, for which it has already been pointed out that the observation angular sector limits the transverse resolution on the image by relation (6.36): $\delta y = \lambda_0/(2\Delta\theta)$. At a 1-cm wavelength ($f_0$ = 30 GHz), a 3-m resolution is thus obtained for an angular extent of 1.7 mrad, that is, at 5 km (a range below which the process would be of no interest whatsoever) a motion of 8.5m. It is therefore necessary to know the motion of the target, with an accuracy much better than the wavelength, over distances of several meters. This can be considered, taking many precautions, when the sole motion is that of the carrier (case of the side-looking mapping radar), but obviously becomes much more difficult when the motion is that of an unknown target, a priori "uncooperative."

More precisely, the motion of the target relative to the radar changes the received signal in two aspects:

1. A shift in range and Doppler (which must be eliminated);
2. An apparent rotation about a reference point—arbitrarily defined on the object—at an angular velocity that must be known (to carry out the Doppler/cross-range scale conversion).

Two methods for motion compensation are conventionally considered [1, 4]: reference to a scatterer, and tracking of the center of gravity. Both processes aim at aligning in range and phase the consecutive impulse responses received by choosing, as center of rotation, either a scatterer observed on every impulse response (usually a strong and nonfluctuating scatterer is preferred) or the center of gravity of these impulse responses. In the first case it is implicitly assumed that the scatterer selected as a center of rotation is steady and unique in the range resolution cell (failing which the alignment would lead to range skips from one scatterer to another). In the second case, the target is assumed to be composed of enough scatterers, sufficiently steady, for the center of gravity of the responses to have a fixed position in relation to the target. A possible technique [4] consists of compensating the successive impulse responses simultaneously in range and phase by correlation of successive *complex* impulse responses; the range shift is estimated by the position of the correlation maximum, and the phase difference between successive impulse responses is estimated by the phase of this correlation maximum.

The apparent rotation rate of the target is generally deduced from assumptions about the motion of the target (uniform straight-line path) combined with azimuth, range, and Doppler measurements delivered by the same radar or by an auxiliary radar. It is important to note that an error in this rotation rate distorts the image but does not degrade the resolution of this image, in contrast to an impulse response alignment error.

The complexity of the practical implementation of these compensations, combined with the wide bandwidth required to obtain satisfactory range resolution, sometimes leads to the adoption of a degraded imaging mode in which the range resolution is abandoned. The radar then delivers "transverse" responses (fine Doppler analysis of the received signal, described in Section 6.3.5), which do not require the previously described compensations, except that the scale of this transverse response is determined by the apparent rotation speed (assumed to be constant) of the target. Although sometimes referred to as inverse synthetic aperture radar (ISAR), this process is conceptually closer to the impulse response process, as seen in Section 6.3.5.

Although some implementations of this type of ISAR have been referred to in the literature [2–4], the field of application remains limited, mainly due to the complexity of the radar (coherent over a wide band) and to the necessary compensation processing, in relation to the quality of the images obtained, which are difficult to interpret.

7.2.3 Automatic Classification of Targets

7.2.3.1 The Problem

The automatic classification approach is a response to a concern generally expressed very pragmatically—for a radar of given characteristics, how are the signals received best exploited for target classification, or again, at the design stages, which radar characteristics could help classify the targets? Generally, in this approach, classification is thus no longer the prime objective of the radar but is rather considered as a potential capability, given characteristics mainly resulting from higher-priority requirements (detection, location) [8]. Adopting this approach, we shall first examine a few typical situations in which a classification can be considered before describing the procedures that can be envisaged.

First, for air targets (aircraft, helicopters), two configurations seem to be of immediate interest: the ground-based air surveillance radar of medium range (from 30 to 50 km), typically using a wavelength of 5.5 cm and an antenna with a gain of 28 dB and a bearing resolution of 2°, rotating in azimuth at 90°/second; the mean power is 0.1 kW, the repetition frequency is 3 kHz, the range resolution is 150m, and the Doppler resolution is 100 Hz. This set of characteristics, resulting from various trade-offs (e.g., power available, desired range, and location accuracy) thus leads to a radar that does not resolve the target in range but whose Doppler resolution can be used to analyze the jet engine modulation lines or the spectral broadening due to the helicopter rotor blades (see Section 6.4).

The second interesting configuration is that of the combat aircraft radar, which is now often of the "multimode" type (with, for example, various repetition frequencies and various transmission frequencies) one mode of which, HPRF is typically the following: wavelength 3 cm, angular resolution (elevation and bearing) 3°, antenna gain 34 dB, azimuth scan speed 180°/second, mean power 2 kW, pulse repetition frequency 150 kHz (highly ambiguous radar in range), range resolution 150m, and Doppler resolution 300 Hz. Here again, the range resolution does not enable analysis of the target, but the velocity resolution enables analysis of the jet engine modulation lines or the helicopter rotor blades. Moreover, the pulse repetition frequency of 150 kHz ensures that these spectra of moving parts will not be folded, that is, that the spectrum width is less than the Doppler ambiguity.

Figure 7.1 shows an example of such spectra observed in X band (λ = 3 cm), with a Doppler resolution of 300 Hz; it can be seen in particular that the fluctuations from one spectrum to the other (the two spectra corre-

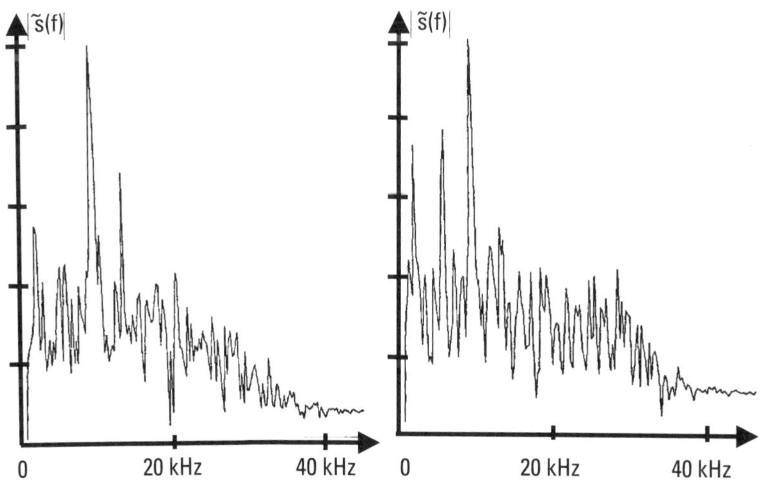

Figure 7.1 Aircraft classification (consecutive spectra with a 3-ms interval) (ONERA measurements).

spond to two consecutive measurements with a 3-ms interval) are significant. Moreover, these spectra are obviously subjected to a global shift in frequency due to the inherent Doppler shift of the aircraft.

Considering next the case of maritime targets (ships), we shall choose the typical example of the airborne maritime surveillance radar operating at a wavelength of 3 cm, with an antenna of 25-dB gain and 3° bearing directivity scanning space by rotation at 120°/second; the peak power of such a radar is 25 kW approximately, its repetition frequency is 1 kHz, and the range resolution is variable between 7.5 and 150m. Such a radar thus delivers impulse responses of the observed ship—as well as a possible transverse response (Doppler) and a frequency response by using frequency agility on transmission.

Figure 7.2 shows, as an example, two ship impulse responses (in modulus), measured for ship aspect angles differing by 1° in azimuth. Here again, substantial fluctuations of the relative levels of the peaks from one response to the next can be observed. Moreover, these impulse responses are obviously subjected to a global shift in range due to the change in the aircraft-ship distance.

Finally, we consider the third situation, that of the ground target (vehicle) detection radars. A typical example of a radar that can be installed in a remotely piloted vehicle (drone) or in a helicopter, could be the following: wavelength 2 cm (frequency 15 GHz), antenna gain 36 dB, bearing resolution 0.5°, mean power 2W, repetition frequency 40 kHz, range resolution 5m, Doppler resolution 600 Hz, and agility over 500 MHz (64 different

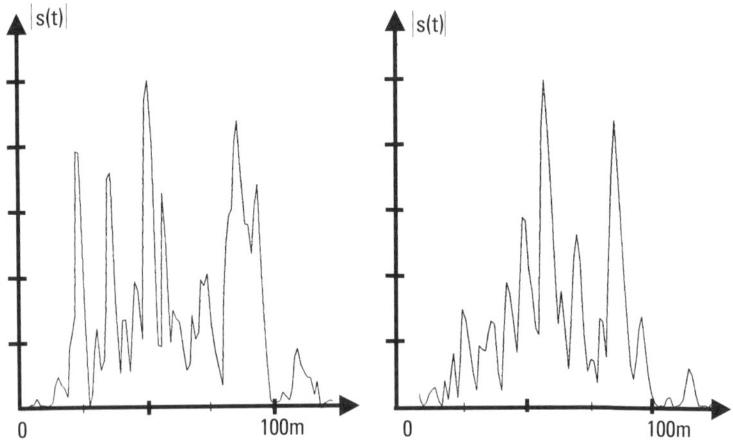

Figure 7.2 Ship classification (impulse responses for two aspect angles).

transmission frequencies). Although such a radar does not have a sufficient resolution in range or in velocity to analyze the target, it can in contrast use its frequency agility as a source of information. The echo received versus frequency is in fact the frequency response $q(f)$ of the target (Section 6.2.3), whose Fourier transform is the impulse response with resolution $c/2\Delta f$, equal to 30 cm here.

Generally, however, the phase of $q(f)$ is difficult to exploit, due to the relative radar-target motion and to the transmitter stability. Hence, the radar only delivers, for one frequency scan (thus every $64/600 \approx 0.1$ second), the square of the modulus of $q(f)$, whose Fourier transform is, according to a conventional theorem, the autocorrelation of the impulse response. This type of frequency agile radar thus enables calculation of the autocorrelation of the target pulse response. We are going to see that this feature can be used for automatic classification.

7.2.3.2 Automatic Classification

An automatic classification—or pattern recognition—procedure generally operates according to the diagram shown in Figure 7.3. A first processing extracts discriminating features (such as signal level or pulse response) from the received signals, then a decision procedure compares these features to features previously recorded in a learning memory (e.g., thresholds or pulse responses of various targets) and displays the result. In this diagram, the aim of the extraction procedure for the features is to produce a "caricature" of the observed target—it carries out a data compression at the same time, which reduces the processing volume required for the decision procedure.

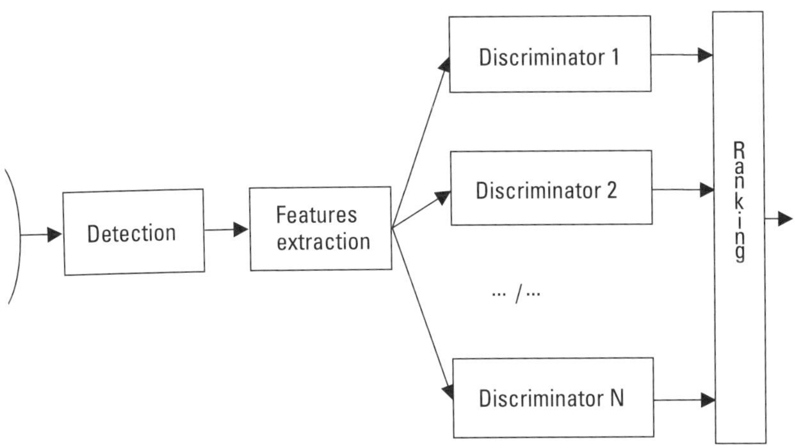

Figure 7.3 Classification procedure.

It is also important to note that the future extraction procedure is typical of the specific radar problem; in contrast, the aim of the decision procedure is to compare vectors with each other, and all pattern recognition applications generally use the same types of procedure such as k nearest neighbors, linear discrimination, and so on, whose principles will be described below in Section 7.2.3.5.

7.2.3.3 Feature Extraction

The examples presented in Section 7.2.3.1 showed the wide diversity of observable situations. They also showed that, in spite of this diversity, there are some general characteristics.

- The signal available is generally one-dimensional (impulse response, Doppler spectrum, frequency response).
- The signal available may have a shift that is variable from one measurement to the next (in time, in frequency, or more generally in both domains at the same time, since the target range and radial velocity are variable with respect to the radar).
- The fluctuating nature of the signals makes it necessary to carry out the recognition on the basis of an average—to be defined—of consecutive signals, in order to improve reliability.

Without analyzing the signals received in greater detail, it is clear that the radar is characterized by the fact that it is desirable to extract from the signals received an invariant feature by *time-shifting and frequency-shifting* the signal. Such a value can then be averaged over the consecutive signals, before being used by the decision procedure (the target mean range and Doppler data are obviously used by the radar elsewhere and could even be used as additional criteria for the final decision). Initially, the extraction of the features depends on the definition of a feature invariant by shifting the received signal in time and frequency.

Such features can obviously be specified in an ad hoc way, such as spacing between the lines of a spectrum, frequency of the "blade flashes" of a helicopter, apparent length of an impulse response, number of local maxima of a response. These features are often useful but rarely sufficient to ensure the desired identification probability. We shall see that the common characteristics of the problems raised by these three examples—air targets characterized by their Doppler signature, maritime targets characterized by their impulse response, and ground targets characterized by their frequency response—enable the definition of a common extraction procedure for the features, based on the "discriminant invariants" definition.

Let $s(t)$ be the complex target echo delivered by the radar. For air targets, $s(t)$ is thus the series of the complex samples received over time, in the target range bin. For the HPRF radar referred to previously, $s(t)$ is for example composed of 500 (complex) echoes, in the target range gate, in response to the 500 pulses transmitted. For the maritime surveillance situation, $s(t)$ is composed of the 50 (complex) echoes received in the 50 range bins (7.5m each) covering the target. In the case of the vehicle detection radar, $s(t)$ is composed of the 64 (complex) echoes received, in the target range bin, in response to the signals transmitted at the 64 consecutive frequencies.

For these three situations, the preceding considerations lead to the determination of a feature, deduced from $s(t)$, invariant by shift of $s(t)$ in time or in frequency.

A systematic method of determining such a feature consists of using the time-frequency representation of this signal [9]. The definition of the time-frequency representation is based on the definition of the "useful part" of the signal $s(t)$, equal to $U(t_1, t_2) = S(t_1)S^*(t_2)$; given for any value of the (t_1, t_2) pair, this useful part characterizes the signal to within an arbitrary phase shift (the signal amplitude is $\sqrt{U(t, t)}$, and its phase is the argument of $\frac{U(t, t_0)}{\sqrt{U(t_0, t_0)U(t, t)}}$). The time-frequency representation (Wigner-Ville representation) is then defined as $K(t, f)$:

$$K(t, f) = \int s\left(t + \frac{u}{2}\right) s^*\left(t - \frac{u}{2}\right) e^{2\pi jfu} du \tag{7.1}$$

The given information of this representation is equivalent to that of the useful part, and consequently to that of the signal, to within an arbitrary phase shift. This representation has several interesting properties; for example, the integral with respect to frequency of $K(t, f)$ is equal to the square of the modulus of the signal $\int K(t, f)\, df = |s(t)|^2$, and similarly in frequency $\int K(t, f)\, dt = |\tilde{s}(f)|^2$, (where $\tilde{s}$ is the Fourier transform of s).

Moreover—and this property is fundamental for our application—a translation of the signal, in time or in frequency (a translation in time by t_0 being equivalent to the multiplication of $\tilde{s}(f)$ by $e^{2\pi jft_0}$) results in the same translation for its time-frequency representation, along the time axis or the frequency axis, depending on the case.

To extract from this function a feature invariant by translations in time and frequency, it is then sufficient to calculate its autocorrelation $H(t, f)$:

$$H(t, f) = \int K^*(t + u, f + v) K(u, v) \, du \, dv \tag{7.2}$$

It is easy to check that this feature is actually invariant by shifting the signal in time and in frequency, and it is also demonstrated that it is none other than the ambiguity function of the received signal, not to be confused with the conventional ambiguity function of the transmitted signal, as defined in (3.21):

$$H(t, f) = \left| \int s(u) s^*(u + t) e^{-2\pi j f u} \, du \right|^2 \tag{7.3}$$

This value is therefore, a discriminant invariant (two-dimensional) of the received signal, and can thus be estimated by averaging the consecutive discriminant invariants calculated on the signals received. As an example, Figure 7.4 shows the ambiguity function obtained by a radar echo from an aircraft. Note the lines, parallel to the time axis, corresponding to the compressor lines discussed previously.

The ambiguity function of the received signal—pulse response, transverse response, or frequency response—must obviously not be confused with the ambiguity function of the radar system used, which can simply be considered as the equipment function limiting the resolution on the received signal $s(t)$.

However, the invariant thus defined has the drawback of being a two-dimensional feature, and consequently rather heavy to handle for the pattern recognition procedures. It is therefore preferable to derive this feature from other invariants that are one-dimensional this time. This can be done simply, at the cost of a loss of information that must be assessed in each case, by projection of $H(t, f)$ parallel to the axes; two one-dimensional discriminant invariants are thus defined:

$$\tilde{A}(f) = \int H(t, f) \, dt \qquad \text{and} \qquad B(t) = \int H(t, f) \, df$$

where the ~ index designates the frequency dependent functions.

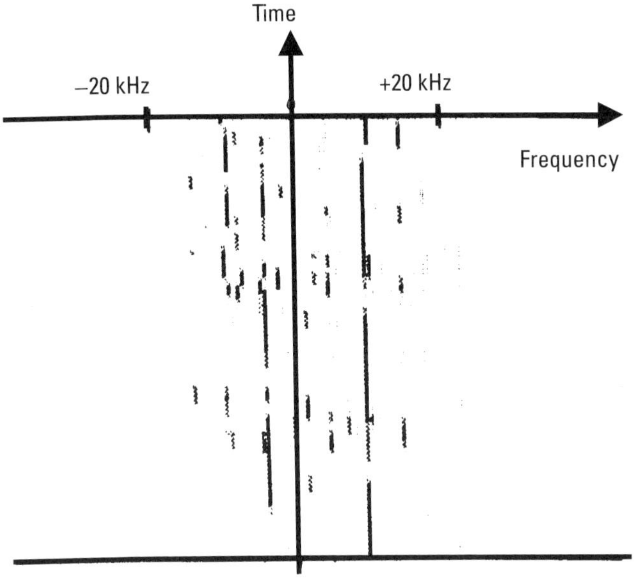

Figure 7.4 Ambiguity function of a radar echo from an aircraft.

Elementary calculations show that these features can also be written, if ⊗ designates the correlation operator:

$$\tilde{A}(f) = |\tilde{s}(f)|^2 \otimes |\tilde{s}(f)|^2 \quad \text{and} \quad B(t) = |s(t)|^2 \otimes |s(t)|^2 \tag{7.4}$$

These two invariants obviously have their corresponding Fourier transforms, which are proved by elementary calculations to be the two sections of $H(t, f)$ by $t = 0$ and $f = 0$:

$$A(t) = H(t, 0) \quad \text{and} \quad \tilde{B}(f) = H(0, f)$$

where

$$A(t) = |s(t) \otimes s(t)|^2 \quad \text{and} \quad \tilde{B}(f) = |\tilde{s}(f) \otimes \tilde{s}(f)|^2 \tag{7.5}$$

Thus, four discriminant invariants are defined: $A(t)$, the square modulus of the autocorrelation of the signal, and $B(t)$, the autocorrelation of the square modulus of the signal, and their respective Fourier transform $\tilde{A}(f)$, autocorrelation of the modulus of the signal spectrum and $\tilde{B}(f)$, the square modulus of the autocorrelation of the signal spectrum. All these invariants involve the autocorrelation operator, which renders invariant to a translation in its domain, and the modulus operator, which renders invariant to a translation in the conjugate domain.

Returning to the radar examples previously mentioned, it is then possible to interpret these discriminant invariants in more concrete terms. In the case of the HPRF radar, $\tilde{A}(f)$ is the autocorrelation of the square modulus of the Doppler spectrum, plotted in Figure 7.5(a), and $\tilde{B}(f)$ is the square modulus of the autocorrelation of the spectrum, plotted in Figure 7.5(b); it can be seen that the jet engine modulation lines appear much more clearly on $\tilde{A}(f)$ than on $\tilde{B}(f)$.

This results from the fact that $\tilde{A}(f)$ is a projection parallel to the time axis of the ambiguity function $H(t, f)$ plotted in Figure 7.4, where the lines appear in the form of traces parallel to the time axis. The feature $\tilde{A}(f)$ thus determined contains most of the information about the jet engine lines and no longer depends on the inherent Doppler shift of the target. It can therefore be averaged over several consecutive observations, in order to constitute the feature for use by the decision procedure.

In the case of the maritime surveillance radar, similarly, $A(t)$ is the autocorrelation of the modulus of the impulse response, plotted in Figure 7.6 for the two successive impulse responses shown on Figure 7.2 (this

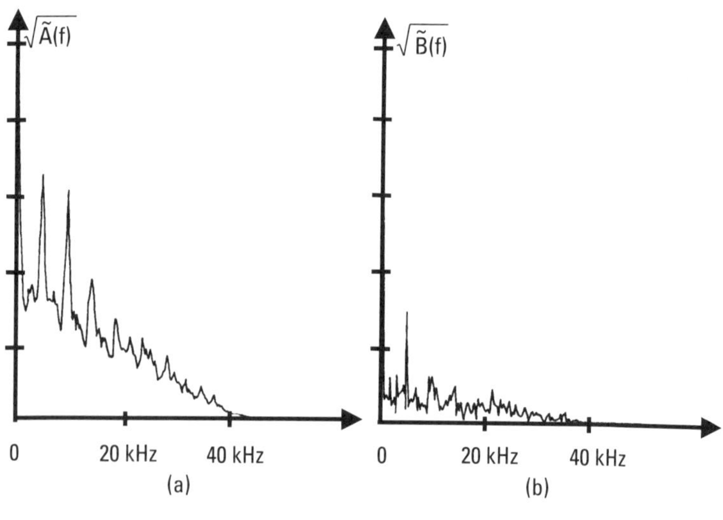

Figure 7.5 Aircraft signature example: (a) $\tilde{A}(f)$, and (b) $\tilde{B}(f)$.

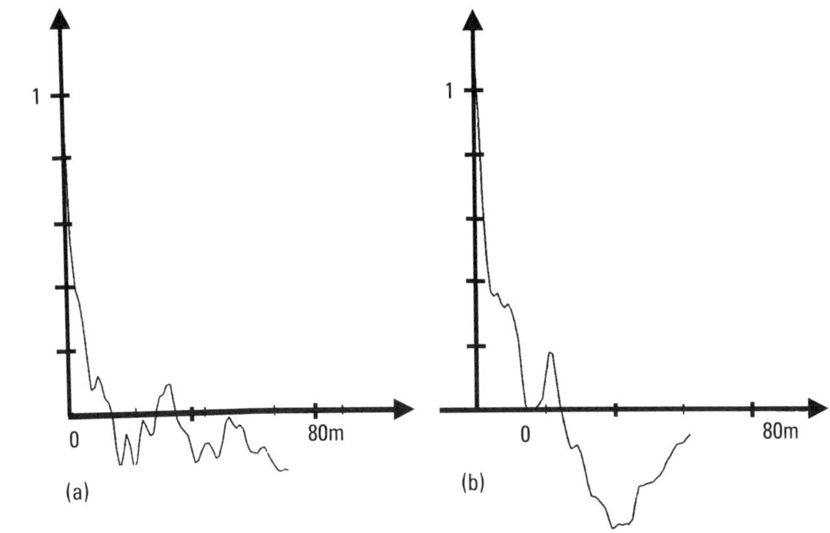

Figure 7.6 Autocorrelation of successive impulse responses of a ship (ONERA measurements) (after preprocessing, such that mean = 0 and maximum =1). (a) $A(t)$, and (b) $B(t)$.

feature has the advantage that it can be delivered by a "noncoherent" radar, since only the modulus of the impulse response is used). $\tilde{B}(f)$ is the modulus of the autocorrelation of the frequency response (which is delivered by some types of "stepped-frequency" radars and has the advantage of not requiring the synthesis of a short pulse).

Finally, for the ground target detection radar, as $s(t)$, the signal received is in this case the frequency response of the target[1], $\tilde{B}(f)$ is the square modulus of the autocorrelation of the target impulse response, and $\tilde{A}(f)$ is the autocorrelation of the square modulus of the impulse response. There again, $\tilde{A}(f)$ can be obtained as a Fourier transform of $A(t)$, which here is the autocorrelation of the square modulus of the target frequency response, and does not require knowledge of the relative phase of the consecutive echoes at the different frequencies.

In all cases, the feature depends on the target aspect angle; learning will thus require measurement of these features for different aspect angles of each type of target. However, inasmuch as the procedure uses a feature

1. We must apologize for this diabolical notation: $s(t)$, the time sequence of received signals, is in this case identical to the frequency response of the target, since the transmitted frequency is changed from pulse to pulse.

averaged over a given angular sector, it will generally be sufficient to learn on a small number of aspect angles—a number obviously depending on the particular case processed but often on the order of ten.

The main purpose of these three examples was to show that the necessity to average the consecutive received signals leads to the use of features of the same type: autocorrelation of the square modulus, or square modulus of the autocorrelation of the signal, in order to eliminate the range-Doppler shift. They also show that, depending on the type of the signal to recognize (harmonic, like the jet engine signal, or pulsed, like the ship signal) one will be led to choose a particular type of invariant: $A(t)$ or $\tilde{A}(f)$ for harmonic signals, $B(t)$ or $\tilde{B}(f)$ for pulse signals. Depending on the type of radar, one invariant or the other will be more or less easily measured; a short-pulse radar, for example, directly delivers a detected pulse response, and so the invariant $B(t)$ is used. It appears that, for a given application, the selection of the appropriate invariant is the outcome of a compromise between the nature of the expected signals and the characteristics of the radar, generally defined by other considerations related to its detection or location performance.

Finally, this invariant definition solves the problem of the relative motion of the carrier and the target, discussed in Section 7.2.2. This compensation, which was necessary for imaging, is replaced here—at the cost of some loss of information—by the adoption of range-Doppler invariants.

This approach has shown, once again, the specific nature of radar problems, characterized by the fundamental importance of the two quantities conjugated by Fourier transform: time and frequency. This particularity, the source of the properties of the ambiguity function (simultaneous *measurement* of time and frequency), here explains the importance of the autocorrelation and the modulus (simultaneous *elimination* of time and frequency).

Finally, note that in this analysis, it has been assumed necessary to eliminate the shifts in time *and* in frequency (range and Doppler). In some situations, one of these shifts can prove negligible.

For example, consider the case of a radar of range resolution ΔR; to record the frequency response of a target, of dimensions smaller than or equal to ΔR (target contained in a range resolution bin), this radar must sample the frequency response with a frequency interval:

$$\delta f \leq \frac{c}{2\Delta R}$$

according to the Fourier transform relationship existing between frequency response and pulse response (see Section 6.3.5).

The effect of the Doppler shift f_d will then be negligible if:

$$f_d << \delta f$$

If v is the radar-target relative radial velocity, the Doppler shift is negligible if:

$$\frac{2v}{\lambda} << \delta f \Leftrightarrow v\Delta R << \frac{c^2}{4f_0}$$

If this relation is true—which is the case in many real situations—it is possible to adopt as invariant the modulus of the frequency response, or the autocorrelation of the complex pulse response, for which only the time shift is eliminated. This approach is described in [10], for example, applied to radar targets recognition and measurement of their aspect angle.

7.2.3.4 Specific Criteria

The approach adopted in Section 7.2.3.3 enables the major part of the useful information to be extracted. However, as is often the case in pattern recognition, it frequently proves useful to supplement the data thus obtained with a few additional features, extracted at little cost from the received signals. By removing the remaining uncertainties between sometimes very different targets, such criteria sometimes considerably improve the final performance. The definition and choice of these features are obviously closely related to the specific problem to be solved, but they are generally part of the toolbox shared by all signal processing experts. We can mention in particular:

- The mean level of the square modulus of the received signal (or, equivalently, of its spectrum);
- The length of the impulse response, or the width of the Doppler spectrum, measured in the quadratic sense according to (2.2) and (3.4) (duration and bandwidth of a signal);
- The number of zero crossings—or crossings of a set threshold—of the signal, or its derivative or the filtered signal. This approach, interesting due to its simplicity of application, is described in detail in [11].

In some situations, such a set of specific criteria, selected according to the expected targets and to the radar (or sonar) used, may prove sufficient to obtain the required performance (discrimination between complex targets and point echoes, for example).

7.2.3.5 Decision Procedures

Once the features have been selected, the appropriate decision procedure remains to be specified. This definition is the subject of pattern recognition theories, which will not be examined here in detail. These theories, now amply described and analyzed in the literature [8, 12], can be used to define procedures for comparison between a pattern composed of a vector of measurements (here, the vector of the consecutive samples composing the chosen feature: mean autocorrelation of the square modulus of the impulse response, for example), and learning patterns (features of the same nature, determined previously on various types of target). Two types of procedure, routinely used, will be described briefly: the *k* nearest neighbors procedure and the linear discrimination procedure.

Once the features have been selected, the decision can be represented as a "boundary" determination problem, illustrated in Figure 7.7 in the case in which the features consist of two values. The features are then represented by a point in a two-dimensional space (generalizing to an n-dimensional space does not raise any particular problem). Similarly, the features measured when learning on objects of different classes to recognize (tankers and aircraft carriers, for example) are represented by points in this same plane space. In this situation, the *k* nearest neighbors procedure simply consists of determining the category to which the target belongs by a majority vote among the *k* nearest neighbors of the measurement point.

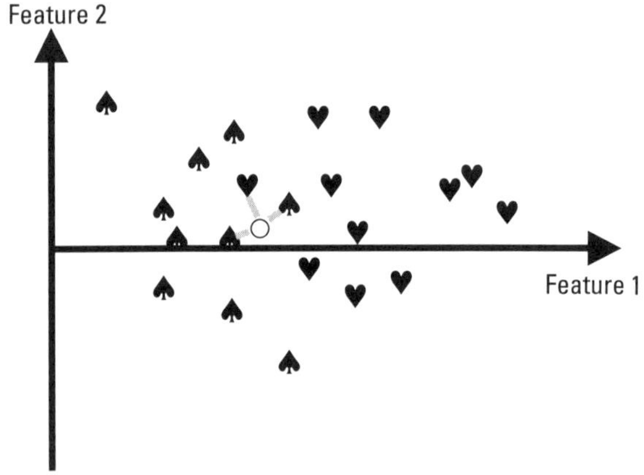

Figure 7.7 The *k* nearest neighbors recognition procedure. ○: measurement to be recognized; ♠: learning, class 1; and ♥: learning, class 2.

Similarly, the linear discrimination procedure, as shown in Figure 7.8, simply consists of determining, when learning, the equation of the straight line "best" separating the two sets of points corresponding to the two classes. Recognition is then carried out by determining on which side of this line the measurement point characterizing the unknown target is located. The straight line separating the sets can be determined, under certain assumptions (Gaussian), from the correlation matrices of the different sets of points, or by more empirical iteration procedures.

This simplistic description of the decision procedures that can be used shows that the global effectiveness will depend, to a nonnegligible extent, on the form of the set of points associated with each class, the best performance being obtained when the sets are rather homogeneous, relatively isotropic, and clearly separated. The form of the set itself depends on the chosen feature—and possibly on normalizations of this feature generally defined empirically. It also depends on the choice of the classes to recognize, some types of target being more naturally associated by the procedure than others. In other words, the choice of feature also influences the definition of the classes. This is an aspect of the well-known fact that, beyond well-established general rules (choice of features resulting from the physics of the problem, choice of decision procedures), the definition of a recognition procedure generally makes ample use of the designers' expertise and intuition. As an example, [10] describes the use of the Chi-2 metric, making it possible to normalize the frequency responses of the target and to select the most discriminating frequencies.

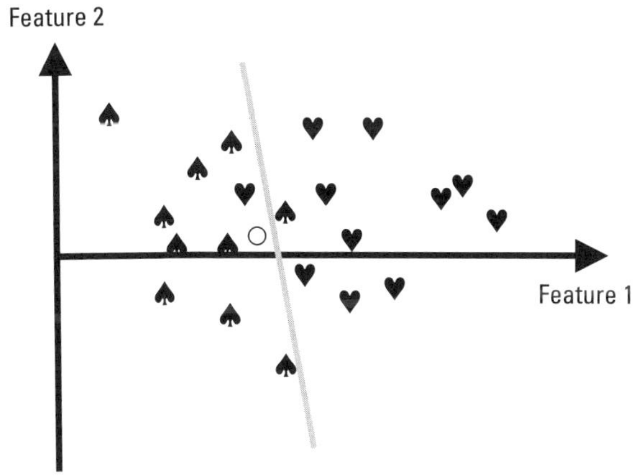

Figure 7.8 Linear discrimination recognition procedure. ○: measurement to be recognized; ♠: learning, class 1; and ♥: learning, class 2.

7.2.3.6 Other Approaches

The approach described above generated four types of features often used in radar pattern recognition, taking into account the problem (encountered in imaging) of the compensation of the target range-Doppler motion. In some cases, other features can obviously prove useful. An example is the use of target resonances, which has been proposed for recognition of targets observed at quite low frequencies (wavelength on the order of the target dimensions). This information has the useful property of being independent of the aspect angle (the received signal levels depend on the aspect angle, but the resonant frequencies do not); although this condition is not absolutely necessary for the implementation of a radar identification procedure, it is however an advantage, making it possible to reduce the size of the learning memory.

In this context, two similar approaches have been suggested [13, 14]. Both use the complex pulse response, whose Fourier transform is modeled by a rational fraction of order n (poles and zeroes) and whose poles constitute the feature.

The first approach determines the radar pulse that must be transmitted to annihilate these resonances (K-pulse), and the second simply determines the location of these poles by processing the impulse response.

Finally, let us mention another approach [15] in which, as with the K-pulse, the radar transmission is adapted (in frequency) to reveal the target characteristics.

7.3 Extended Target Tracking

We have seen in Section 3.4 that the optimum receiver for estimating the angular position, range, and velocity of a point target breaks down into an angular processing and a range-velocity processing. This enables the tracking processor to be defined in the simplified case when the transmitted signal is a pure wave $e^{2\pi j f_0 t}$. In this section we shall describe the behavior of this point target tracking processor in the situation where the target is not a point scatterer and outline a method that improves this behavior.

7.3.1 The Glint Phenomenon

According to Section 3.4, the optimum receiver in the case of a point target observed in the presence of additive white noise forms, from the signals recorded at a time t_0 on the antenna, a sum channel Σ and a difference channel Δ (derived from the previous one with respect to θ). The target direction is then estimated by the relation:

$$\hat{\theta} = -\theta_0 \frac{\mathrm{Re}(\Sigma\Delta^*)}{|\Sigma|^2} \tag{7.6}$$

For a point target, the sum and difference channels are given by:

$$\Sigma = \alpha D_\Sigma(\theta) + n_\Sigma$$
$$\Delta = \alpha D_\Delta(\theta) + n_\Delta$$

n_Σ and n_Δ are independent samples of Gaussian white noise, α is the target complex backscattering coefficient, and $D_\Sigma(\theta)$ and $D_\Delta(\theta)$ are the amplitude patterns of the antenna sum and difference channels [where $D_\Delta(\theta) = \theta_a D'_\Sigma(\theta)$, and θ_a is D_Σ mainlobe width]. In the case of a linear antenna, composed of sensors along axis Ox, over a length L, the sum pattern is:

$$D_\Sigma(\theta) = \int_{-\frac{L}{2}}^{+\frac{L}{2}} a(x) \exp\left[2\pi j \frac{x}{\lambda} \sin\theta\right] dx \tag{7.7}$$

where $a(x)$ is an aperture tapering, enabling apodisation of the antenna pattern.

Now let us consider the case of an extended target (see Figure 7.9) illuminated by a plane wave in the direction AO. The current point M of the antenna receives an echo coming from direction OM. If $q_b(\varphi)$ is the bistatic complex backscattering coefficient (illumination in direction AO, reception in direction OM) of the target, we obtain:

$$\Sigma(\theta) = \int_{-\frac{1}{2}}^{+\frac{1}{2}} a(x) q_b(\varphi) \exp\left[2\pi j \frac{x}{\lambda} \cos\alpha\right] dx$$

Elementary geometry considerations show that for small x/R:

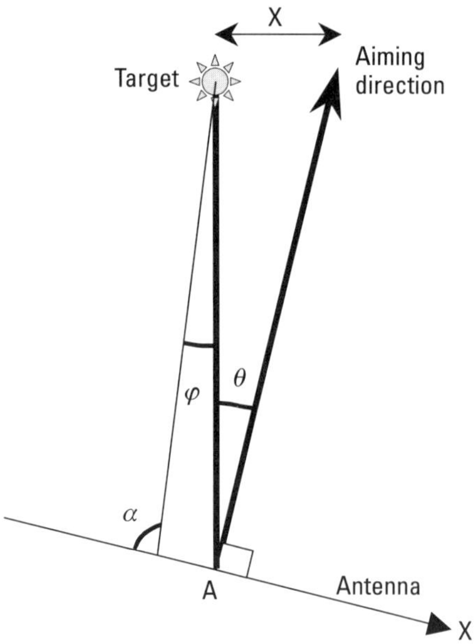

Figure 7.9 Tracking of an extended target.

$$\varphi \simeq \frac{x}{R}\cos\theta \qquad \text{and} \qquad \cos\alpha \simeq \sin\theta$$

$$\Rightarrow \Sigma(\theta) = \int_{-\frac{L}{2}}^{+\frac{L}{2}} a(x)\, q_b[(x/R)\cos\theta]\exp\left[2\pi j\frac{x}{\lambda}\sin\theta\right]dx$$

When the bistatic angle is small, the bistatic backscattering coefficient is shown to be equal to the monostatic coefficient for an observation along the bistatism angle bisector, that is, here:

$$q_b\left(\frac{x}{R}\cos\theta\right) = q\left(\frac{x}{2R}\cos\theta\right) \tag{7.8}$$

q: monostatic backscattering coefficient.

The demonstration is straightforward in the framework of multiple-scatterers approximation, where (6.34) giving the monostatic coefficient is written, in the bistatic case:

$$q_b(\vec{k}) = \int I(\vec{r}\,')e^{j(\vec{k}_i - \vec{k}_d)\vec{r}\,'}\,d\vec{r}\,' \tag{7.9}$$

where $\vec{k}_i$ is the incident wave vector, and $\vec{k}_d$ the diffracted wave vector. Equation (7.8) then results from $\vec{k}_i - \vec{k}_d = 2\vec{k}$ where $\vec{k}$ is the wave vector along the bisector of $(\vec{k}_i, \vec{k}_d)$.

If, moreover, as shown in Figure 7.10, the beamwidth of the antenna diagram is, at range R, significantly more extended than the target, that is if[2]:

$$\frac{\lambda}{2L} R >> D \tag{7.10}$$

where D is the transverse dimension of the target, it is then also true that, $\dfrac{\lambda}{2D} R >> L$, that is, that the main lobe of the target backscattering pattern (of width $X/2D$) is, at range R, much greater than the dimension L of the antenna. It is then possible to consider that $q(\varphi)$ varies slowly along the antenna, and to carry out a limited expansion of $q(\varphi)$, in the neighborhood of φ_0.

Hence

$$\Sigma(\theta) = q(\varphi_0)D_\Sigma(\theta) + \left.\frac{dq(\varphi)}{d\varphi}\right|_{\varphi_0} \frac{\cos\theta}{2R} \int_{-\frac{L}{2}}^{+\frac{L}{2}} xa(x)\exp\left(2\pi j\frac{x}{\lambda}\sin\theta\right)dx$$

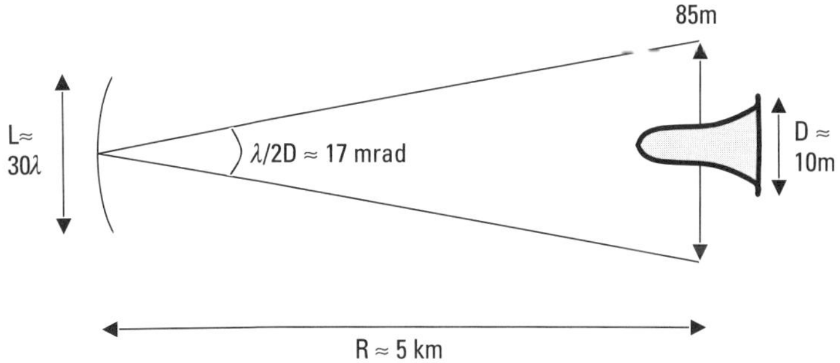

Figure 7.10 Tracking: typical configuration.

2. The beamwidth is assumed equal to $\lambda/2D$ in order to take into account the two-way antenna diagram.

or

$$\Sigma(\theta) = q(\varphi_0)D_\Sigma(\theta) + \left.\frac{dq(\varphi)}{d\varphi}\right|_{\varphi_0} \frac{\lambda}{4\pi jR}\frac{dD_\Sigma(\theta)}{d\theta} \quad (7.11)$$

This equation thus gives the expression for the signal delivered at any time by an antenna of pattern $D_\Sigma(\theta)$, when a target with backscattering coefficient $q(\varphi)$ is observed at long range, as in (7.10).

The difference channel Δ being, in the absence of noise, proportional to the derivative of the sum channel with respect to θ:

$$\Delta(\theta) = \theta_a \frac{d\Sigma}{d\theta}$$

we obtain

$$\Delta(\theta) = \theta_a q(\varphi_0)\frac{dD_\Sigma(\theta)}{d\theta} + \theta_a \left.\frac{dq(\varphi)}{d\varphi}\right|_{\varphi_0} \frac{\lambda}{4\pi jR}\frac{d^2D_\Sigma(\theta)}{d\theta^2} \quad (7.12)$$

Equations (7.11) and (7.12) thus describe the information available on the antenna at a given time. Due to the long-range approximation [enabling the limited expansion of $q(\varphi)$ in the neighborhood of φ], any other data collected over the extent of the antenna would in fact be redundant with Σ and Δ, since the field varies approximately linearly on the antenna, Σ and its derivative Δ summarize all the available information.

If the optimum receiver in the presence of a point target is then used in this extended target situation, the antenna is then servo controlled in a direction such that $\mathrm{Re}(\Sigma\Delta^*) = 0$. The resulting direction $\hat{\theta}$, in the case of a parabolic amplitude pattern (which amounts to applying the approximation that, in tracking phase, the direction pointed at remains close to the pattern maximum, that is, that the tracking is locked on target), verifies:

$$\mathrm{Re}(\Sigma\Delta^*) = \theta \Leftrightarrow \mathrm{Re}\left[q\left(1 - \frac{\theta^2}{2\theta_a^2}\right) - \frac{\lambda}{4\pi jR} q'_\varphi \frac{\theta}{\theta_a^2}\right] \quad (7.13)$$
$$\left[-q^* \frac{\theta}{\theta_a} + \frac{\lambda}{4\pi jR} q'_\varphi \frac{1}{\theta_a}\right] = 0$$

assuming

$$q(\theta) = A(\varphi)\exp[j\Omega(\varphi)] \Rightarrow q'_{\varphi} = \left(\frac{A'}{A} + j\Omega'\right)q$$

For $\dfrac{\theta}{\theta_a} << 1$, we obtain:

$$\hat{\theta} = -\frac{\lambda}{4\pi r}\frac{d\Omega}{d\varphi} \tag{7.14}$$

The correct direction being $\theta = 0$, there is thus a pointing angular error proportional to the derivative of the phase of the backscattering coefficient with respect to the aspect angle, and inversely proportional to the range R from the target to the radar. This tracking error, observed in the absence of noise when the target is not a point, is referred to as glint.

The physical reason for this tracking error is related to the nature of the wave backscattered by the target. For a point target, the backscattered wave is spherical, and consequently equivalent to a planar wavefront at the radar range (far field). To locate the target, the sole information available to the radar is the orientation of the wavefront at the place where it is located, and consequently the estimated target direction is perpendicular to this wavefront—which maximizes $|\Sigma(\theta)|^2$, and hence annulates $\mathrm{Re}(\Sigma\Delta^*)$.

When, on the contrary, the target is extended, the radiated wave is only approximately spherical. The wavefront has minor undulations that distort the angular location carried out by the radar—which cannot, at a given time, do anything other than line up in the wavefront direction, maximizing $|\Sigma(\theta)|^2$. As the target moves, and thus as its aspect angle changes, the direction of the wavefront will also change, with the passage of the undulations of the wave front on the antenna.

This phenomenon clearly appears in Figure 7.11, illustrating the angular glint of a target composed of a pair of scatterers of unequal amplitude, rotating about the median point of the segment connecting the two points. The aspect angle φ is indicated on the abscissa, and the angle error given by (7.6) is indicated on the ordinate. Two superimposed phenomena can then be observed: a slow sinusoidal fluctuation on the one hand, representing the evolution of the center of gravity of the scatterers and, on the other hand, very sharp peaks corresponding to the passage of the undulations of the wavefront, caused by interference between the scatterers (each peak corresponds to a minimum RCS, because the phase of q rotates faster when its modulus approaches zero).

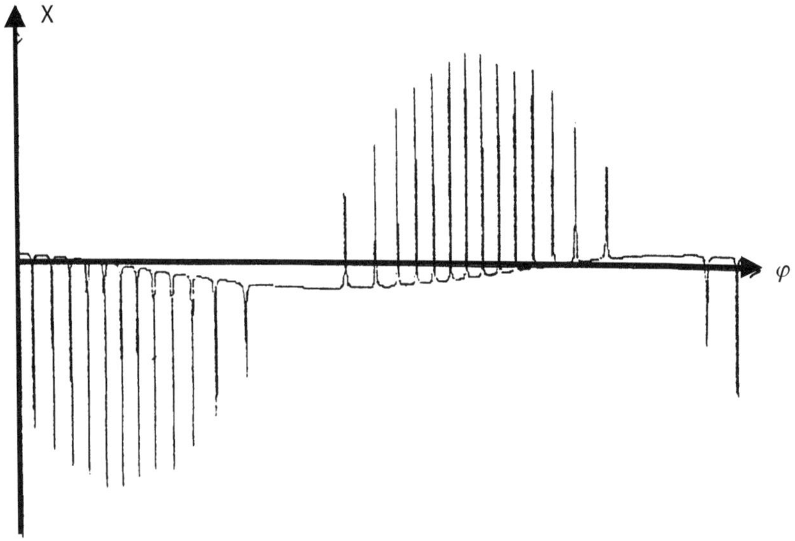

Figure 7.11 Point pair glint.

From the point of view of tracking effectiveness, these two phenomena must be distinguished because they affect the performance of the system differently; fast fluctuations are partly filtered by the later servo controls and filtering of the system, whereas slow fluctuations are not. Although this distinction is straightforward in the case of the pair of points, it is no longer obvious for a real target, for which the barycenter of the scatterers is not fixed with respect to the target (because of the angular anisotropy of the scatterers). In this case, the fluctuations of the barycenter and the fluctuations caused by the interference can occur at similar frequencies. It is then useful to carry out a spectral analysis of the statistical characteristics of these glint fluctuations. To this end, we shall use a suitable tool, namely the statistical spectral representation [16].

Let us consider the glint fluctuation signal (for a regular rotation of the target) plotted in Figure 7.12. Let s be a fixed threshold corresponding to a maximum permissible error, and d the time during which this threshold is exceeded, the unavailability period in other words; if d is expressed as a percentage of the total observation time, $d(s)$ is the distribution function—empirical—of the glint signal assumed to be random. Now, considering the same glint signal filtered with a time constant τ, it is similarly possible to plot the unavailability duration $d(s, \tau)$, as a function of the threshold s

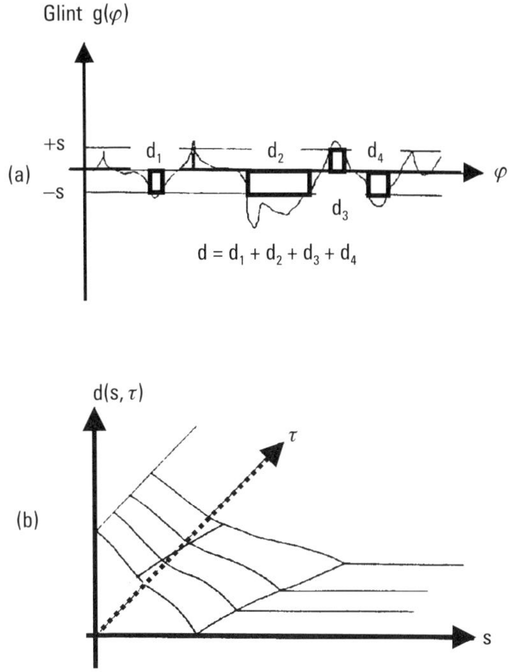

Figure 7.12 Statistical spectral representation: (a) signal, and (b) SSR.

specifying this unavailability and the time constant of the filtering applied (the term *time constant* must be understood here in the broad sense because it is related to the angular sector over which the integration is performed). This surface $d(s, \tau)$, which we will refer to as the statistical spectral representation (SSR), from the mathematical point of view, is just the empirical distribution function of the filtered random process, as a function of the filtering time constant τ.

From the user's point of view, this function has the interesting property of representing the target characteristics independently of the system observing it and of the tracking scenario. For a given observation system and a given target motion scenario, the user knows the time constant τ and can therefore read, on the SSR, the corresponding glint distribution function and hence deduce the mean and standard deviation of these tracking fluctuations. Other uses can also be considered. For a limit value of d, for example, beyond which the system is said to be unavailable, the evolution of the unavailability duration as a function of the usable time constant (section at a given s) can be read on the SSR.

Finally, note that depending on the application considered, various similar representations can be deduced from this SSR. For example, by derivation with respect to s, the probability density of the filtered signal as a function of the filtering time constant can be obtained. This representation can certainly be used profitably to analyze the spectral content of random colored signals, and even to synthesize the random processes of any statistical and spectral characteristics.

As an illustration, Figure 7.13 shows the results obtained with a real aircraft target observed at 3 GHz. The glint fluctuations are plotted in Figure 7.13(a) as a function of the aspect angle, and the SSR of these glint fluctuations is plotted in Figure 7.13(b).

One can note in particular a rather marked plateau effect, corresponding to the threshold values lying between the maxima of the highest glint peaks and all the maxima of the mean amplitude peaks. The unavailability duration then slowly varies with the threshold level because the signal slope is very steep for strong and narrow peaks. This example (corresponding to a phenomenon frequently observed in practice) shows that the SSR reveals a signal characteristic that is both important for the user and difficult to reveal with conventional means.

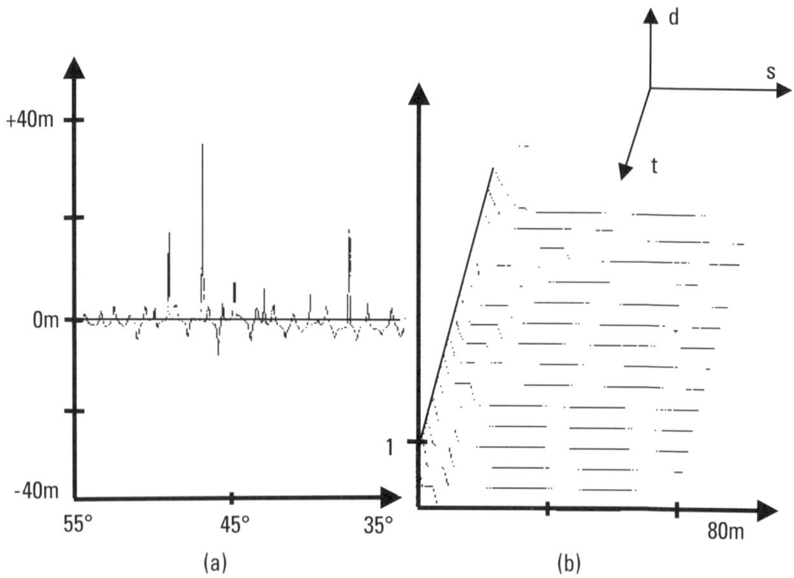

Figure 7.13 Analysis of the glint of an aircraft: (a) glint; and (b) SSR.

7.3.2 Glint Modeling

If the target is modeled by the random model described in Section 6.2.3 [$q(\varphi)$: random process], the glint probability density is, according to (7.14), that of the derivative of the phase of the corresponding Gaussian process (that is, to the nearest 2π, that of the instantaneous frequency of this process).

Defining $q(\varphi)$:

$$q(\varphi) + A(\varphi)e^{j\Omega(\varphi)} \Rightarrow q'(\varphi) = q(\varphi)\left(\frac{A'(\varphi)}{A(\varphi)} + j\Omega'(\varphi)\right)$$

$$\Rightarrow \Omega'(\varphi) = \operatorname{Im}\frac{q'(\varphi)}{q(\varphi)} \tag{7.15}$$

It is shown in Appendix 7B that if $S(f)$ is the power spectral density of $q(\varphi)$, the probability density of $\Omega'(\varphi)$ is a Student law:

$$p_{\Omega'}(\Omega') = \frac{1}{4\pi\Delta f}\left[1 + \frac{(\Omega' - 2\pi f_0)^2}{(2\pi\Delta f)^2}\right]^{-3/2}$$

f_0 being the mean frequency and Δf the mean-square bandwidth of the power spectrum of $q(\varphi)$:

$$f_0 = \frac{\int fS(f)\,df}{\int S(f)\,df}$$

$$(\Delta f)^2 = \frac{\int (f - f_0)^2 S(f)\,df}{\int S(f)\,df}$$

Taking into account (6.45), previously established:

$$f_0 = \frac{2}{\lambda}y_0$$

$$\Delta f = \frac{2}{\lambda}\Delta y$$

where y_0 and Δy respectively denote the mean transverse abscissa (transverse coordinate of the center of gravity of the scatterers distribution) and the transverse extent (transverse width of the scatterers distribution) of the target.

Returning to the definition of glint g, (7.14):

$$g = \frac{\lambda}{4\pi R} \Omega'_\varphi$$

the glint probability density is obtained:

$$p_g(g) = \frac{R}{2\Delta y\left[1 + \frac{(g - y_0/R)^2}{(\Delta y/R)^2}\right]^{3/2}} \tag{7.16}$$

The glint therefore follows a Student law, of mean value y_0/R, which means that, on average, the radar is pointed to the centroïd of the target scatterers.

The glint standard deviation is infinite (divergent integral)—in practice, the filtering imposed by the tracking loop bandwidth leads to a finite standard deviation. However, the dispersion of g about its mean value can still be characterized by the quantity $E|g - E(g)|$. We obtain:

$$E|g - E(g)| = \Delta y/R \tag{7.17}$$

Here again, the interpretation of the results obtained is simple: The glint "dispersion" (measured by the mean of the absolute value of the deviation from the mean) is proportional to the width of the scatterers transverse distribution.

Finally, it is worth noting that the glint probability density depends on the characteristics of the target by two parameters only:

- y_0, the barycenter of the scatterers;
- Δy, the width of the scatterers transverse distribution.

In particular, neither their arrangement in depth nor their relative phases has any influence on the glint probability density.

This glint model, based on the stochastic modeling of the targets described in Section 6.2.3, generally leads to a rather exact description of the behavior of the tracking radars and proves in any case more accurate than that obtained by the models of a few scatterers sometimes used.

7.3.3 Antiglint Processing

Various techniques have been suggested to reduce this tracking error caused by the target extension. One way or the other, they generally amount to the integration of the received signals in order to filter this error. Indeed, the received signals instantaneously comply with the system of two complex equations, (7.18) and (7.19), with five real unknown quantities: q, q' and θ:

$$\Sigma = qD + kq'_{\varphi} D'_{\theta} \tag{7.18}$$

$$\Delta = \theta_a qD'_{\theta} + k\theta_a q'_{\varphi} D'_{\theta} \tag{7.19}$$

A way of solving the problem consists of applying the Σ and Δ measurements to consecutive time samples. For example, by eliminating q'_{φ} between the two equations, then differentiating the equation obtained with respect to time in order to get an additional equation, and eliminating q, the following relation is obtained:

$$(\Sigma\theta + \Delta\theta_a)\left(-\frac{\varphi}{k} + \frac{\dot{\theta}}{\theta_a}\right) = \dot{\Sigma} - \theta\dot{\Delta} \quad (\dot{\Sigma}\text{: time derivative of } \Sigma) \tag{7.20}$$

This equation system relating θ, $\dot{\theta}$, and φ is the base of some techniques of antiglint processing; it consists of two real equations with three unknown variables (θ, $\dot{\theta}$, φ). Elimination of $\dot{\varphi}$ between these two equations gives:

$$-\theta^2 \mathrm{Re}(\dot{\Delta}\Sigma^*) + \theta \mathrm{Re}(\dot{\Sigma}\Sigma^* - \theta_a^2 \dot{\Delta}\Delta^*) + \theta_a^2 \mathrm{Re}(\dot{\Sigma}\Delta^*) = \frac{\dot{\theta}}{\theta_a^2} |\Sigma\theta + \Delta\theta_a^2|^2$$

Taking into account (7.18) and (7.19), it can be shown that the second member of this equation is on order k^2 and is therefore negligible, so that the equation giving the ecartometry θ is written:

$$-\theta^2 \mathrm{Re}(\dot{\Delta}\Sigma^*) + \theta \mathrm{Re}(\dot{\Sigma}\Sigma^* - \theta_a^2 \dot{\Delta}\Delta^*) + \theta_a^2 \mathrm{Re}(\dot{\Sigma}\Delta^*) = 0 \tag{7.21}$$

Similar results can be obtained with a simplified relation, neglecting terms in θ^2:

$$\theta = -\theta_a^2 \frac{\mathrm{Re}(\dot{\Sigma}\Delta^*)}{\mathrm{Re}(\dot{\Sigma}\Sigma^*)} \tag{7.22}$$

Figure 7.14 shows the result of this ecartometry, applied to the tracking of a target made of three scatterers (overall dimension: 3m), at 5 km, with an antenna beamwidth of 3°. Conventional ecartometry (top) is compared with the ecartometry given by (7.22) (bottom), showing a significant reduction in target glint (reduced to very short spikes).

Another approach, which was described in Section 5.5.5, consists of representing the extended target as a point target to which colored noise is added (correlated between sensors). It has been shown that in this case, the optimum processing leads to the estimator $\hat{\theta}$:

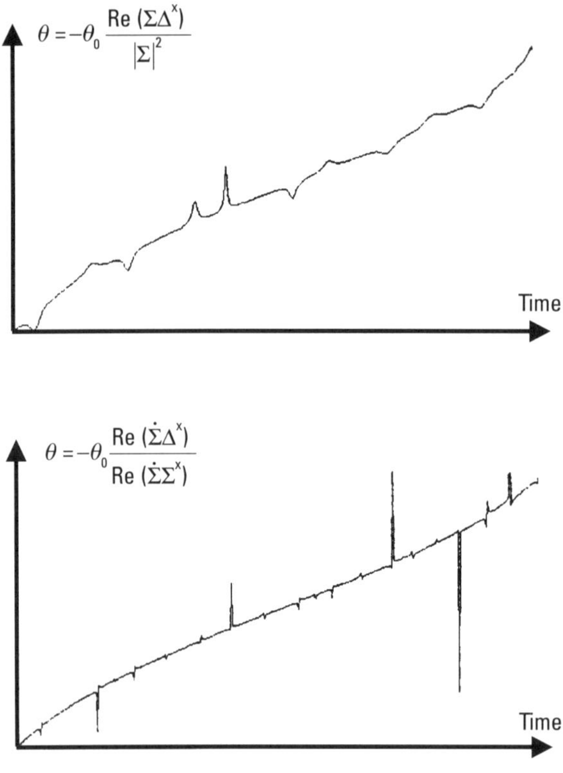

Figure 7.14 Antiglint processing.

$$\hat{\theta} = -\theta_a \frac{\text{ReE}(\Sigma\Delta^*)}{\text{E}|\Sigma|^2} \tag{7.23}$$

which separately averages the numerator and denominator of the usual angle-error measurement (this technique can also be used in frequency agility, averaging being carried out on the measurements made at the consecutive transmitted frequencies). This extremely simple procedure is also highly effective when the integration time is long enough.

Of course, this brief outline only gives a very general idea of the principles used to reduce the glint effect. In order to analyze the phenomenon in further detail, the Doppler-range processing must also be taken into account because, in the case of an extended target, this information is no longer independent of the angular information. This Doppler-range processing can for example be used to resolve the different scatterers of the target, as seen in Section 6.3.5 (pulse response and transverse response) and thus to proceed from the situation of an extended target in a range-Doppler bin to a situation of quasi-point targets distributed over several range-Doppler bins.

7.4 Target Extraction from Clutter

In theory, the clutter can be considered as an external interference of a random nature, and thus taken into account by the optimum processing in colored noise described in Chapters 4 and 5.

More particularly, an adaptive receiver capable of detecting targets in the presence of clutter by time filtering was described in Section 5.4.3; similarly, adaptive receivers using spatial filtering (sensor array) were described in Section 5.5. However, sometimes this optimum processing cannot be used, either because the radar does not have the required characteristics (pulse coherence, array antenna), or for simplicity. In such a situation, the receiver used is the optimum receiver in the presence of white noise—or a good approximation of this receiver—followed by a postdetection processing device which takes the clutter into account.

In this section we shall examine a typical example of such an approach: the use of frequency agility. Then we shall take polarization into account, as an additional element making it possible to discriminate the target from the clutter.

7.4.1 Use of Frequency Agility After Detection

For example, consider the case of a radar scanning the ground in search of fixed point targets. Let us assume that the receiver of this radar first carries out the optimum processing, with the assumption of white noise: matched filtering, sampling, and Doppler filtering, then quadrature envelope detection; the result of such processing is, for zero Doppler, a range profile, for a given direction of observation, composed of a real signal that is a function of the range $r(d)$ (the same type of processing could be carried out on the series of observations at constant range, as a function of the observation bearing). Strictly speaking, the problem is here again two dimensional (angle-range), but we reduce it to a one-dimensional problem in order to simplify the operation of the radar; each direction of observation is thus processed independently of the neighboring directions.

In the presence of clutter echoes, this range response of course has several local maxima, representing the various echoes from the clutter at the various ranges. These echoes will be modeled as random, following a log-normal law (the logarithm of the clutter RCS is therefore assumed to be normally distributed). Although it is obviously not universal, this model often provides a reasonable approach to reality—while still remaining analytically tractable.

In order to improve the detection of a possible target, it is obviously desirable to reduce these fluctuations due to the clutter. A means of achieving this is the use of frequency agility, that is, illuminating every plot of land successively at different frequencies and averaging the obtained results. If the range resolution is ΔR, the echoes received at frequencies separated by $c/2\Delta R$ can be considered independent [sampling criterion of a frequency response, for a target of depth ΔR, as indicated by (6.37)]. A frequency scan over a band Δf thus gives of $2\Delta f \Delta R/c$ samples, assumed to be independent.

In order to model detection, it is necessary to make an assumption on the law of clutter echo fluctuation (magnitude) as a function of the transmitted frequency. We shall assume that this law is a Rayleigh distribution (or exponential distribution for the RCS), which amounts to making the assumption of a large number of scatterers in a radar resolution bin. This Rayleigh distribution is characterized by the parameter z, itself obeying a log-normal law, according to the preceding considerations.

If y is the RCS of an area of clutter, the probability density of y is thus given by:

$$h(y) = \int_z p(y/z) g(z)\, dz \tag{7.24}$$

where

$$p(y/z) = \frac{1}{z} e^{-y/z} \tag{7.25}$$

(RCS fluctuations following an exponential distribution with frequency),

and

$$g(z) = \frac{1}{\sigma_z \sqrt{2\pi z}} \exp\left[-\frac{\ln^2(z/m_z)}{2\sigma_z^2} \right] \tag{7.26}$$

(log-normal fluctuations with the spatial variable z).

Moreover, we assume that the detection chain includes a logarithmic amplifier, which is often the case with this type of radar, for better CFAR operation in clutter (see Section 3.3.8); the output variable is then $u = \ln z$. Let us calculate the standard deviation σ_u.

The mean value is:

$$E(u) = \iint \ln y p(y/z) g(z)\, dy\, dz$$

and hence the variance:

$$\sigma_u^2 = \iint (\ln y)^2 \frac{1}{z} e^{-y/z} g(z)\, dy\, dz - [E(u)]^2$$

Defining $y' = y/z$ and $z' = z$, we obtain: $dy'\, dz' = (1/z)\, dy\, dz$

$$\Rightarrow \sigma_u^2 = \sigma_1^2 + \sigma_z^2 \tag{7.27}$$

in which σ_1^2 is the variance of the logarithm of an exponential variable, and σ_z^2 the variance of the log-normal variable.

It is shown in Appendix 7A that σ_1^2 is a constant depending in particular on the selected logarithm base. Indeed, this important result is one of the justifications for the use of logarithmic amplifier receivers. When observing a target—or clutter—composed of a large number of scatterers, the standard deviation of the fluctuations at the output does not depend on the target RCS level.

As a result, the variance σ_u^2 of the output fluctuations for this model of clutter (Rayleigh in frequency, log-normal in space) is:

$$\Rightarrow \sigma_u^2 = \sigma_z^2 + k \tag{7.28}$$

It thus appears that the spatial and frequency variances are added in this model.

Let us therefore assume that the receiver (after quadratic detection) averages consecutive received samples, at the different frequencies used, and for every resolution cell of the radar:

$$v = \frac{1}{N}\sum_{i+1}^{N} \ln(y_i \text{ if } z)$$

z: parameter of the Rayleigh distribution for the resolution element observed.

$$m_v = m_u = \mathbf{E}[\ln(y \text{ if } z)] = \int \ln y p(y \text{ if } z)\,dy$$

$$= \int (\ln y)\frac{1}{z}e^{-y/z}dy = \int (\ln zy')e^{-y'}dy' \text{ with } y' = \frac{y}{z}$$

Let

$$m_u = \ln z - C = m_v \text{ where } C \text{ is Euler's constant} \tag{7.29}$$

Furthermore, it has been established that the variance of $\ln y/z$ is k (variance of the logarithm of a Rayleigh variable). Consequently, the variance of v is:

$$\sigma_v^2 = \frac{k}{N} \tag{7.30}$$

N being the number of independent frequencies used.

To calculate the total variance, we shall note that the variable v, conditional to z, has a normal distribution $\mathbf{N}(v/z)$ (as the sum of a large number of independent identically distributed variables):

$$\mathbf{N}(v/z) = \frac{1}{\sigma_v\sqrt{2\pi}}e^{-\frac{(v-m_v)^2}{2\sigma_v^2}}$$

The total variance is then given by:

$$\sigma^2 = \iint \frac{v^2}{\sigma_v\sqrt{2\pi}} e^{-\frac{(v - m_v)^2}{2\sigma_v^2}} g(z)\, dz\, dv$$

$$- \left[\iint \frac{v^2}{\sigma_v\sqrt{2\pi}} e^{-\frac{(v - m_v)^2}{2\sigma_v^2}} g(z)\, dz\, dv\right]^2$$

Defining $t = \dfrac{v - m_v}{\sigma_v}$, obtain:

$$\sigma^2 = \iint (\sigma_v t + m_v)^2 \frac{e^{-t^2/2}}{\sqrt{2\pi}} g(z)\, dz\, dt$$

$$- \left[\iint (\sigma_v t + m_v) \frac{e^{-t^2/2}}{\sqrt{2\pi}} g(z)\, dz\, dt\right]^2$$

$$\sigma^2 = \sigma_v^2 + \int m_v^2 g(z)\, dz - \left[\int m_v g(z)\, dz\right]^2$$

with $m_v = \log z - C = m_u$

$$\sigma^2 = \sigma_v^2 + \int (\ln z)^2 g(z)\, dz - \left[\int \ln z\, g(z)\, dz\right]^2$$

$$\sigma^2 = \sigma_z^2 + \frac{k}{N} \tag{7.31}$$

This important result shows that, in this model and for this logarithmic receiver, the spatial and frequency variances are added, and that frequency agility enables the frequency variance to be divided by the number of independent frequencies used, $2\Delta f \Delta R / c$, while leaving the spatial variance unchanged. This result corresponds to a physical reality more generally than the simple field of validity of the clutter model used here: the use of frequency agility indeed reduces the variance of the clutter fluctuations, but a residual variance remains that only a spatial average can reduce.

7.4.2 Use of Polarimetry [17]

The survey of the polarization properties of the wave backscattered by a body showed (see Section 6.2.1) that these properties contain certain information on the nature of the target. It is then natural to try to exploit those properties to discriminate between the target echoes and the background clutter. We are going to see that such an exploitation can be done essentially on the basis of the statistical characteristics of the polarization of the received wave.

7.4.2.1 Definitions

The polarization of the wave received from a moving target varies over time because of two phenomena:

1. The evolution over time of the target aspect angle changes the relative phase shift between scatterers, in a given resolution cell, and thus changes the phase difference between the two polarized components of the wave, and hence its polarization.
2. This change in the aspect angle also modifies the basic backscattering matrices of the various reflectors (since these matrices depend on the aspect angle, as established in Section 6.2.1), and hence also the global backscattering matrix of the target.

The same phenomena occur on reflection from the clutter, at different scales. For these reasons, the wave received has a random nature inasmuch as its components in a given base—for example *H* and *V* linearly—have a power and a correlation that change with the geometrical configuration. If *x* and *y* are the two components of the wave received, the correlation matrix **J** of this wave is then defined by:

$$\mathbf{J} = \begin{pmatrix} J_{xx} & J_{xy} \\ J_{yx} & J_{yy} \end{pmatrix} \text{ with: } J_{xx} = \mathrm{E}|x|^2;\ J_{yy} = \mathrm{E}|y|^2;\ J_{xy} = \mathrm{E}(xy^*);\ J_{yx} = \mathrm{E}(yx^*) \tag{7.32}$$

The energy of the received wave is $J_{xx} + J_{yy}$; the trace of matrix **J** (invariant by any basis change, for example by changing from the linear polarization basis to the circular polarization basis) and the correlation coefficient between the two components is:

$$\mu_{xy} = \frac{J_{xy}}{\sqrt{J_{xx} J_{yy}}} \tag{7.33}$$

The analysis of such waves, known as "quasi-monochromatic," was carried out by Born and Wolf [18], who defined in particular the notion of an unpolarized wave. An unpolarized wave is a wave such that the intensity of its component in any direction orthogonal to the propagation direction is the same, and this intensity is not affected by the delay of one of the components relative to the other. It is easy to show that these conditions imply:

$$J_{xx} = J_{yy}\,;\; J_{xy} = J_{yx}^{*} = 0$$

A fully polarized wave, on the contrary, is defined as a wave such that $\det \mathbf{J} = 0$. This is in particular the case with a pure, monochromatic wave, such as the one described in Section 6.2, since if a_x and φ_x, a_y, and φ_y are respectively the amplitudes and phases of the x and y components of such a wave, $\mathbf{J}$ is given by:

$$\mathbf{J} = \begin{pmatrix} a_x^2 & a_x a_y e^{j(\varphi_x - \varphi_y)} \\ a_x a_y e^{-j(\varphi_x - \varphi_y)} & a_y^2 \end{pmatrix} \Rightarrow \det \mathbf{J} = 0 \qquad (7.34)$$

$\det \mathbf{J}$ being invariant by coordinate change, the "polarized" nature thus defined is indeed characteristic of the wave itself. The correlation coefficient of such a wave is simply given according to (7.31):

$$\mu_{xy} = e^{2j(\varphi_x - \varphi_y)} \qquad (7.35)$$

and consequently has a modulus of 1.

The modulus of the correlation coefficient between components is thus maximum for a fully polarized wave and zero for an unpolarized wave. This quantity could thus characterize the degree of polarization of the wave. However, it has an important drawback: it depends on the choice of the basic polarizations (circular, linear) defining the components x and y.

To define the degree of polarization of a wave intrinsically, it is first shown that any wave, of correlation matrix $\mathbf{J}$, can be broken down in a unique way into a sum of two independent waves of matrices $\mathbf{J}_1$ and $\mathbf{J}_2$:

$$\mathbf{J} = \mathbf{J}_1 + \mathbf{J}_2 \qquad (7.36)$$

where $\mathbf{J}_1$: correlation matrix of an unpolarized wave:

$$\mathbf{J}_1 = \begin{pmatrix} A & 0 \\ 0 & A \end{pmatrix} \qquad (7.37)$$

and $\mathbf{J}_2$: correlation matrix of a fully polarized wave:

$$\mathbf{J}_2 = \begin{pmatrix} B & D \\ D^* & C \end{pmatrix}, \text{ with } BC - |D|^2 = 0 \tag{7.38}$$

The solution of the system formed by (7.34) to (7.36) actually leads to the single solution:

$$\begin{aligned} A &= \frac{1}{2}(J_{xx} + J_{yy}) - \frac{1}{2}\sqrt{(J_{xx} + J_{yy})^2 - 4 \det J} \\ B &= \frac{1}{2}(J_{xx} - J_{yy}) + \frac{1}{2}\sqrt{(J_{xx} + J_{yy})^2 - 4 \det J} \\ C &= \frac{1}{2}(J_{yy} - J_{xx}) + \frac{1}{2}\sqrt{(J_{xx} + J_{yy})^2 - 4 \det J} \\ D &= J_{xy} \end{aligned} \tag{7.39}$$

It is then natural to define the *degree of polarization* of the wave as the ratio of the intensity of the polarized wave to the total intensity:

$$p = \frac{\text{Trace }(\mathbf{J}_2)}{\text{Trace }(\mathbf{J})} = \sqrt{1 - \frac{4 \det \mathbf{J}}{(J_{xx} + J_{yy})^2}} = \sqrt{1 - \frac{4(J_{xx}J_{yy} - |J_{xy}|^2)}{(J_{xx} + J_{yy})^2}} \tag{7.40}$$

This quantity is intrinsic [because $\text{Tr}(\mathbf{J}_2)$ and $\text{Tr}(\mathbf{J})$ are invariant by change of coordinates] and included between 0 (in the case of an unpolarized wave) and 1 (in the case of a polarized wave).

This degree of polarization is simply related to the average Stokes parameters of the wave (defined in Section 6.2.1 in the case of a monochromatic wave):

$$\mathbf{g} = \begin{bmatrix} g_0 \\ g_1 \\ g_2 \\ g_3 \end{bmatrix} = \begin{bmatrix} \text{E}|H|^2 + \text{E}|V|^2 \\ 2\,\text{Im}[\text{E}(HV^*)] \\ \text{E}|H|^2 - \text{E}|V|^2 \\ 2\,\text{Re}[\text{E}(HV^*)] \end{bmatrix}$$

where $g_0^2 = g_1^2 + g_2^2 + g_3^2$ for a polarized wave (Section 6.2.1).

By definition of the unpolarized wave, such a wave is characterized by $g_1 = g_2 = g_3 = 0$. The intensity of the polarized part of the wave is therefore:

$$\sqrt{g_1^2 + g_2^2 + g_3^2}$$

and the degree of polarization is then written, in terms of mean Stokes parameters:

$$p = \frac{\sqrt{g_1^2 + g_2^2 + g_3^2}}{g_0} \tag{7.41}$$

This parameter is a fundamental characteristic of the waves received from fluctuating targets or clutters. For a given illumination time, the estimation of this parameter is a means of discriminating between fluctuating clutter and stationary targets, for example.

7.4.2.2 Properties of Targets and Backgrounds [18]

Given the extreme variability of the characteristics of the targets and of the backgrounds, it is difficult to summarize all their polarimetric properties. Some of them are however worth noting:

- For targets, the intensity of the received signal in cross-linear polarization (modulus of term S_{HV}) is generally lower than the intensity of the received signal on the copolarization (8 dB lower is generally an appropriate order of magnitude). This can be explained simply by the fact that the depolarizing elements are generally ridges or wires of lower RCS on average than that of nondepolarizing elements. In circular polarization, the cross-polarization echo is higher than the echo in copolarization (more single reflections than double ones), by a few decibels approximately. Moreover, the degree of polarization of targets is generally high but depends on the duration used for averaging [18].
- For clutter, the relations between the echoes received in cross polarization and echoes received in copolarization are, in order of magnitude, similar to those described for the targets. For sea clutter, the term S_{HH} is generally somewhat lower than the term S_{VV}. For rain, finally, the echo in cross circular polarization is much higher than the echo in copolarization; this is one of the most widely used properties of polarized radar returns.

These very general comments must, of course, be expressed with slight differences for each particular case, taking into account the durations of observation, the radar resolution cells, and the particular types of target and background.

7.4.2.3 Processing of Polarization

From a theoretical point of view, the radar can control the polarization of the wave it transmits, and receive on two polarization channels to analyze the wave received. By combining these transmitted and received polarizations with different waveforms, with different reception processing (white noise, colored noise), for various types of targets, it is possible to imagine a wide variety of different operating principles, some of which are only being studied at present. Instead of listing all the types of processing described in the literature [18], we shall describe two significant examples: adaptive canceling in polarization, and polarization encoding.

Adaptive cancelling in polarization [19] aims at improving the target/clutter contrast by using polarization analysis on reception only, in order not to make the radar excessively complex (this analysis already implies that the receiver has two parallel analysis channels).

The block diagram is given in Figure 7.15. We can recognize the diagram of the adaptive antenna by subtraction of colored noise, if s_2 is a signal incorporating a noise reference, and if the gain W is given by $W = \dfrac{\mathrm{E}(s_1 s_2^*)}{\mathrm{E}|s_1|^2}$ or, in adaptive mode:

$$W_n = W_{n-1} + 2\mu x^*(n) s_2(n)$$

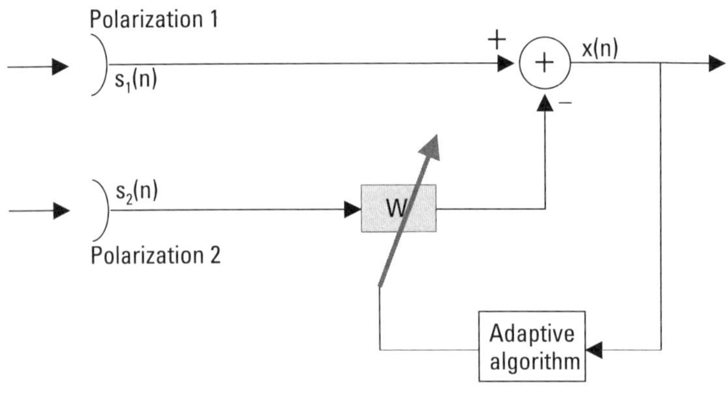

Figure 7.15 Adaptive canceling in polarization.

In comparison to the assumptions with which the adaptive antenna was constructed in Chapter 5, this process seems to be effective when:

- Channel s_2 is a good noise reference, that is, the target RCS in crossed polarization is very low in relation to that in transmission polarization;
- The noise can be considered stationary and uncorrelated with the signal received from the target.

It can be seen that this is essentially an empirical adaptation of a known device (adaptive antenna), mainly justified by analogy; this device however has the advantage of simplicity, and has been used for the canceling of rain clutter. Adaptation is then carried out from one range cell to the next, on a pointing direction of the ground radar antenna, and increases the target/clutter contrast by several decibels. Similar processes can be imagined, on the basis of optimum processing defined in Chapter 5, for various models of targets and backgrounds.

In polarization encoding [20], polarization is used at transmission and reception in order to encode the transmitted pulse in the same way as by phase encoding (Chapter 3). The wave transmitted is thus composed of a series of waves polarized according to a given code, in right-hand or left-hand circular polarization, for example. If, for this principle analysis, only zero-Doppler targets are considered, the receiver performs the correlation between the vector (two polarization components) composing the transmitted wave and the vector composing the received wave. Qualitatively, this correlation has a sharp maximum at the target range if the target is a highly polarized or little polarized reflector. If, on the contrary, the received signal is of an intermediate polarization (right-hand and left-hand components of equivalent intensities), the result of the correlation will show no clear maximum.

According to this principle, the radar benefits from the "pulse compression" only on targets returning a right- or left-hand circularly polarized echo. On more complex targets, the echo received will not be compressed, and hence not detected. This principle therefore improves the accuracy of the range measurement of simple objects, while increasing their contrast with more complex ambient echoes. This improvement is, however, associated with a considerable complexity of the radar at both transmission and reception.

7.4.2.4 Comment

The use of polarization can also be associated with the target recognition procedures described in Section 7.1.3, in order to discriminate between stationary targets and ground clutter and therefore to improve the detection probability of stationary targets at a given false alarm probability. However, it is reasonable to expect recognition rates of only approximately 90% in the best cases—such a rate is an excellent performance for a pattern recognition procedure—which would enable rejection of nine false alarms out of 10 due to clutter, that is, a gain of an order of magnitude in the probability of false alarm due to clutter.

7.4.3 Ambiguous Airborne Radars

In the case of airborne radars, spurious echoes returned by the ground limit the detection of possible air targets flying at low altitude and thus observable by looking downward. We are going to see that the attempt to define a system capable of performing this "look-down detection" function leads to the definition of radars that are ambiguous both in range and in velocity. The analysis will be based on the example of the fighter aircraft radar trying to detect air targets in the front sector (see [1, 21]).

7.4.3.1 Clutter and Ambiguity [1]

Let us consider the case of the radar shown schematically in Figure 7.16. Whatever the type of antenna (parabolic reflector, slotted array, electronic scanning array), its pattern—defined in Section 3.5.2 as the response of the antenna to a point target in direction θ relative to the antenna line of sight—is not perfect and in particular includes sidelobes through which the spurious echoes arrive, originating from directions other than the pointed direction.

Thus, for a given aiming direction and a given range-velocity bin, the level of the received signal (after optimum white noise processing: correlation

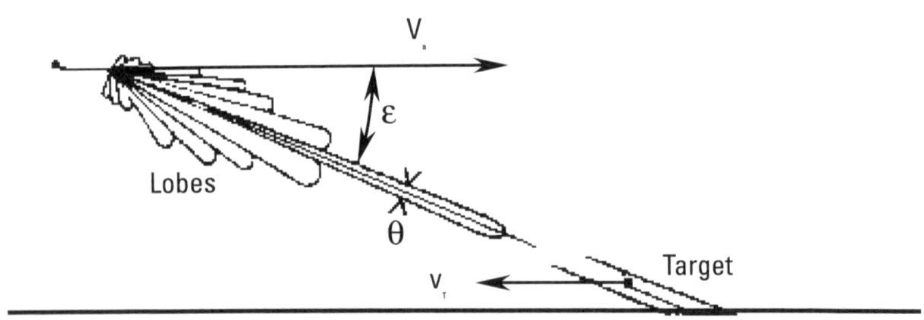

Figure 7.16 Airborne radar.

with the replica corresponding to the range-Doppler bin under consideration, quadratic detection, and normalization with respect to the replica power) results from the phase and amplitude addition of the echoes received from the angle-range-Doppler bin actually observed with the echoes received from the other directions, at the same range and same Doppler, weighted by the antenna sidelobes corresponding to these directions. (The range-Doppler ambiguity function is assumed to be perfect; that is, it has only one peak at $\tau = \nu = 0$, in order to simplify the following reasonings.) The spurious echoes coming from other directions will of course modify the result, and sometimes exceed the detection threshold when no target is present in the antenna main lobe. Let us examine the distribution of these spurious contributions due to clutter, for a given aiming direction of the antenna, versus range and Doppler.

In range first (see Figure 7.17), the echo received from the clutter by the main lobe is of course rather strong (range $h/\sin\epsilon$). As the range increases, the echo received by the sidelobes eventually decreases because incidence becomes grazing (refer to Section 6.5.1).

As the range decreases, the echo received from the clutter increases because the angle of incidence increases (see the sine law of Section 6.4.1), but it is seen through sidelobes of decreasing level. The level observed remains—very roughly—constant up to a range h, for which the ground is observed vertically. The related echo, even seen through weak sidelobes, is generally strong (altitude return). At ranges shorter than h, obviously no echo can originate from the ground clutter.

In Doppler, for an aircraft flying horizontally at velocity v, the echoes due to ground clutter appear as shown in Figure 7.18 at Doppler values

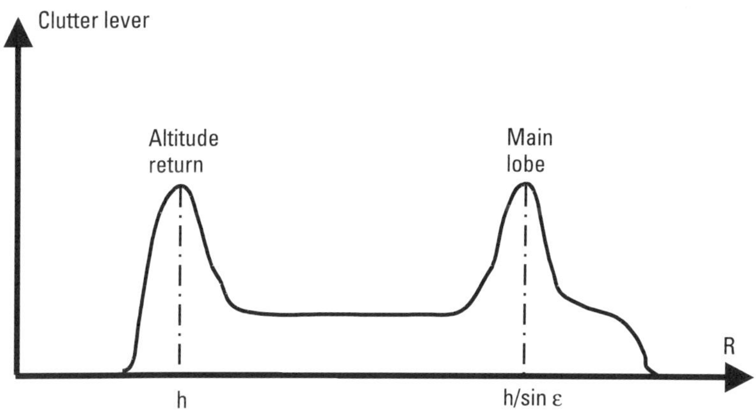

Figure 7.17 Clutter distribution in range.

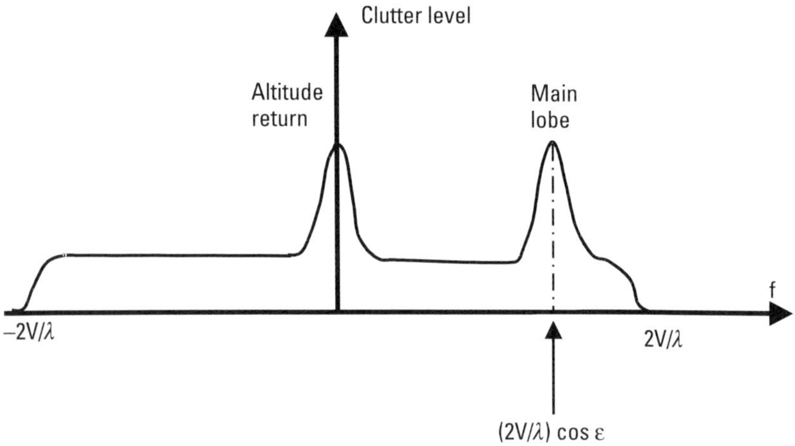

Figure 7.18 Clutter distribution in Doppler.

ranging from $-2v/\lambda$ (echoes coming from the rear horizon, hence weak) and $+2v/\lambda$ (echoes coming from front horizon). The same argument as that applied in the case of the range response shows that the clutter echoes will be much stronger at the Doppler shift corresponding to the intersection of the line of sight and the ground [Doppler $(2v/\lambda)\cos\epsilon$], and at zero Doppler corresponding to the altitude return.

Finally, the Doppler-range map of the "average" clutter as displayed by the radar, after passing through the antenna, is shown in Figure 7.18 for a given antenna pointing direction (side-looking on this figure). The echoes coming through the main lobe (here at zero Doppler) and the first sidelobes, as well as the altitude return, are clearly apparent. The problem faced by the radar is to detect the possible targets, superimposed on this Doppler range map, at a priori any ranges and Doppler shifts. This situation is obviously quite different from the case of the point target in the presence of white noise, examined in Chapter 3, which explains why specific processing techniques must be implemented.

The conventional method of countering these clutter echoes consists of distinguishing them from targets by their immobility. For a ground-based radar, these echoes are mainly observed in the neighborhood of zero Doppler (fixed echoes); for a mobile radar, their Doppler can be known if the carrier velocity is known or measured. However, this principle does not make it possible to suppress the echoes received by the sidelobes. Moreover, the phenomenon is further complicated by the fact that the radar generally transmits pulse trains, at a given repetition frequency. It has been seen in

Section 3.3.7 that the use of this type of a signal results in a range ambiguity D_a and a velocity ambiguity V_a such that $D_a V_a = c\lambda/4$.

Because of the radar ambiguities, the clutter map shown in Figure 7.19 actually seems to be folded back on itself; this folding consists of superimposing consecutive portions of range extent D_a and velocity extent V_a. The result of this folding is shown in Figure 7.20, for three different medium pulse repetition frequencies. The folding leads to two major consequences:

1. The Doppler shift and range of the possible targets are now only measured to within a certain ambiguity; we shall see that this indeterminacy can be removed by the transmission of consecutive trains at different repetition frequencies.

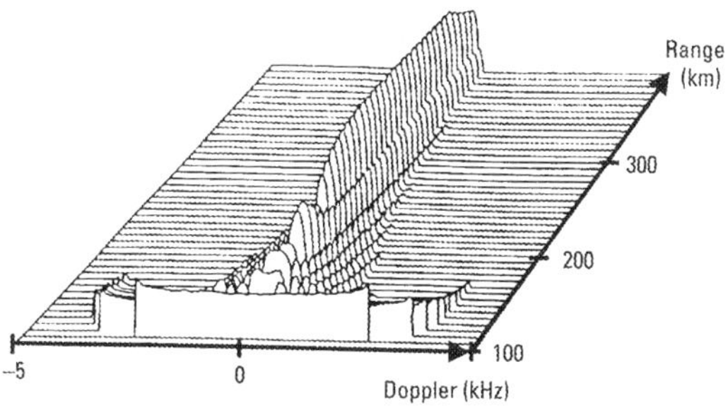

Figure 7.19 Range Doppler clutter. © 1985 IEEE [22].

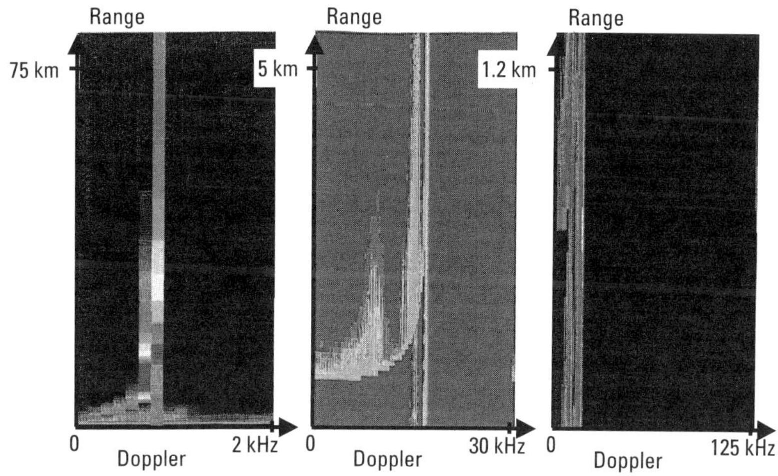

Figure 7.20 Clutter after folding.

2. The proportion of range-Doppler space polluted by the clutter is larger. From this point of view, it seems in particular that the Doppler ambiguity is more of a hindrance than the range ambiguity, since this proportion of polluted domain increases faster when folding occurs in Doppler, according to the appearance of the map in Figure 7.19. This explains why airborne downward-looking radars adopt rather high pulse repetition frequencies in order not to be ambiguous in Doppler shift—at the cost of a near-total ambiguity in range—or at least in order to be more ambiguous in range than in Doppler shift.

In practice, there are three types of airborne radars.

1. The LPRF radars are unambiguous in range (or of ambiguity D_a greater than the effective range), but highly ambiguous in Doppler shift.
2. The MPRF radars, with few ambiguities in range and Doppler. At 10 GHz, a frequency often used for this type of radar because it is a good trade-off between low attenuation by the atmosphere and sufficient directivity of the antenna (of dimensions limited by the diameter of the fuselage), these radars may use, for example, a repetition frequency close to 40 kHz (range ambiguity: D_a = 3.7 km, velocity ambiguity: V_a = 600 m/s) (see [1, 21]);
3. The HPRF radars are unambiguous in Doppler shift but highly ambiguous in range.

Before examining how the range or velocity ambiguities of these radars can be removed, it is advisable to question the necessity of using ambiguous radars. In fact, this ambiguity comes from the choice of a transmission signal composed of pulses repeated at regular intervals. A priori it would also be possible to consider the use of more elaborate signals, based for example on irregularly spaced pulse trains or on more complex modulations. The ambiguity function of such signals does not have these regular spikes every D_a in range and V_a in velocity, but rather the appearance of an upside-down thumbtack: a main peak at $\tau = \nu = 0$, surrounded by a low level, or pedestal. There are generally two main reasons for avoiding the use of this type of transmission signal for airborne radars:

1. Their processing is more complex, because the irregular spacing of the pulses prohibits (or complicates) the use of fast Fourier transform algorithms for Doppler filtering.

2. Above all, and more fundamentally, the pedestal of the ambiguity function [imposed by the invariance relationship of the internal volume to the radar ambiguity function, see (3.29)] is finally more of a hindrance, in relation to clutter, than the local increases in the ambiguity function of a periodic pulse train (see [1] for a demonstration). The clutter echoes, distributed by nature in range and Doppler, accumulate via the pedestal of this instrument function so that they eventually appear at all range and Doppler shifts. This near-uniform clutter distribution is thus a greater hindrance than its occurrence at certain well-defined ranges and velocities.

7.4.3.2 Ambiguity Removal

A technique often used to remove range and velocity ambiguities consists of transmitting consecutive pulse trains at different repetition frequencies (Figure 7.21).

The signals received for a repetition frequency F_r are ambiguous: ambiguity $c/(2F_r)$ in range and $(\lambda/2)F_r$ in velocity. Varying the repetition fre-

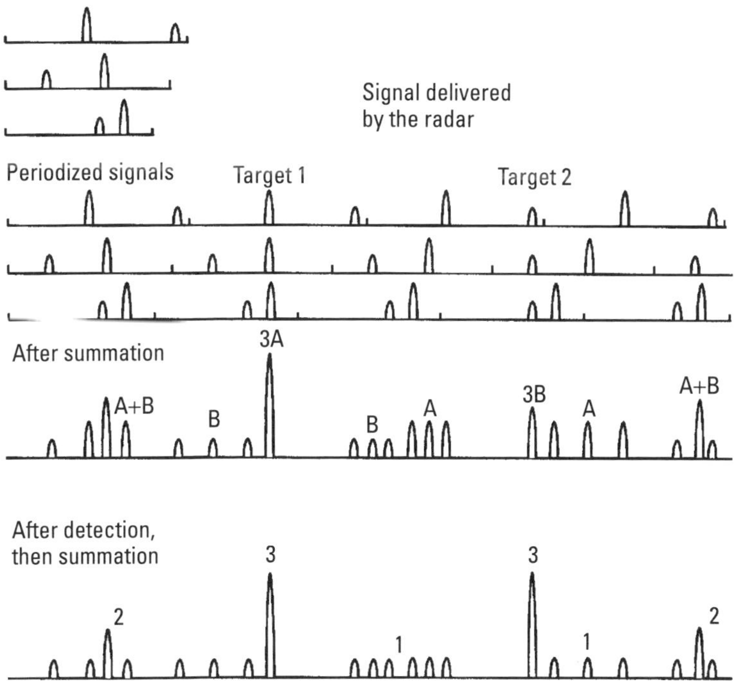

Figure 7.21 Ambiguity resolution by coincidence.

quency amounts to varying the ambiguous range and velocity, which enables the real range and distance to be determined as follows (for simplicity, we consider the case of the range measurement, a similar argument can be applied for velocity measurement).

Upon receiving the first train of pulses (repetition frequency F_r), the position of the possible targets can be measured to within the ambiguity D_a. The received signal in range can therefore be periodized, or unfolded, as shown in Figure 7.21. Similarly, the reception of the second pulse train (repetition frequency F_r', range ambiguity D_a') enables the construction of the periodized signal shown in the figure, and the same again for a third train at frequency F_r''.

It is clear that, if the target range has not significantly changed during the transmission-reception time of these three trains, its position on the three periodized signals must remain unchanged. Consequently, by summing these three signals, the echoes due to the target will accumulate for $D = D_t$, the actual range of the target. This processing by periodization followed by summation consequently enables the ambiguity to be removed, in principle.

In practice, however, several targets (or clutter echoes) can be observed simultaneously, as shown in the figure. If their level differs by more than a factor 3, which can obviously be the case in practice (see Chapter 6), the contrast between real and apparent targets due to summing of two ambiguities will prove insufficient, even negative; it is therefore necessary to previously "threshold" the signals received for each pulse train before carrying out a detection by coincidence. Figure 7.21 shows that the result obtained is appreciably improved (at the cost, however, of a theoretical loss in signal-to-noise ratio, since the thresholding carried out before summation somewhat degrades the equivalent signal-to-noise ratio). This example illustrates well how taking into account a realistic configuration of multiple targets leads to deviation from the optimum processing in the presence of a single target.

The same processing principle—very convenient for the digital processors used by these types of radars—applies to the elimination of ambiguities in velocity. One must however note that the velocity ambiguity [$V_a = (\lambda/2)F_r$] can also be removed by varying the wavelength, from one pulse train to the next. In practice, the few consecutive trains are transmitted at wavelength/repetition frequency pairs selected in order to optimize detection, taking into account the clutter distribution in range and Doppler for each train [1].

7.4.3.3 Discussion

This paragraph devoted to airborne radars does not, of course, cover the whole subject. More particularly, there has been no discussion of the respective advantages and drawbacks of HPRF, LPRF, and MPRF radars in the various configurations of use, such as upward-looking, downward-looking, obstacle avoidance, or target recognition, taking into account the various existing technological limitations, including dynamic ranges of the processed signals, coherence of the transmitter/receiver, and spurious reflections on the front radome protecting the radar. This analysis (which can be found in [1]) does not fall within the scope of our discussion, the main purpose of which is to show how the taking into account of clutter echoes leads to modification of the optimum processing in white noise, while keeping the upstream part of the receiver. A more satisfying approach, intellectually, would consist of taking into consideration the nature of the signals received from the clutter further upstream in order to optimize the entire processing and not only the final detection. This approach, based on the principles developed in Chapter 5, is only now starting to be used for certain specific radars, as illustrated in Section 5.4.3.

7.4.4 Airborne Detection of Slow Moving Targets [23], [24]

For slow target detection with airborne array radars, the ground clutter Doppler spreading due to the platform motion is a limiting factor, especially for long wavelength radars for which the antenna directivity is poor and the antenna sidelobes are high. In this paragraph, two procedures for slow target detection (Doppler-angle imaging, or space-time processing, and motion compensation) will be described and shown to be equivalent [25–28]. Furthermore, a real-time procedure for antenna array calibration using clutter returns will be defined; this procedure can be combined with the detection techniques to increase subclutter visibility in the presence of high clutter returns.

7.4.4.1 Doppler-Angle Imaging (Space-Time Processing)

For an airborne radar, the velocity of clutter returns can take any value between $-v$ and $+v$ (Doppler between $-2v/\lambda$ and $+2v/\lambda$), if v is the platform speed. When seen through a directive antenna of length L, whose 3-dB resolution is $\theta_a = \lambda/L$ (radians), their extent in velocity is on the order of $v\theta_a$ (3-dB extent), that is, 10 m/s for a 3° beamwidth radar on board of an airborne platform with speed 200 m/s. For slow target detection, with velocity in the order of a few meters per second, it is therefore necessary to reduce or analyze this spectral widening of clutter returns.

Let us consider a digital beamforming airborne radar with a quasi-omnidirectional transmitting aerial E, and a linear array of receiving aerials C_i fitted on the flank of the platform. For this description, it will be assumed that the array alignment is parallel to the speed vector and that the spacing between adjacent aerials is D. It will be further assumed in Section 7.4.4.2 that within a pulse repetition period Δt, the aerial C_i takes the place of C_{i+1} (zero drift angle, perfectly known ground speed)[3].

In the following, x_{ik} will be the complex signal (after amplitude-phase demodulation) received by the ith aerial at the kth pulse repetition interval, in range gate number R. By Fourier transform of this sequence for $k = 0, 1, \ldots, M-1$, one gets the Doppler samples corresponding to the same range gate R (for this aerial number i):

$$\tilde{x}_{il} = \sum_{k=0}^{M-1} p_k x_{ik} e^{-2\pi j \frac{kl}{M}} \tag{7.42}$$

where

p_k is a tapering time window for Doppler sidelobes reduction;
The actual Doppler frequency (filter number l) is:

$$\frac{l}{MT_r} = \frac{lF_r}{M} \pmod{F_r}.$$

For ground clutter returns, it is known that if v is the platform speed, there exists a relation between the Doppler shift f_d of the clutter returns and their look direction θ (for the azimuth angle in the plane defined by the array and the look direction, see Figure 7.22):

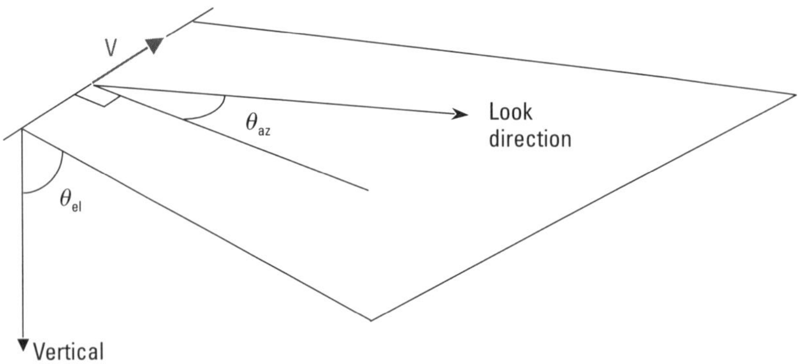

Figure 7.22 Airborne antenna.

3. The calibration technique described hereafter can also be extended to the case of unknown drift and speed [29].

$$f_d = \frac{2v}{\lambda}\sin\theta \tag{7.43}$$

that is

$$\frac{lF_r}{M} = \frac{2v}{\lambda}\sin\theta_1 \,(\text{mod } F_r) \Rightarrow l = \frac{2v}{\lambda}\sin\theta_l \frac{M}{F_r}\,(\text{mod } F_r) \tag{7.44}$$

For these clutter returns, the received samples $\tilde{x}_{il}$ at the output of the Doppler filter associated with direction θ_l (corresponding to Doppler filter number l) for the aerial number i can be written as:

$$\tilde{x}_{il} = \sum_{k=0}^{M-1} p_k x_{ik} e^{-\frac{4\pi jv}{\lambda}\frac{k\sin\theta_l}{F_r}} \tag{7.45}$$

For instance, for a platform speed of 200 m/s, 100-km range, 1.5-m wavelength, Doppler analysis of a 1-s duration pulse train gives a 400-m synthetic antenna leading to an angular resolution of 1.5/400 radians, that is, 3.75 mrad (about 0.25°), or 375m at 100 km. If the repetition frequency is 500 Hz, this Doppler analysis is obtained through a Fourier transform of the 500 successive samples in a range gate.

This angular resolution will determine the angular accuracy of the calibration method described in Section 7.4.4.4.

When this Doppler analysis has been performed on each sensor C_i, it is therefore possible, through angular digital beamforming (classical or adaptive), to coherently sum these N signals in order to draw a Doppler-angle image of the returns, for the range gate number R. If $\tilde{F}_{\theta l}$ is the signal corresponding to direction θ and Doppler l, $\tilde{F}_{\theta l}$ is classically obtained through spatial Fourier transform of the samples $\tilde{x}_{il}$ received on the elementary aerials $i = 1, 2 \ldots, N$:

$$\tilde{F}_{\theta l} = \sum_{i=0}^{N-1} w_i a_{i\theta} \tilde{x}_{il} e^{-\frac{2\pi j}{\lambda} iD\sin\theta} \tag{7.46}$$

w_i being a tapering aperture windowing for angular sidelobe reduction.

This operation, known as space-time processing, results in an image shown in Figure 7.23 for a typical situation, with angular resolution θ_a and velocity resolution δ_v. The gray zone along the diagonal represents "ideal" clutter returns, that is, for an infinite resolution in angle and Doppler, their

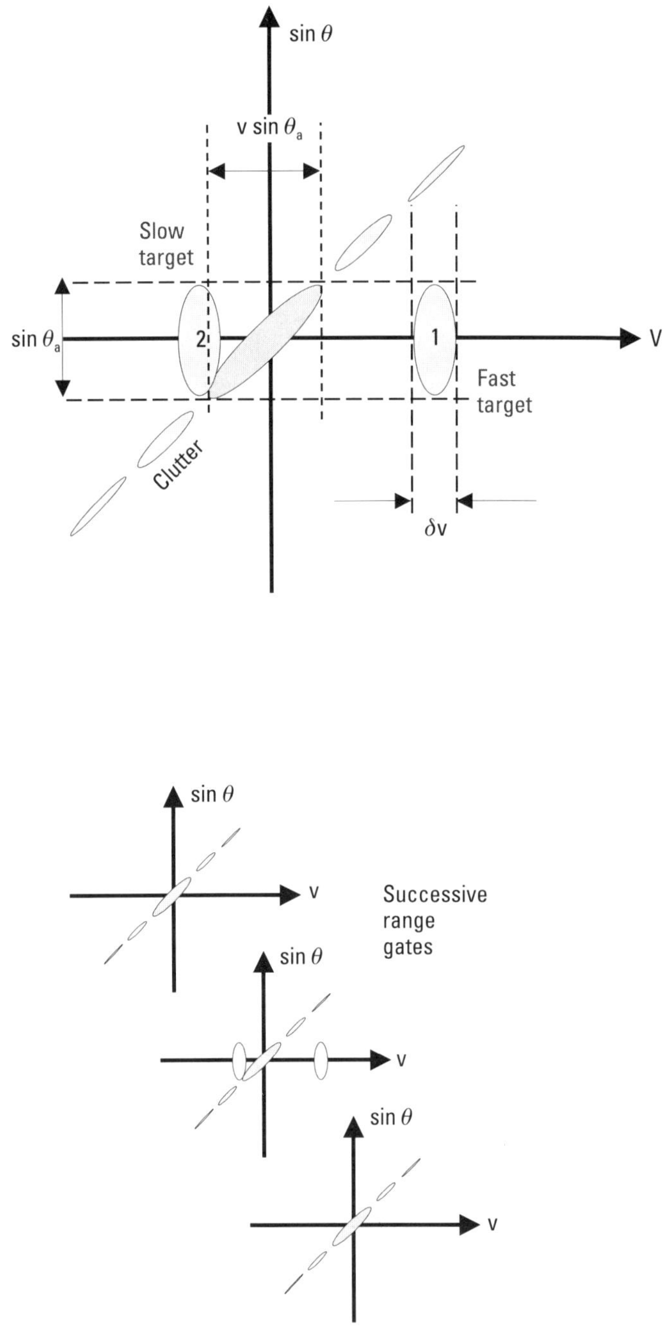

Figure 7.23 Doppler-angle imaging.

spreading parallel to the velocity axis is only due to their own motion of clutter elements (typically about 1 m/s). As seen by the radar, this clutter knife-edge spreads because of the limited velocity resolution δ_v (spreading parallel to the vertical axis), and because of the limited angular resolution θ_a (spreading parallel to the horizontal axis).

Target and clutter returns coming from a specific angular bin, for example bin $\theta = 0$, are separated by Doppler analysis (see target 1) if the target velocity is sufficient that is, if the condition $|v_{\text{target}}| > |v \sin\theta_a|$ is valid, and inversely, in a given Doppler gate, angular analysis allows the moving target number 2 to be separated from clutter returns if angular resolution is sufficient.

Clutter suppression can be achieved in this Doppler-angle image by defining a contour inside which every return will be assumed to be a clutter return, or by defining a detection threshold locally adapted to the neighboring (in Doppler, angle, and possibly range) clutter levels.

Depending on the targets coherence time compared to the integration time necessary for clutter Doppler-angle imaging, it may be appropriate to implement afterwards a specific target detection scheme based on noncoherent integration taking into account the Doppler-spreading of accelerating targets.

7.4.4.2 Detection Through Motion Compensation

For optimized detection of targets whose Doppler is near to the Doppler of clutter, it may seem preferable to artificially stabilize the receiving array by using the successive receivers instead of a hypothetical fixed receiver in space. In this section, the design principle of such a technique will be summarized, before comparing it with the Doppler-angle imaging technique.

The motion compensation technique involves three steps.

First, new sample sequences X_{ik} of M samples received from the virtual nonmoving aerials are formed from sequences x_{ik} (extracted along the diagonals as represented in Figure 7.24) in each range gate:

$$X_{ik} = x_{i+k,k} \tag{7.47}$$

for $k = 0, 1, 2, \ldots, M-1$, and for $i = 0, 1, 2, \ldots, N-M$, ($M-1$ aerials are "lost" for directivity, in order to compensate for the platform motion during a burst of M pulses).

The second step is digital beamforming in direction θ_p, classically obtained through coherent summation of the samples X_{ik}:

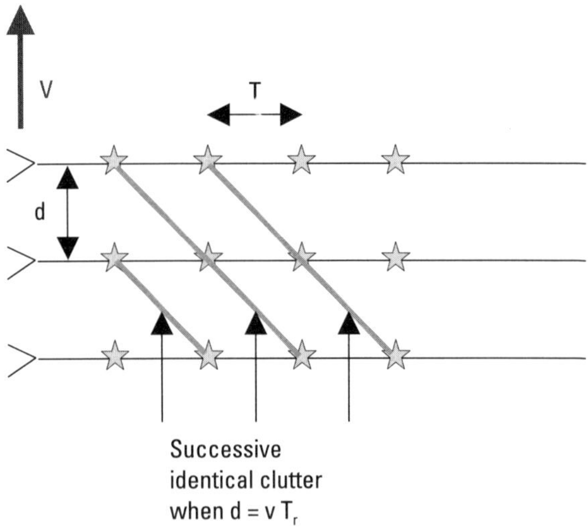

Figure 7.24 Motion compensation for an antenna array.

$$F_{pk} = \sum^{N-M} w_i' X_{ik} e^{-\frac{2\pi j}{\lambda} iD \sin\theta_p} \qquad (7.48)$$

w_i' being a tapering window for angular sidelobe reduction.

The third step is Doppler analysis in each direction θ_p, for each range gate R, through Fourier transform of the M samples F_{pk} ($k = 0, 1, \ldots, M - 1$):

$$\tilde{F}_{pl} = \sum_{k=0}^{M-1} p_k' F_{pk} e^{-2\pi j \frac{kl}{M}} \qquad (7.49)$$

Target extraction is then performed through the same steps as for ground radars (e.g., suppression of Doppler filters corresponding to slow returns, and CFAR thresholding for the other filters).

7.4.4.3 Equivalence Between Space-Time Processing and Motion Compensation [28]

It is interesting to compare the two procedures described in Sections 7.4.4.1 and 7.4.4.2. Such a comparison was drawn in [24–26] for adaptive processing. We show here that for each motion compensation processing, there exists an equivalent Doppler-imaging processing, so that the two procedures are equivalent. For this purpose, we will compare Doppler-angle imaging with $N - M + 1$ sensors with compensation processing.

According to (7.42), (7.45) and (7.49), the output of these processings are, for direction θ and Doppler number 1:

Without compensation:

$$\bar{F}_{\theta l} = \sum_{\substack{i=0,\ldots,N-M \\ k=0,\ldots,M-1}} p_k w_i x_{ik} e^{-\frac{2\pi}{\lambda} iD \sin\theta} e^{-2\pi j \frac{kl}{M}} \tag{7.50}$$

With compensation:

$$\bar{F}'_{\theta l} = \sum_{\substack{i=0,\ldots,N-M \\ k=0,\ldots,M-1}} p'_k w'_i x_{i+kk} e^{-\frac{2\pi}{\lambda} iD \sin\theta} e^{-2\pi j \frac{kl}{M}} \tag{7.51}$$

Taking into account the Doppler transformation involved in the compensation method, the quantities to be compared are $\bar{F}_{\theta l}$ and $\bar{F}'_{\theta, l-l_c}$ where l_c is the number of the filter polluted by clutter, for a look-direction θ:

$$l_c = \frac{2v}{\lambda} (\sin\theta) \frac{M}{F_r} = \frac{2D}{\lambda} (\sin\theta)\ M$$

Identity between the two processing techniques will be proven if, for given i_0 and k_0, weights applied to $x_{i_0 k_0}$ can be made equal:

$$p_{k_0} w_{i_0} = p'_{k_0} w'_{i_0 - k_0} \tag{7.52}$$

$$p_{k_0} w_{i_0} = 0 \text{ for } (i_0,\ k_0) \text{ such that } i_0 < k_0$$

This equality is valid only if Doppler-angle imaging is modified in order to take into account a weighting in k (time) depending on i (space) and the look direction θ, that is, a really two-dimensional Doppler-angle processing with appropriate steering vector. Under this condition, the two processing techniques can be made equivalent by suitable choice of the weights.

In practice, the weights will be computed through an adaptive two-dimensional algorithm (for example, sample-matrix inversion, the sample correlation matrix being evaluated in adjacent range gates, or preferably by averaging in the time domain [27] in order to take into account the clutter reflectors that are in the range bin under test) so that Doppler-angle imaging will be equivalent to motion compensation technique and will be easier to calibrate when the platform is drifting [29].

7.4.4.4 Calibration on Clutter Returns [29]

This schematic reasoning makes obvious the requirement for identical aerial diagrams (in amplitude and phase, for each direction) so as to ensure that angular sidelobes—(for Doppler-angle imaging) or Doppler sidelobes—(for motion compensation) do not hide small slow targets.

In the real world, the elementary diagrams are not identical—especially for low-frequency radars—for different reasons: receiver imperfections, coupling between adjacent antennas, and interaction between antennas and neighboring structures (wings). For these reasons, it becomes necessary to calibrate the received signals for each receiving direction.

Since the Doppler of the clutter returns depends only on their angle of arrival, the calibration coefficients can be computed using the clutter returns. More precisely, a clutter element at distance R and with Doppler f_d necessarily comes from a direction θ_s in elevation (equal to h/R, for a given platform altitude h) and θ_g in azimuth, represented in Figure 7.22. Because this echo is simultaneously received by every elementary aerial with the same Doppler, a comparison between the corresponding signals makes it possible to compensate for the discrepancies between the different aerials' diagrams.

The calibration procedure consists of comparing the signal received from clutter, by successive aerials number i, in range gate R and for direction θ_1. If the calibration were perfect, these signals would be equal apart from a known phase shift depending on direction. More precisely, if $\bar{X}_{il}$ is the calibrated signal corresponding to $\bar{x}_{il}$, the following relation must be valid for clutter returns:

$$\bar{X}_{il} = \bar{X}_{i+1l}\, e^{\frac{2\pi j}{\lambda} D \sin \theta_l} \tag{7.53}$$

For this relation to be valid, the signal $\bar{x}_{il}$ must be calibrated by the calibration coefficient a_{i1}, the first aerial being chosen as a reference:

$$\begin{aligned} \bar{X}_{1l} &= \bar{x}_{1l} \\ \bar{X}_{2l} &= a_{2l}\bar{x}_{2l} \\ \bar{X}_{il} &= a_{il}\bar{x}_{il} \end{aligned} \tag{7.54}$$

with

$$a_{il} = \frac{\overline{\bar{x}_{il} x^{*}_{i+1l}}}{\left|\bar{x}_{i+1l}\right|^2} e^{-\frac{2\pi j}{\lambda} D \sin \theta_1} \tag{7.55}$$

the averaging indicated as an upper bar being implemented on several adjacent range gates so that elevation does not noticeably vary. This averaging (obtained through least-square estimation) reduces the perturbation induced by possible target echoes in the clutter returns used for calibration.

After calibration, the procedures summarized in (7.50) and (7.51) become:

Without compensation

$$\tilde{F}_{\theta l} = \sum_{\substack{i=0,\dots,N-M \\ k=0,\dots,M-1}} p_k w_i a_{i\theta} x_{ik} e^{-\frac{2\pi j}{\lambda} iD \sin\theta} e^{-2\pi j \frac{kl}{M}} \tag{7.56}$$

With compensation

$$\tilde{F}_{\theta l} = \sum_{\substack{i=0,\dots,N-M \\ k=0,\dots,M-1}} p_k' w_i' a_{i+k\theta} x_{i+kk} e^{-\frac{2\pi}{\lambda} iD \sin\theta} e^{-2\pi j \frac{kl}{M}} \tag{7.57}$$

As an illustration, a simulation of the Doppler-angle imaging procedure (sample matrix inversion, no noise-only reference) is presented in Figure 7.25, for an array composed with five subarrays of nine elements each, and a transmitted pulse train of 128 pulses. The platform speed is 80 m-s^{-1}, and the wavelength is 3 cm. Clutter to noise ration is 16.5 dB per sensor and per pulse, and one target is present.

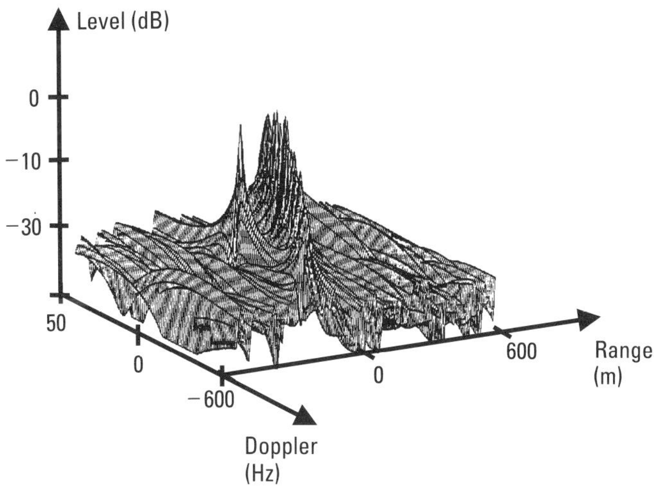

Figure 7.25 Doppler-angle imaging, 45 elements array with 5 subarrays.

The amplitude errors are between –4 dB and +1.2 dB, and phase errors are between –4.4° and +19.6°. The loss of energy on the clutter knife-edge and on the target, due to calibration errors, is clearly visible in Figure 7.26, and the calibration restores the correct amplitudes of clutter and target (Figure 7.27).

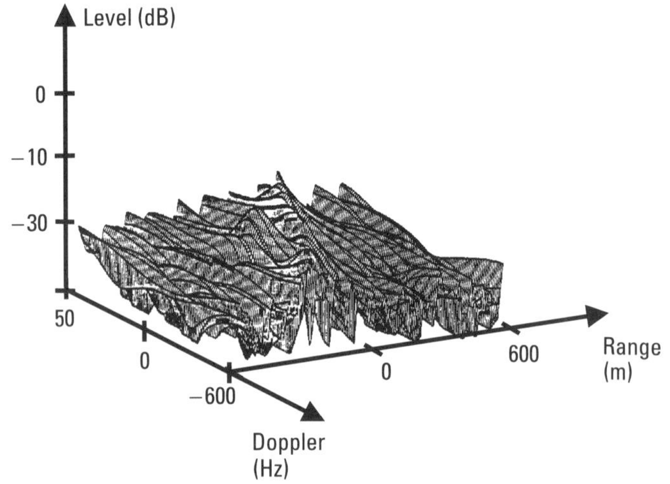

Figure 7.26 Doppler-angle imaging, with calibration errors.

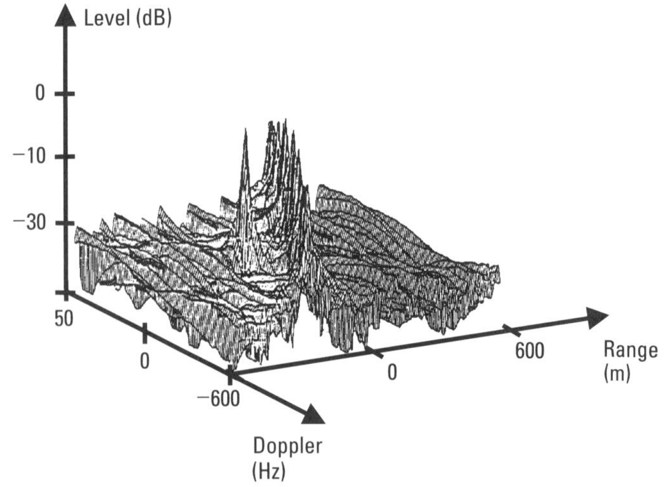

Figure 7.27 Doppler-angle imaging, result after calibration.

7.5 Low-Frequency Radars

7.5.1 Scope

Recent evolutions in operational scenarios tend to favor the use of low-frequency radars, that is, operating below 1 GHz. Such major evaluations are: antiradiation missile threat, increased helicopter threat, and above all, the irruption of stealthy targets (in conventional radar band). In this context, low-frequency radar has definite advantages but also limitations that have to be considered, if only to compare the relative advantages of the various low-frequency bands: HF band (3–30 MHz), VHF band (30–300 MHz) and UHF band (300–1,000 MHz).

The purpose of this section is to review the main properties of low-frequency radars, first from the standpoint of the physical phenomena involved (retrodiffusion of targets, passive and active stealth, propagation, possibility of target recognition) then from the standpoint of the radar design in itself (e.g., available technologies, exploration of space, and system integration).

7.5.2 Targets

At low frequencies, wavelengths are no longer negligible compared to target sizes; therefore, the signature results from a complex interaction between the incident electromagnetic wave and all of the target component elements. The radar signatures obtained in this way are very specific but can be reliably modeled with numerical methods. Conversely, measurement of characteristics is difficult, essentially because of the particularities of propagation in these frequency bands. This more "global" interaction between the target and the electromagnetic wave also considerably reduces the efficiency of the available stealth processes, an essential argument in the favor of the use of these low-frequency bands.

7.5.2.1 Modeling

The modeling methods that can now be used differ depending on whether the observation frequency is less than or greater than approximately 300 MHz.

For frequencies less than 300 MHz (HF and VHF bands), the so-called exact methods that numerically solve Maxwell equations at the target surface (methods of finite elements or moments) apply with today's data processing means and make it possible to calculate, with a good approximation, the RCS values of the targets, as long as they are not penetrable, that is, as long as they can be assimilated to an external surface (with a given surface impedance condition). An example is given in the HF band (decametric waves) in Figure 7.28, representing the HF bistatic RCS of an aircraft near the crossing point on a link between transmitter and receiver (therefore near the forward scatter, i.e., for a bistatic angle of 180°) as a function of time. Figure 7.28 shows the results of a comparison between modeling and measurement. Good agreement is observed between modeling and measurement, as in other comparisons made in the high VHF band (metric waves), for instance on missile signatures.

For frequencies higher than around 300 MHz, the use of exact methods becomes difficult because of the number of unknowns to be calculated. It is then preferable to use hybrid methods combining asymptotic methods (valid for high frequencies) and exact methods (to take into account the finest details of the structure). This area is currently being developed in various directions such as:

- Direct hybridization between the areas processed using the moments method and those processed by the physical theory of diffraction (with an intermediate zone that allows for coupling between these zones);

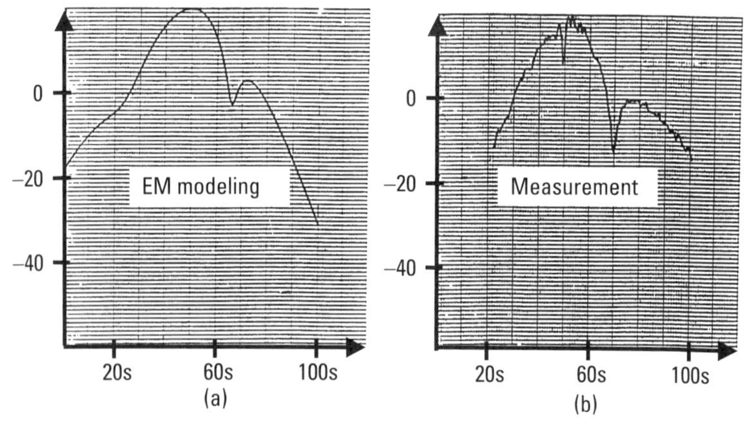

Figure 7.28 HF bistatic RCS of an aircraft: (a) modeling; and (b) measurement.

- More direct digital hybridization by the manipulation of the hollow matrix of couplings between currents on edges or by iterative resolution from a starting solution obtained by the asymptotic method;
- Segmentation of the object into parts treated separately by the method of moments, with additional unknowns permitting the chaining together of the successive segments.

7.5.2.2 Measurements

RCS measurement in the HF, VHF, and UHF bands is often a difficult matter because the effects of propagation interfere with any measurement made on the ground. A variety of solutions have been used so far.

- Measurements are performed in an anechoic chamber on real objects or on models; when the quality of the anechoic chamber allows this kind of measurement, and when the dimension of the object to be measured is modest (small missiles), this technique gives good accuracy. The problems of near fields are then less crucial than in conventional radar bands (distance of nearby field with respect to an object having dimension D, observed at wavelength λ: $2D^2/\lambda$, that is, 18m on a target measuring 3m at 300 MHz). As soon as target dimensions become larger, it becomes necessary to use measurements on models (as long as the target is conductive) or to operate outdoors (measurements in a chamber on "pieces" of targets are generally barely representative of reality because of the overall character of target-wave interaction).
- Measurements are made on the field with "deconvolution" of propagation effects—and first, deconvolution of the effect of multipath, described in Section 6.5.2. This type of measurement takes into consideration real targets such as aircraft, in spite of the inaccuracy due to the modeling of propagation effects.
- In-flight measurements: This technique overcomes to a large extent the interference caused by propagation phenomena, as long as the effects of clutter echoes are suitably processed.

7.5.2.3 Stealth

Reducing the RCS of a target depends on three essential techniques, which may be combined: optimized shapes, the use of absorbing materials, and the implementation of active modulation processes (neutrodyning and RCS modulation).

Stealth Shapes

The basic principle of stealth by shaping consists of reflecting the energy emitted by the radar in a different direction. To do this, the aircraft consists of facets (F 117) or of more rounded shapes (B2, F22) equipped for conventional radar bands with some directivity. For instance, a facet having a dimension $L = 2\text{m}$ has directivity $\lambda/2L$ (in radians), that is, 1.4° at 3 GHz and 14° at 300 MHz; the effect of the shaping is far less important at low frequencies, and the angular sectors where some discretion can be obtained will be far smaller. In the low-VHF and HF bands, this shape effect becomes even less efficient because the wavelength becomes comparable to the dimension of the target; backscattering is then necessarily nondirective.

Absorbing Materials

To absorb electromagnetic energy near a conducting body, it is possible to use the electric field (absorption by Joule effect) or the magnetic field; the latter solution can be used to design thin, absorbent materials because the magnetic field is maximum near the surface of a conductor, whereas the electrical field is zero on that surface.

Accordingly, absorbing materials at frequencies less than 1 GHz are either thick (thickness comparable to the wavelength so that the wave can penetrate the material and be absorbed in it) or are thin (a few millimeters for magnetic materials) but heavy (around 15 kg/m^2 of overlaid surface). These characteristics of thickness or weight are all the more detrimental in that the RCS of the target is not due, in these frequency bands, to few specific points but rather to the overall shape meaning that it is necessary to cover large surfaces (counted in several square meters). Absorption by passive materials, as things stand today, therefore appears to be difficult to use in practice, at least on nonspecialized aircraft.

Another solution could consist of developing thin absorbing materials that could be adapted to the actual radar wavelength by electronic means. If a system of this type could be produced, it would be complex in any case (one would need to measure the direction and frequency of the radar by means of a listening system that itself would need to be discreet) and would be vulnerable to low-cost countermeasures such as, for instance, frequency agility[4]; in addition, it would be inefficient with respect to bistatic or multistatic radar.

4. In Section 7.5.3 we will see that this frequency agility capability will in any case generally be required in order to alleviate problems due to propagation or multipath.

Modulation

The concept of RCS modulation involves two different principles. Neutrodyning consists of transmitting by an antenna a signal equal and opposite to the signal returned by the object, thus implying that the latter signal has to be known, in amplitude and in phase, for different directions. RCS modulation consists of dispersing the echo returned in the Doppler cells and, if necessary, the radar range.

The first limitation on *neutrodyning* is linked with the need to know the direction, the frequency, and the polarization of the radar signal with sufficient accuracy. Indeed, if we attempt to send a signal equal to and opposite of the echo returned by the cell, it is essential to know the phase of that echo with accuracy of around 6° (1/10 radian) to reduce the total echo by 20 dB. Now, the echo phase returned by the target varies by 360° when the backscattering angle varies by $\lambda/2L$ (in radians), where L is the transverse dimension of the target; at 300 MHz, for a 10-m target, the returned echo phase goes through 360° for an angular presentation variation of around 3°. Accordingly, the direction of the radar with respect to the target must be known at around $3/360 \times 6 = 0.05°$. Such accuracy is certainly difficult to attain, in particular onboard an aircraft.

In terms of frequency (the phase of the echo returned by the target varies by 360° when the frequency varies by $c/2L$, where L is the depth of the target), the same thinking leads to frequency measurement accuracy of 250 kHz, which is difficult to reach instantaneously (during a pulse duration) and which is meaningless as soon as the instantaneous band width of the radar exceeds that value.

To summarize, neutrodyning seems hardly realistic for VHF radars (and even less so at higher frequencies for the same reasons) as far as aircraft type targets are concerned. At lower frequencies, it becomes more difficult to implement because of the complex coupling that occurs between the transmitting and receiving antennas and the target itself.

RCS modulation, in theory, does away with knowledge of the precise characteristics of the radar because it consists of dispersing energy by amplitude and phase modulation of the received signal. This technique becomes all the more efficient as the processing gain (Doppler and range) of the radar increases. In Section 7.5.6, we will show that low-frequency radars have this very property to compensate for the low gains of the antenna. The difficulty of applying these processes comes from the need to modulate the reflectivity of the scattering zones; it involves coupling modulating elements with bright zones and therefore designing modulators that are strongly coupled with the surrounding surface currents, that is, antennas that are strongly coupled with the surrounding structure.

7.5.2.4 Orders of Magnitude

Various measurements and modeling carried out so far suggest that the RCS of weapon-carrying aircraft, using horizontal polarization, have an order of magnitude of several tens of square meters in metric bands (300 MHz) and a hundred or so square meters in decametric bands (30 MHz). With vertical polarization, the RCS of aircraft or missile type targets in the decametric band are substantially smaller (around $5m^2$) and slightly less in the metric band (a few square meters).

From the standpoint of the radar, it is important to take into account the fluctuations of the target (for $P_d = 0.9$ and $P_{fa} = 10^{-8}$, the necessary signal-to-noise ratio is 8 dB higher if the target fluctuates according to a Swerling I law, as compared to a nonfluctuating target). For the VHF-UHF bands, fluctuations of an aircraft-type target are illustrated in Figure 7.29 (aircraft and missile observed in linear polarization HH at 270 MHz and 3 GHz); fluctuations are clearly visible, with lobe widths, at 270 MHz, of about 5° (approximately $\lambda/2L$ where L is the transverse dimension of the target).

For a stealthy aircraft, considering the limitations given above, it is realistic to consider an RCS of $0.1m^2$, not fluctuating (or $1m^2$ fluctuating) as a minimum.

For missiles, it is worth noting that the smaller the missile, the less its RCS will fluctuate in the metric band; therefore, the targets with the smallest RCS can be considered as nonfluctuating. In practice, measurement cam-

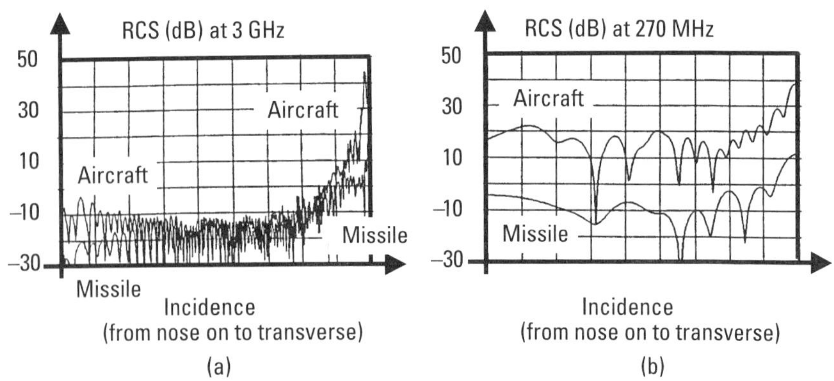

Figure 7.29 RCS of an aircraft and a missile at (a) 3 GHz, and (b) 270 MHz.

paigns and modeling demonstrate that an RCS of 0.1m^2 (nonfluctuating) is a minimum for missiles observed in the metric band.

For helicopters, the RCS of the blades in VHF and UHF can be assimilated to a diffracting edge having an RCS equal to L^2/π where L denotes the length—typically 8m^2 for the VHF and UHF bands.

In the HF band, fluctuations are almost nonexistent, as shown in Figure 7.30, representing the RCS of an aircraft with horizontal polarization at 15 MHz.

For missiles in the HF band, the RCS decreases substantially in these frequency bands (Rayleigh zone). An RCS of $-30\ \text{dB/m}^2$ (nonfluctuating) is quite possible in the HF band for an antiship missile [in this frequency band, the RCS of the missiles obeys a Rayleigh law: $\sigma = (4/\pi)k^4F^2V^2$, where $k = 2\pi/\lambda$ and V is the volume of the target and F a shape factor of around 0.5].

7.5.3 Propagation [30]

Propagation at low frequencies incorporates a number of particularities that have to be taken into consideration at the radar design stage: multipath effects, surface wave effects, and ionospheric reflection in HF and diffraction by masks and propagation in HF and VHF by lateral waves. These phenom-

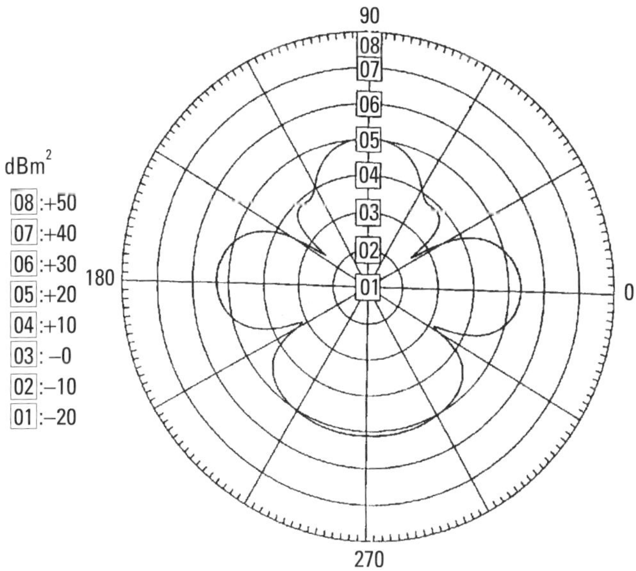

Figure 7.30 RCS of an aircraft at 15 MHz (horizontal polarization).

ena are also put to good use in the basic operation of the radar, in particular at the lower frequencies: transhorizon radar by sky wave (ionospheric reflection) or by surface wave, VHF antihelicopter surveillance radar (diffraction by masks and lateral waves), for instance. Conversely, the phenomena of atmospheric attenuation encountered in conventional radar bands (e.g., molecular attenuation, rain, fog, and insects) can be considered nonexistent in these frequency bands.

7.5.3.1 Multipath

When a radar target is observed at low grazing angles, the multipath phenomenon, described in Section 6.5.2, may appear: the target is illuminated by a direct radar-target ray and by a ray reflected on the surface of the land (or water). These two rays, which interfere constructively or destructively, depending upon their differential length, produce amplification or attenuation of the field received by the target. Similarly, the echo returned to the radar propagates simultaneously along these two paths, with the same interference effect on arrival.

If the surface is considered flat, this phenomenon can be interpreted as an image effect; the received complex signal is then multiplied by a complex "propagation factor" F_c [(6.61)]:

$$F_c = 1 + \rho \exp\left[-j\left(\varphi + \frac{2\pi\delta}{\lambda}\right)\right]$$

where φ is the phase shift during reflection on the surface, ρ the amplitude of the reflection coefficient on the surface and δ the difference in length between the direct path and reflected path. The magnitude of this factor varies between 0 (destructive interference) and 2 (for a coefficient reflection of 1 on the ground with constructive interference).

It was shown in Section 6.5.2 that for low incidence angles, the phase shift φ is generally around π and coefficient ρ around 1 (the tangential component of the field on the surface is null) so that, at a given range R, interference between target and image is destructive at zero altitude and only becomes constructive again when the target altitude is:

$$h_t = \frac{R\lambda}{4h_r} \quad (h_r\text{: radar altitude})$$

and then becomes destructive again at twice that altitude, and so on. This is the lobing phenomenon, referred to in this way because it produces on a received echo an effect similar to what an antenna pattern having successive lobes in elevation would produce if the lobes' width were:

$$\theta = \lambda / 4h_r$$

Because the interference is destructive at a zero altitude, it appears that targets moving at very low altitudes are difficult to detect by radars using low frequencies. For instance, at 300 MHz, for a 15-m radar antenna altitude, radar range is far below an elevation of 1/(4 × 15) radians, that is, around 1° (target altitude 85m at 5 km).

In this region, the propagation factor F_c is more or less equal to:

$$F_c = 2\left|\sin\left(2\pi\frac{h_r h_t}{R\lambda}\right)\right| \simeq 4\pi\frac{h_r h_t}{R\lambda}$$

that is, two-way round-trip power attenuation in λ^{-4} at a given altitude (the angular altitude of this zone in elevation varies in λ) and in R^{-4}—total attenuation in R^{-8}. It can be seen that detection on flat ground (without microrelief) at very low altitudes is particularly difficult at low frequencies[5] without recourse to specific propagation modes such as surface waves or diffraction by microrelief (see Section 7.5.3.4).

This attenuation phenomenon is associated with a glint phenomenon because the target, as seen at low elevation by the radar, in reality comprises the real target and its image in the ground. As an initial approximation, the radar therefore observes a target consisting of the real target and its image on the ground, and the elevation angle measurement is perturbed by glint effects. These effects generate more considerable errors when the target and its image are added in phase opposition, that is, for minimum RCS of the object comprising the target and its image.

These low-elevation effects make it particularly difficult, if not impossible, to measure elevation by a low-frequency radar for low-altitude targets. They also make it necessary to use *frequency agility* to ensure detection in spite of the lobing effects (by moving the interference lobes).

5. This phenomena is, however, not so detrimental as it might appear, except over water. For surveillance radars, the range is generally sufficient for very low altitude targets to be masked by microrelief rather than by a lobing effect. For example, for a 30-km range, a radar at a height of 15m above the average level will detect targets above altitudes of 250m. The limitation will generally come from the microrelief, as dealt with in the following section, rather than from the lobing effect.

7.5.3.2 Ionospheric Reflection [31]

The existence of ionized layers in the high atmosphere (between 90 and 500 km) causes the refracting of the HF waves, put to good use by over-the-horizon backscattering radars (OTH-B) to detect radars beyond the optical horizon at distances counted in the thousands of kilometers. The refraction properties of the ionospheric medium vary, in particular depending on solar activity (on the scale of the day, the year, and the decade) so that the measurements made have to be corrected and the radar parameters optimized (waveform and transmission frequency) according to the state of the ionosphere, itself modeled or measured.

7.5.3.3 Surface Wave

Analysis of propagation over the sea demonstrates that for vertical polarization, the decrease of the electric field with range is much lower than for horizontal polarization. Figure 7.31 illustrates this phenomenon for a transmitter at an altitude of 20m and a receiver at an altitude of 20m (one-way attenuation above free-space loss).

For this configuration, the propagation losses are 2 dB at 40 km and 4 dB at 80-km range with vertical polarization against 48 dB at 40 km and 57 dB at 80-km range for horizontal polarization (depending on the state of the sea surface, a few decibels should be added to these losses). The existence

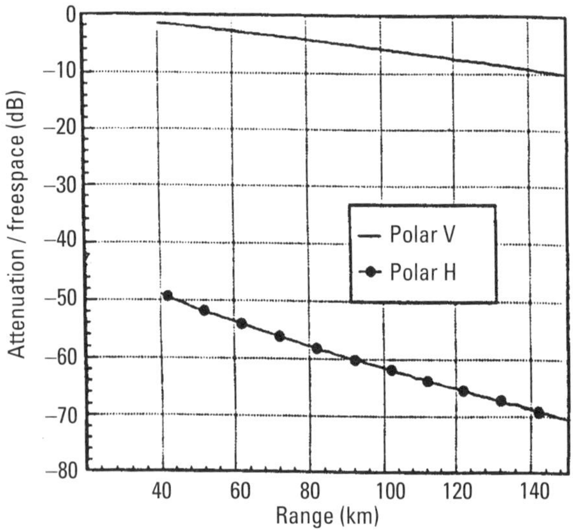

Figure 7.31 Propagation above the sea at 10 MHz; transmitter height: 20m; receiver height: 20m.

of this surface wave[6] with vertical polarization (whatever the atmospheric conditions) means that targets can be detected at very low altitudes beyond the horizon.

7.5.3.4 Diffraction on Masks

When the radar and target are not in direct line of sight, a good approximation often consists of considering that the target is illuminated by diffraction from the upper edge of the mask; the attenuation corresponding to a radar-target path (one-way path) is then plotted in Figure 7.32 (additional attenuation with respect to standard atmospheric attenuation), extracted from [30]. It is 6 dB when the ridge is flush with the target-radar path and it obeys, with good approximation, the following law:

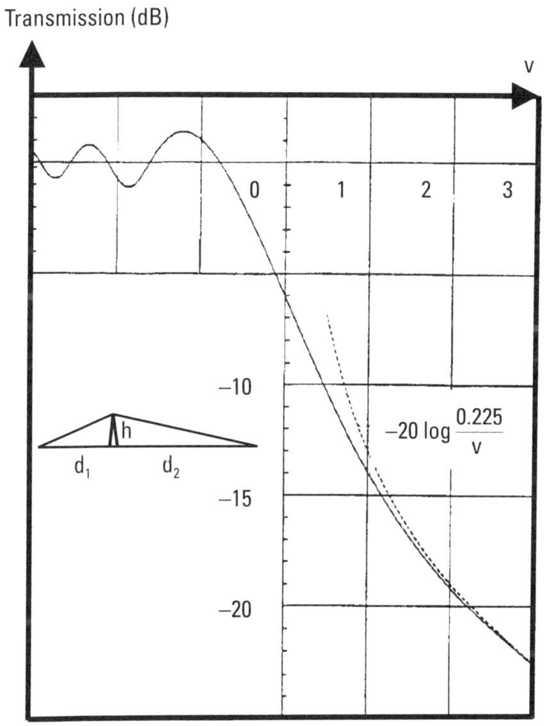

Figure 7.32 Diffraction by an edge [30].

6. Although widely used, this terminology is not correct and a spherical diffraction wave would be more appropriate.

$$A\ (\text{dB}) = -6.9 - 20\log\left[\sqrt{(v - 0.1)^2 + 1} + v - 0.1\right]$$

where v is given by:

$$v = h\sqrt{\frac{2}{\lambda}\left(\frac{1}{d_1} + \frac{1}{d_2}\right)}$$

(In this expression, h is the height of the mask, d_1 the distance from the radar to the mask and d_2 the distance from the mask to the target.)

In particular, it appears that the round-trip power attenuation on a target is proportional to λ^{-2}, favoring the use of low frequencies. For $d_1 = 1$ km, $d_2 = 10$ km, and $h = 30$m (i.e., the mask exceeds the radar-target link by 30m); the attenuation caused by diffraction is 16 dB, that is, 32 dB two-way at 300 MHz. This phenomenon has been experimentally confirmed and it allows detection of targets with a high RCS (helicopters) at large distances behind vegetation masks.

Various other propagation modes also increase the range of the radars in the low frequencies, including lateral waves propagating with horizontal polarization at the interface between the two dielectrics consisting of the vegetation and the atmosphere. It thus appears that for frequencies less than 300 or 400 MHz (but rarely beyond them), the radar has a significant detection capacity behind vegetation masks.

7.5.4 Clutter and Interference

In general, the lower the frequency the more the ground clutter becomes immobile because mobile clutter elements are small and their RCS become smaller at low frequencies (in particular in the HF band, where ground clutter can be considered as a fixed echo). The level of diffused clutter echoes tends to decrease with the frequency (more or less proportionally), but, conversely, the level of the point-like echoes tends to increase.

As an illustration, Figure 7.33, taken from [32], represents the variation of the RCS density of UHF (425 MHz) land clutter and sea clutter echoes.

The jammings affecting radars are of various origins but are often not intentional. In particular, for HF, the radar noise factor is due to external noise, a good approximation of which is given by the formula (in a calm environment):

$$F\ (\text{dB}) = 56 - 25.3\log f\,(\text{MHz})$$

which, for instance, leads to a noise factor of 18.6 dB at 30 MHz.

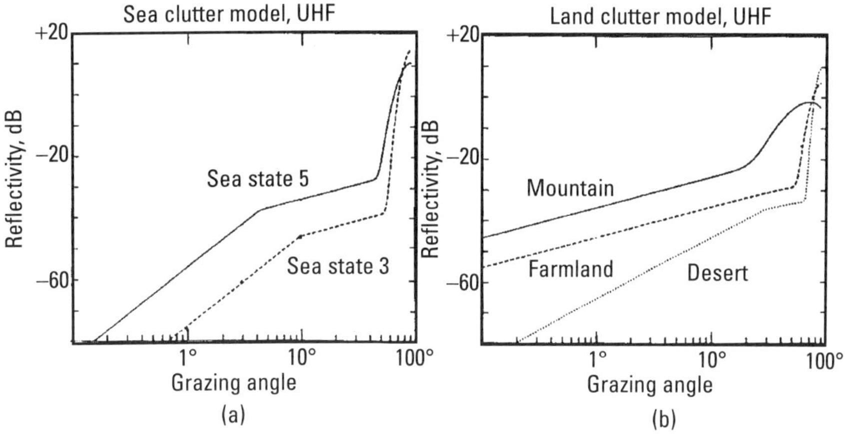

Figure 7.33 Clutter RCS density: (a) sea; (b) ground. (*From:* [32].)

Furthermore, concerning ground radars, it should be noted that the multipath effect tends to considerably attenuate the hindrance caused by jammers (whether intentional or not) located on the ground. Nevertheless, it is possible for helicopter or aircraft carried jammers to radiate effective powers to about several tens of kilowatts.

7.5.5 Target Recognition

Various discrimination procedures between airborne targets can be used by radars operating on the metric or decametric bands. These procedures are based on the use of signatures or features that can be extracted from the radar signal: RCS, frequency response (or its Fourier transform, the impulse response), the angular response (or its Fourier transform, the transverse response), the response in angle and frequency (holographic imaging), kinematics, analysis of moving parts, and resonance.

To implement these procedures, it is generally necessary to utilize specific operating modes for the radar. The purpose of this section is not to describe all the recognition procedures that can be used by low-frequency radars but only to review the constraints imposed upon a low-frequency radar by the use of each of these possible features.

7.5.5.1 Utilization Constraints

Utilization constraints inherent to this frequency band are mainly linked with the effects due to propagation, limitations caused by spectral occupancy, and the required observation times.

Propagation

We have shown that propagation plays an important role for ground radars in the metric and decametric bands, whether enabling the detection of targets masked by vegetation or modifying visibility on flat ground by low elevation effects (lobing effect). For discrimination function, the signature of the target is combined with this propagation in a manner that is predictable to some extent in a given plot of land. This means that the radar has to use some amount of frequency diversity to detect targets and to modify propagation effects in order to compensate for the effects on the observed signatures.

For airborne radars, propagation phenomena are generally negligible except for some configurations with low grazing angles, at long or medium ranges.

Bandwidth

The legislation normally rules out the use of large bandwidths. In practice, only broken bands with a width of a few megahertz, themselves at intervals of a few megahertz, are available in the metric bands.

Waveform

Analysis of the target signature generally imposes certain particular constraints on the transmitted waveform: frequency agility, repetition frequency, and observation time. These constraints will be described in Section 7.5.5.2 for each characteristic magnitude.

7.5.5.2 Discriminating Features

In this section, each of the different features is examined in order to identify the difficult points that its use could raise.

RCS

Measuring the RCS requires calibration of the radar, which is difficult because of propagation. If the land can be modeled, some estimation of the RCS is possible with an accuracy that is yet to be evaluated (a few decibels for an accurately known land).

Frequency Response and Pulse Response

The target frequency response, or its Fourier transform, the impulse response, can be obtained by a frequency scan (synthetic band technique) using a Δf band in excess of 75 MHz (range resolution of 2m). The number of frequencies to be transmitted equals the ratio between the range resolution of the target and the final desired resolution; typically, it means transmitting 60 or so bursts at intervals of approximately 1.25 MHz, if possible taking into account the frequency occupation plans. A considerable limitation of the radar system will derive from the time needed for this waveform, because each burst itself will last between 10 and 50 ms (requirement for Doppler integration, in particular to limit the peak power to be transmitted); the total observation time will thus be included between 0.6 and 3 seconds.

Angular Response or Transverse Response (One-Dimensional ISAR)

This technique is not particularly demanding for the radar system because a fixed frequency is used. To obtain a transverse resolution δy on the target, it is necessary to observe through an angular sector $\Delta\theta$ with sampling every $\delta\theta$ [see (6.36) and (6.37)]:

$$\delta\theta = \lambda/(2L),\ \Delta\theta = \lambda/(2\delta y)$$

Typically, to obtain transverse resolution of $\delta y = 2.5$m on a 15-m target, it is necessary to observe, in the metric band, an angular sector $\Delta\theta = 12°$ sampled every 2°. The operation condition here is a certain amount of stability in propagation conditions (or good compensation of their effects), including multipath, during this observation, which can last approximately 10 seconds.

Application of these techniques in the decametric band, to estimate the dimensions of the targets, has also been experimentally demonstrated.

Response in Angle and Frequency, Holography (or Two-Dimensional ISAR)

Because this technique combines the two previous features (see Section 7.2.2), implementation requires both good compensation of effects due to propagation and the transmission of a synthetic band waveform. Furthermore, it requires that the aircraft path be known (or compensated) to within $\lambda/4$, a major constraint (all the more so in that the efficiency of the motion compensation processes that can be used for ISAR is not guaranteed in these frequency bands where the image is highly diffuse).

Kinematics

Use of the target kinematics for target recognition imposes great accuracy upon the measurements made, in particular in terms of angles. Recourse to an ecartometry function is then necessary whether it is obtained by a monopulse type system or by a high-resolution analysis method (adaptive beamforming is also required as a countermeasure against jamming). Acceleration measurements can be made by means of a Doppler extent index without any major modification to the waveform used.

Moving Parts

The existence of moving parts[7] appears at the output of Doppler analysis without recourse to any specific waveform; therefore, it does not impose any particular constraint upon the radar system other than a repetition frequency higher than the kHz or so in the metric band to detect helicopter rotor blade flashes and the use of horizontal polarization to see the main rotor blades.

Resonance (see Section 7.2.3.6)

The constraints imposed by the use of this technique are essentially the same as those imposed by frequency response analysis: compensation of propagation effects and observation times included between 0.6 and 3 seconds.

7.5.5.3 Synthesis

This brief overview demonstrates that there is no doubt that holographic imaging (or two-dimensional ISAR) imposes the severest constraints for relatively limited efficiency in these frequency bands (very diffuse image that is difficult to interpret). Conversely, these features, whose use seems less constraining, appear to be an analysis of angular fluctuations (one-dimensional ISAR) and analysis of the kinematics and of moving parts. Feasibility of a reliable measurement of a frequency response is an important point to demonstrate to obtain precise discrimination between targets in the metric bands (measurement of target depth and resonance analysis).

7.5.6 System Aspects

The use of low frequencies has a certain number of consequences on a radar system, in particular on the modes of space scanning or on integration in detection and tracking systems. Furthermore, a brief run-through of the existing technologies reveals that there are some consequences that apply to the definition of the radar system and its performance.

A basic property of low-frequency radars is a high illumination duration, with coherent integration (typically 100 ms for VHF or 1 second for HF); this is due to the combination of several factors, including:

- Low antenna directivity for given dimensions, meaning long durations of possible target presence in the radar beam;

7. Essentially propeller blades, because turbine blades are barely visible in the low-frequency bands (wavelength larger than the air intakes' diameter).

- Slow fluctuation of the target due to the small dimensions, expressed in wavelengths, which allows a high coherent integration time;
- The need to compensate for the low transmission gain by a high Doppler integration duration-rarely less than 0.1 second in metric or 1 second in decametric bands.

This high illumination duration drives some essential properties of low-frequency radars, in terms of elimination of clutter echoes and exploration of space.

7.5.6.1 Technologies

Low-frequency transmitters can usually be designed from solid-state technology, making it possible to transmit a few continuous kilowatts in the HF and VHF bands. In addition, the quality of master oscillators and transmitters means that a low phase noise can be obtained, typically:

- At 300 MHz: −135 dBc/Hz, at 1,000 Hz of carrier and −105 dBc/Hz at 100 Hz;
- At 30 MHz: −130 dBc/Hz, at 100 Hz of carrier, −90 dBc/Hz at 1 Hz.

This spectral quality ensures a high clutter-attenuation factor. Contrary to the case of conventional radar bands, the detection capacity of stealth targets by metric pulse radars is not limited by the spectral purity of the transmitted signals.[8]

Antenna dimensions are an essential limitative factor of such radars. Generally, antennas comprise arrays of elementary sources (e.g., dipoles, patches, or slots). The size of each radiating feed is generally no less than $\lambda/2$ (if we rule out, at least in the immediate future, excessive adaptation losses or recourse to supraconductor elements), and thickness can vary from $\lambda/5$ for broadband sources (30% bandwidth) to $\lambda/50$ for narrowband sources (2% instantaneous bandwidth), that are tunable if necessary. Dimensional constraints more particularly limit the possibilities of installation on the aircraft and, in general, prohibit the use of frequencies lower than 300 MHz or so. Furthermore, the coupling existing between radiating sources and the nearby environment means that recourse to automatic calibration techniques

8. Furthermore, the long integration duration also tends to improve the detectability of weak targets by proportionally reducing the residual phase noise in each Doppler filter.

on external sources or clutter echoes is essential, in particular for airborne radars (see Section 7.4.4).

7.5.6.2 Exploration of Space

As mentioned at the beginning of this chapter, the diverse constraints specific to low frequencies favor the use of long observation times, themselves leading to adoption of space exploration based on digital beamforming in order to preserve a sufficiently high refresh rate (generally in excess of 1 Hz). Beamforming is made possible by the relatively low number of sources required (generally less than 30).

In the higher part of the VHF and UHF bands, to preserve the power budget,[9] it is generally necessary to combine beamforming and scanning (electronic or mechanical). For the VHF and HF bands, beamforming is generally a natural solution.

7.5.6.3 Waveforms

The waveforms used are generally made up of coherent bursts (or noncoherent for long range modes where clutter is not present beyond the transition range) with a large duty factor, typically included between 0.1 for monostatic radars and 1 for bistatic or multistatic radars,[10] and feature a low-repetition frequency to benefit from Doppler range nonambiguity. This means using pulse compression, at least in an acquisition or confirmation mode.

It can be seen that low-frequency radars generally use high processing gains. For instance, the use of 10 sources, a coherent burst of 500 pulses, and a pulse compression rate of 20, gives a total processing gain equal to 50 dB. The resultant complex processing is nevertheless not excessive because the low angular resolution, and often also the low range resolution, means a number of cells processed being 10 or 100 times lower than in the case of conventional radars.

But when a low-frequency radar has a resolution equal to or greater than conventional radars, and when volumetric surveillance is provided, the required processing capacities may become considerable (for instance, in the case of the large air defense VHF radar RIAS [33]).

---

9. Beamforming is not equivalent to sequential scanning from the power budget standpoint unless the low transmission gain (necessary to illuminate space extensively) can be compensated by longer integration times, which must be compatible with fluctuations and motions of the target.
10. The use of a continuous wave (not pulsed) by a monostatic radar is generally impossible at low frequencies because the necessary decoupling between the transmission antenna and the reception antenna cannot be obtained due to the low directivities of the antennas.

7.5.6.4 System Integration

All of these particularities generally mean entrusting a warning role to low-frequency radars (antistealth warning, invulnerability to antiradiation missiles at least for HF and VHF). Often, this means that they must be coupled with another sensor for acquisition or tracking, although this necessity is not absolute—for example, when angular location accuracy of approximately a degree is sufficient to react (self-protection to trigger chaff, for instance) or when a sufficient area is available for the antenna (such as air defense by RIAS [33]).

A variety of coupled sensor systems can be considered, for instance using optronics for short range, acoustics for very short range, or obviously, conventional radars. An attractive system consists of a dual-band or multiband radar in which the antenna integrates the sources of the various radar bands and the receiver simultaneously processes the different bands (integrated arrays and reception). Concepts such as this are now particularly relevant because all-digital reception is becoming accessible (existence of analog-digital coders operating on 8 bits at 4 GHz), limiting the width of the processed band to the capacities of the antenna (barely limited because these are source arrays) and to limits imposed by frequency occupation and tolerances.

7.6 Wideband Radars

The derivation of the received signal for an active radar or sonar system, in (1.2), assumed that the target motion during the observation time T was smaller than the range resolution of the radar, determined by its bandwidth B:

$$vT << \frac{c}{2B} \tag{7.58}$$

When this condition is not met, the implementation of the optimum signal processing has to be changed to take into account the range migration effect, or in other words, to take into account the fact that the Doppler effect can no longer be considered as a mere translation in frequency; the frequency shift is proportional to frequency, or in simpler words, the Doppler effect is a compression (or an expansion) of the signal. In this section, some properties of such high-range resolution radar will be analyzed, and the optimum processing of a sequence of wideband pulses will be defined.

7.6.1 Wideband and Ultra-Wideband Radars [6, 34, 35]

A wideband radar is defined as a radar such that the condition (7.58) is not met. Two classes of such radars can be distinguished: ultra-wideband radars, for which the relative bandwidth is large (typically larger than 25%: $2(F_{max} - F_{min})/(F_{max} + F_{min}) > 1/4$), and wideband radars, for which the relative bandwidth is lower than 25%. Though somewhat arbitrary, this distinction actually corresponds to different types of radar:

- Wideband radars, with carriers between 3 and 30 GHz and range resolution between 1m and 10 cm, corresponding to a pulse duration (usually obtained through pulse compression) on the order of 10 carrier periods;
- Ultra-wideband radars, for which the pulse duration is in the order of one or two carrier periods, and usually using frequencies between 100 MHz and 1 GHz.

The combination of low-frequency properties (see Section 7.5) with high-range resolution gives ultra-wideband radars unique capabilities for camouflaged or stealth targets detection (foliage penetration) or subsurface objects detection (such as pipes or mines). A typical wideband signal is shown in Figure 7.34. The pulse has zero mean (necessary condition for radiation) and only one oscillation with one nanosecond overall duration. The derivative of a Gaussian is a good approximation of such pulses, with the advantage

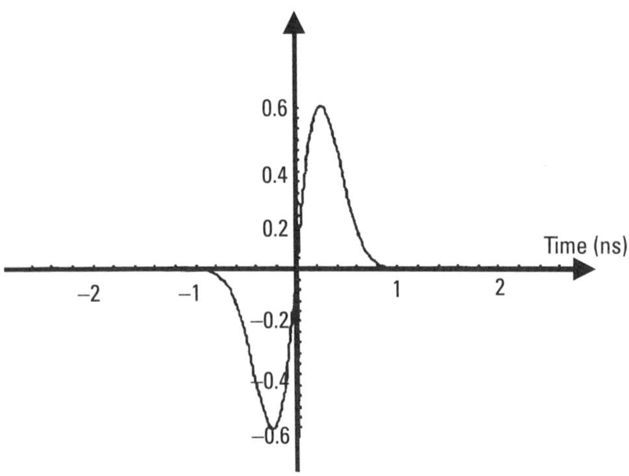

Figure 7.34 Ultra-wideband pulse.

that the spectrum has the same analytical form, standard deviation in time, and frequency being related by: $\sigma_f = \dfrac{1}{2\pi\sigma_t}$.

Reflection on a metallic target changes the sign of this pulse; on reception, the radar correlates this pulse with replicas of the expected signal (Doppler effect is negligible during the pulse duration), and the correlator output is shown in Figure 7.35.

An interesting property of this signal is that no phase is present in the received real signal, so that:

- There is no need for quadratic detection on reception;
- The output noise is Gaussian (and not exponential as in the conventional radar situation).

This property, added to the specificities of low-frequency radars described in Section 7.5 gives ultra-wideband radars some advantages for low RCS target detection.

On the other hand, wideband radars have no special properties except for their specific ambiguity function, which can turn out to be a significant advantage for detection in a hostile environment (i.e., clutter, countermeasures). The ambiguity function of such radars will be analyzed in Section 7.2.1.

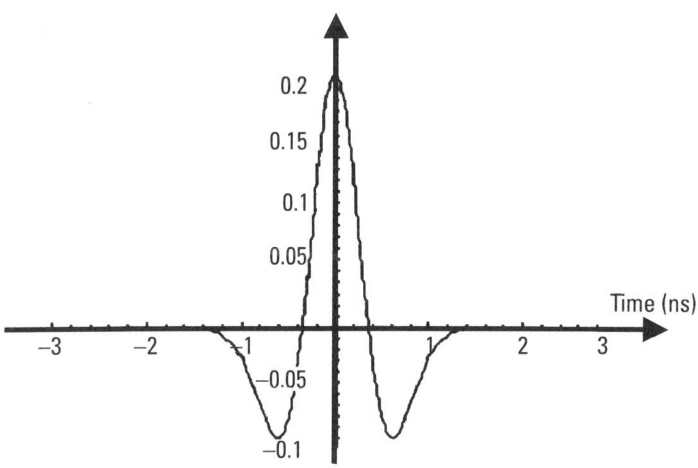

Figure 7.35 Correlator output.

7.6.2 Wideband Radar Processing

Consider a radar transmitting M identical short pulses with repetition period T_r (repetition frequency F_r). The optimum processing will be defined by Fourier decomposition of the received signal in adjacent narrowband channels, in order to show the equivalence between such narrowband processing followed by coherent summation and direct temporal processing (compensation of range migration followed by coherent summation). Time will be denoted by t (N received samples for each transmitted pulse), frequency channel by f (t and f being conjugated by Fourier transform), pulse number by r, and Doppler frequency by ν (r and ν being conjugated by Fourier transform).

If x_{rt} is the complex received sample, after demodulation by the central frequency F_0, the frequency response of the received signal is y_{rf}:

$$y_{rf} = \sum_{t=0}^{N-1} x_{rt} e^{-2\pi j \frac{tf}{N}} \tag{7.59}$$

Doppler analysis in each sub-band f gives:

$$z_{\nu f} = \sum_{r=0}^{M-1} y_{rf} e^{-2\pi j \frac{r\nu}{M}} \tag{7.60}$$

The correspondence between Doppler ν and velocity v is:

$$\frac{2v}{\lambda} = \frac{\nu}{M} F_r \left(\text{mod}\, \frac{F_r}{M}\right) \quad \text{with } \lambda = \frac{c}{F_0 + f\delta F}$$

$$\Rightarrow \nu = \frac{M(F_0 + f\delta F)}{F_r} \frac{2v}{c} \,(\text{mod}\, M) \tag{7.61}$$

It is then possible to define the range/velocity map of the M received signals, through coherent summation of the signals, from the different sub-bands f, corresponding to the same velocity ν:

$$T_{vt} = \sum_f z_{\nu f} e^{2\pi j \frac{tf}{N}} \text{ for } \nu = \frac{M(F_0 + f\delta F)}{F_r} \frac{2v}{c} \,(\text{mod}\, M) \tag{7.62}$$

$$T_{vt} = \sum_{f,r} y_{rf} e^{2\pi j \frac{tf}{N}} e^{2\pi j \frac{r(F_0 + f\delta f)}{F_r} \frac{2v}{c}} \tag{7.63}$$

which is written, using relation (7.59):

$$T_{vt} = \sum_{t',f,r} x_{rt'} e^{-2\pi j \frac{t'f}{N}} e^{2\pi j \frac{tf}{N}} e^{-2\pi j \frac{r(F_0 + f\delta f)}{F_r} \frac{2v}{c}} \tag{7.64}$$

or finally

$$T_{vt} = \sum_{t',f,r} x_{rt'} \underset{\text{A}}{\underset{\uparrow}{e^{-2\pi j \frac{rF_0}{F_r} \frac{2v}{c}}}} \underset{\text{B}}{\underset{\uparrow}{e^{2\pi j \frac{f}{N}}}} \left[t - t' - \underset{\text{C}}{\underset{\uparrow}{\frac{r\delta f}{F_r} \frac{2v}{c} N}} \right] \tag{7.65}$$

A: coherent Doppler summation at frequency F_0;
B: pulse reconstruction;
C: compensation of differential Doppler in the whole bandwidth.

Integrating on f and t', (7.63) is written, in time domain:

$$T_{vt} = \sum_r x_{r,\, t - \frac{r\delta F}{F_r} \frac{2v}{c} N} \, e^{-2\pi j r \frac{F_0}{F_r} \frac{2v}{c}} \tag{7.66}$$

But, if Δf is the total bandwidth and δR is the radar range resolution, we have:

$$\frac{2\delta F}{c} N = \frac{2\Delta F}{c} = \frac{1}{\delta R}$$

$$T_{vt} = \sum_r x_{r,\, t - r\frac{vT_r}{\delta R}} \, e^{-2\pi j r \frac{F_0}{F_r} \frac{2v}{c}} \tag{7.67}$$

The range/velocity map is obtained in the time domain by coherent summation, taking into account the migration effect from pulse to pulse (Figure 7.36), and the Doppler phase shift at frequency F_0. The processing by sub-bands is therefore equivalent to direct time-domain processing.

This wideband processing gives an ambiguity function, Figure 7.37(b), different from that obtained when range migration is negligible, Figure 7.37(a). The periodic spikes in velocity resulting from velocity ambiguity are notably reduced in Figure 7.37(b), corresponding to a pulse train of 30 pulses and a range migration of one range bin after $N/4$ pulses for $V = V_a$ [ambiguous velocity, $V_a = \lambda/(2T_r)$].

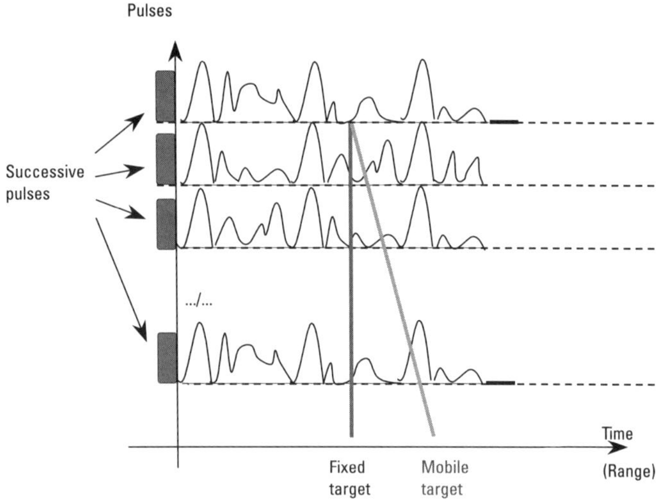

Figure 7.36 Range migration compensation.

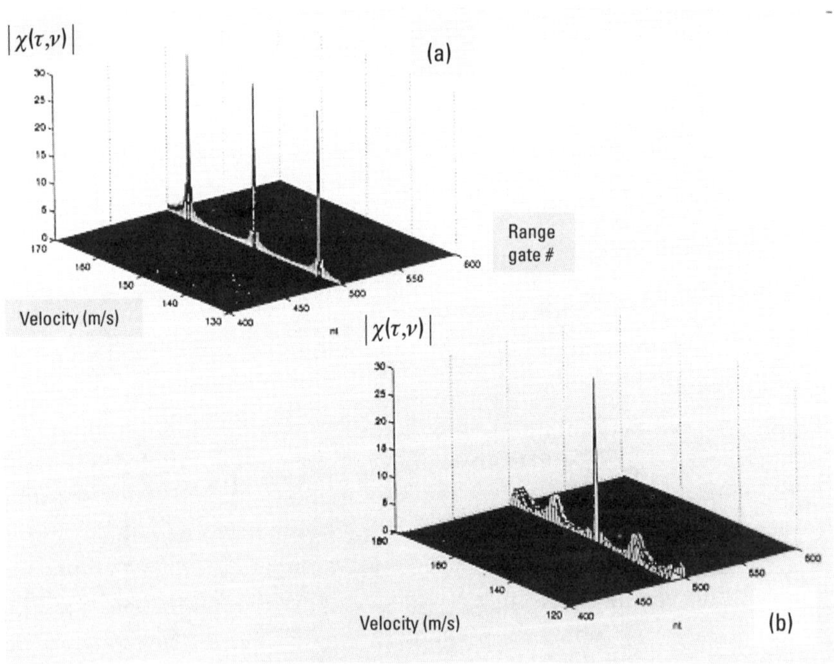

Figure 7.37 (a) Narrowband ambiguity function, with pulse duration = 1 μs; and (b) wideband ambiguity function, with pulse duration = 1 ns; F_0 = 10 GHz, M = 30 pulses, T_r = 1 ms, λ = 3 cm.

This description of the ambiguity function of a wideband pulse train can be completed, if necessary, by taking into account the Doppler effect in the elementary pulse itself. A detailed analysis of such ambiguity functions can be found in [36], taking advantage of the Mellin transform properties for fast computation.

Careful analysis of the performances of such wideband coherent waveforms shows an increase in performance (primarily due to the longer integration time) but also various system benefits [37]:

- Simultaneous long-range visibility of approaching *and receding* targets allows the same waveform to be used for search and track of all targets.
- Classification capability (high-range resolution, range Doppler imaging of moving parts): The waveform provides images of targets in the range-Doppler plane, again with no specific mode and no requirement for increased time on target; for example, main and tail rotors can be easily separated in the signature of stationary helicopters, thus providing clear differentiation of these targets from ground vehicles, for example.
- ECCM capability and discretion (spread spectrum property): Range-Doppler gate pull-off requires reception and storing of a significant part of the pulse burst in order to analyze it and be able to reproduce the range-walk phenomena and the corresponding Doppler compression. That leads to hundreds of megabits fast digital radar frequency memories requirements, with adequate signal analysis and filtering on reception.
- Angular *resolution* of closely spaced targets (monopulse wideband separation): When two extended targets are present *in the same angular beam,* the monopulse signals in the successive high resolution range gates will give the information that two targets are present, and their respective angular location.
- Reduction of multipaths will be essentially separated by range resolution and incoherently integrated in the m/n extractor.
- Simultaneous air-air, air-ground SAR, and air-ground GMTI operation, with a single waveform.

7.7 Periodic Signal Passive Listening

As mentioned in Section 6.5, a problem frequently encountered in passive listening consists of determining the presence, in the received signal, of a periodic component and in estimating this period. To describe the optimum processing for carrying out this operation, we will follow the approach described in [38].

Let us assume that the signal originates from one sensor only. The received signal $r(t)$ is written:

$$r(t) = s(t) + n(t) \tag{7.68}$$

where $n(t)$ is a Gaussian white noise with a limited bandwidth of correlation function (refer to Chapter 2). $B(t) = N_0 \operatorname{sinc}(2F_{\max} t)$ and $s(t)$ is the unknown periodic signal to be determined, with period P also unknown.

As in Chapter 2, let us sample this received signal at frequency $2F_{\max}$. Let r_i, s_i, and n_i be the samples thus obtained. The maximum likelihood criterion leads to determining s_i such that:

$$\Lambda = \prod_i \frac{1}{\pi N_0} \exp\left[-(r_i^* - s_i^*)\frac{1}{N_0}(r_i - s_i)\right]$$

is a maximum with respect to all s_i samples, that is to say by going to the limit value:

$$\min_{/s} \frac{1}{T} \int |t(t) - s(t)|^2 dt \tag{7.69}$$

(This transformation only demonstrates the equivalence between maximum likelihood and least squares method in the case of an additive Gaussian noise.)

As $s(t)$ is periodic, it can be decomposed into the Fourier series:

$$s(t) = \sum_k \xi_k \exp\left(2\pi j \frac{kt}{P}\right) \tag{7.70}$$

where ξ_k and P are the unknown quantities of the problem.

The relation (7.41) is then written:

$$\begin{aligned} \mathrm{Min}_{/(\xi_k, P)} \frac{1}{T} \int \Bigg[|r(t)|^2 - 2\sum_k \mathrm{Re}\left(r^*(t)\xi_k \exp\left(2\pi j \frac{kt}{P}\right)\right) \\ + \sum_{k,l}\left(\xi_k \xi_l^* \exp\left(2\pi j \frac{(k-1)t}{P}\right)\right)\Bigg] dt \end{aligned} \tag{7.71}$$

$$\mathrm{Min}_{/(\xi_k, P)} \left[-2\sum_{k} \mathrm{Re}\left(\tilde{r}^*\left(\frac{k}{P}\right)\xi_k \right) + \sum_{k,l} \left(\xi_k \xi_l^* \,\mathrm{sinc}\frac{(k-1)T}{P} \right) \right] \tag{7.72}$$

where

$$\tilde{r}(k/P) = \frac{1}{T}\int r(t)\exp\left(-2\pi j\frac{kt}{P}\right)dt$$

[line at frequency k/P of the Fourier transform of $r(t)$].

If we assume that:

$$T >> P \tag{7.73}$$

(long time of observation with respect to the period of the signals searched), which is often confirmed for passive listening applications, only the diagonal terms remain in the second term of (7.43):

$$\mathrm{Min}_{/(\xi_k, P)} - 2\sum_{k} \mathrm{Re}\left[\xi_k \tilde{r}^*\left(\frac{k}{P}\right) \right] + \sum_{k} |\xi_k|^2 \tag{7.74}$$

Minimizing, with respect to k, by canceling the derivatives in regard to the real part and to the imaginary part of k, the estimate $\hat{\xi}_k$ of ξ_k is obtained:

$$\hat{\xi}_k = \tilde{r}\left(\frac{k}{P}\right) \tag{7.75}$$

Substituting this estimate into (7.72), the estimation rule for P is obtained as:

$$\mathrm{Max}_{/P} \sum_{k} \left| \tilde{r}\left(\frac{k}{P}\right) \right|^2 \tag{7.76}$$

The optimum estimation procedure for period P and the form of the signal (or of its Fourier components ξ_k) hence consists of:

- Calculating the Fourier transform of the received signal; if the signal is composed of the sample series r_n, the samples $\tilde{r}_k$ of the Fourier transform are thus obtained;
- Calculating what can be called the harmonic signature of the received signal: R_F;

$$R_F = \sum_k |\tilde{r}_{k \times F}|^2 \qquad \text{for } F = 1, 2, \text{etc.} \tag{7.77}$$

- Finding R_F maximum with respect to F: this maximum is obtained for $\hat{F} = \frac{1}{\hat{P}}$, the signal estimated frequency;
- The estimate of the signal Fourier series is then obtained, according to (7.47), by:

$$\xi_k = \tilde{r}\left(\frac{k}{P}\right) \tag{7.78}$$

This approach actually corresponds to a completely intuitive approach, consisting of observing that a signal of period P has a spectrum of lines spaced at $1/P$ intervals (in frequency). Consequently, the best method of determining whether P_0 is the period of the received signal is to sum all the components of the spectrum of the received signal spaced at $1/P_0$ intervals (all the "harmonics"), and to compare the result thus obtained with the results obtained for the other values of P (as the signal is unknown, the phase of each line is unknown, hence the summing can be carried out on the line amplitudes only). The previous demonstration simply proved the optimum character of this approach when the unknown period is small in relation to the time of observation T. Reference [38] analyzes in detail the performance of this procedure of estimation and gives a time-based version of it (by folding back the signals received) generally more expensive in terms of computation time but sometimes more appropriate for a specific processing architecture.

Finally, note that these procedures could be extended to the case of colored noise by using the whitening approach described in Chapters 4 and 5; without going into further details of the calculations, it is easy to imagine that similar procedures could be thus obtained, for which the RF harmonic signature would be a weighted sum of the successive harmonics.

This kind of procedure is particularly suitable for classification of signals received by passive sensors, in radio listening, or sonar applications.

References

[1] Lacomme, P., et al., *Air and Spaceborne Radar Systems,* Norwich, NY: William Andrew Scitech Publishing, 2001.

[2] Wehner, D. R., *High Resolution Radar,* 2nd ed., Norwood, MA: Artech House, 1995.

[3] Pricket, M. J., and C. C. Chen, "Principles of Inverse Synthetic Aperture Radar Imaging," *IEEE EASCON Conf.,* 1980.

[4] "Special Issue on Motion in Signals and Images," *Traitement du Signal,* GRETSI, 1996.

[5] Vignaud, L., "Imagerie Micro-Ondes de Scènes Instationnaires," doctoral thesis, Université de Paris, April 6, 1996.

[6] Astanin, L. Y., et al., *Radar Target Characteristics: Measurements and Applications,* Boca Raton, FL: CRC Press, 1993.

[7] Zhu, Z. D., X. Qiu, and Z. S. She, "ISAR Motion Compensation Using Modified Doppler Centroid Tracking Method," Trans. *Nanjing University of Aeronautics & Astronautics,* Vol. 12, No. 2, Dec. 1995.

[8] Nebabin, V. G. *Methods and Techniques of Radar Recognition,* Norwood, MA: Artech House, 1995.

[9] Bertrand, P., "Représentation des Signaux dans le Plan Temps-Fréquence," *La Recherche Aérospatiale,* Vol. 1983-1, 1983 (English and French versions), pp. 1–12.

[10] Le Chevalier, F., G. Bobillot, and C. Fugier-Garrel, "Radar Target and Aspect Angle Identification," *IEEE International Joint Conference on Pattern Recognition,* Kyoto, Japan, Nov. 1978.

[11] Kedam, B., "Spectral Analysis and Discrimination by Zero-Crossings," *Proc. IEEE,* 74, No. 11, Nov. 1986.

[12] Fukunaga, K., *Introduction to Statistical Pattern Recognition,* 2nd ed., New York: Academic Press, 1990.

[13] Kennaugh, E. M., "Opening Remarks," *IEEE Trans. AP,* AP 29, No. 2, March 1981, pp. 190–192.

[14] Moffatt, D. L., and C. Y. Lai, "Natural Resonance Estimation," *IEEE Trans. IM,* IM 34, No. 4, Dec. 1985, pp. 547–549.

[15] Gjessing, D. I., J. Hjelmstad, and T. Lund, "A Multifrequency Adaptive Radar for Detection and Identification of Objects," *IEEE Trans. AP,* AP 30, No. 3, May 1982, pp. 366–372.

[16] Le Chevalier, F., "Signatures des Cibles et des Fonds et Traitement des Signaux Radar," *La Recherche Aérospatiale,* No. 1986-1, 1986, pp. 1–13.

[17] Born, M., and E. Wolf, *Principles of Optics,* Pergamon Press, 1965.

[18] Giuli, D., "Polarization Diversity on Radars," *IEEE Proc.*, Vol. 74, No. 2, Feb. 1986, pp. 245–269.

[19] Nathanson, F. E., "Adaptive Circular Polarization," *IEEE Conf. Radar 75,* Arlington, VA, April 1975.

[20] Erfurd, R., M. Cohen, and E. Sjober, "Intrapulse Polarization Agile Radar," *Microwave Systems News,* Vol. 14, Feb. 1984, pp. 67–76.

[21] Ringel, M. B., D. H. Mooney, and W. H. Long, "F 16 Pulse-Doppler Radar Performance," *IEEE Trans. AES,* 19, No. 1, Jan. 1983, pp. 147–158.

[22] Clarke, J., "Airborne Early Warning Radars," *IEEE Proc.*, Vol. 73, No. 2, Feb. 1985, pp. 312–324.

[23] Klemm, R., *Space-Time Adaptive Processing: Principles and Applications,* IEE Radar series No. 9, Herts, U.K.: IEE, 1998.

[24] Liao, G., Z. Bao, and Y. Zhang, "Doppler Prefiltering Based Adaptive Array Processing in Airborne Radars," *International Conference on Radar,* Paris, May 1994.

[25] Richardson, P. G., "Analysis of the Adaptive Space-Time Processing Technique for Airborne Radar," *IEE Proceedings Radar, Sonar, Navigation,* Vol. 141, No. 4, Aug. 1994.

[26] Klemm, R., "Adaptive Airborne MTI with Two-Dimensional Motion Compensation," *IEE Proceedings F,* Vol. 136, No. 6, 1991.

[27] Liu, Q. G., and Y. N. Peng, "Short-Time Processor in Airborne Phased Array Radars," *IEEE Trans. AES,* Vol. 32, No. 2, April 1996.

[28] Le Chevalier, F., L. Savy, and F. Durnez, "Clutter Calibration for Space-Time Adaptive MTI Radars," *IEEE International Radar Conference,* Beijing, Oct. 1996.

[29] Le Chevalier, F., and H. Poinsard, *Calibration d'un Radar Aéroporté sur Fouillis de Sol,* Demande de Brevet No. 9506529.

[30] Boithias, L., *Radio Wave Propagation,* New York: McGraw-Hill, 1987.

[31] Gouttelard, C., "The Nostradamus Project, French OTH-B Radar Design Studies," *AGARD CP 486.*

[32] Morchin, W. C., *Airborne Early Warning Radar,* Norwood, MA: Artech House, 1990.

[33] Dorey, J., G. Garnier, and G. Auvray, "RIAS, Radar à Impulsion et Antenne Synthétique," *IEEE International Radar Conférence,* Paris, 1989.

[34] Iverson, D. E., "Coherent Processing of Ultra-Wideband Signals," *IEE Proc. Radar, Sonar, Navig.,* Vol. 141, No. 3, June 1994.

[35] Skolnik, M., G. Andrews, and J. P. Hansen, "Ultra-Wideband Microwave Radar Conceptual Design," *IEEE AES Systems Magazine,* Oct. 1995.

[36] Bertrand, J., P. Bertrand, and J. P. Ovarlez, "Compression d'Impulsion en Large Bande," *12ème Colloque GRETSI,* Juan-Les-Pins, June 1989.

[37] Le Chevalier, F., and A. Becker, "Wideband Coherent Airborne Radar Systems," *International Radar Conf.,* ICR 2001, Beijing, Oct. 2001.

[38] Kincaid, T. G., and H. J. Scudder, "Estimation of Periodic Signals in Noise," *J.A.S.A.,* 46, No. 6, 1969, pp. 1397–1403.

Appendix 7A: Variance of a Rayleigh Variable at the Output of a Logarithmic Amplifier

Let x be a random variable with Rayleigh distribution, and let:

$$r = \ln x; \; y = 20 \log_{10} x = 20r(\log_{10} e)$$

The moments of order 1 and 2 are respectively given by:

$$m_1 = E(r) = \int_0^{\infty} (\ln x) \frac{x}{\sigma^2} e^{-x^2/2\sigma^2} dx$$

$$m_2 = E(r^2) = \int_0^{\infty} (\ln x)^2 \frac{x}{\sigma^2} e^{-x^2/2\sigma^2} dx$$

Let

$$u = \frac{x^2}{2\sigma^2}$$

$$m_1 = \frac{1}{2} \int_0^{\infty} (\ln u + \ln 2\sigma^2) e^{-u} du$$

$$m_2 = \frac{1}{4} \int_0^{\infty} (\ln u + \ln 2\sigma^2)^2 e^{-u} du$$

Now the gamma function is defined by:

$$\Gamma(x+1) = \int_0^{\infty} u^x e^{-u} du$$

$$\Gamma'(x+1) = \int_0^{\infty} \ln u \, u^x e^{-u} du = \psi(x+1)\Gamma(x+1)$$

where

$$\psi(x+1) = \frac{\Gamma'(x+1)}{\Gamma(x+1)}$$

$$\Gamma''(x+1) = \int_0^\infty (\ln u)^2 u^x e^{-u} du$$

hence

$$\int_0^\infty \ln u e^{-u} du = \Gamma'(1) = \psi(1) = -C \text{ (Euler constant)}$$

$$\int_0^\infty (\ln u)^2 e^{-u} du = \Gamma''(1) = (\psi'\Gamma + \psi\Gamma')(1) = \psi'(1) + \psi^2(1) = \frac{\pi^2}{6} + C^2$$

$$\Rightarrow \begin{cases} m_1 = \frac{1}{2}(\ln 2\sigma^2 - C) \\ m_2 = \frac{1}{4}\left[\frac{\pi^2}{6} + (C - \ln 2\sigma^2)^2\right] \end{cases}$$

$$\Rightarrow \text{Var}(r) = m_2 - m_1^2 = \frac{\pi^2}{24} \text{ which is independent of } \sigma$$

$$\Rightarrow \text{Var}(y) = (20\log_{10}e)^2 \frac{\pi^2}{24} = 31{,}025$$

Appendix 7B: Probability Density of the Instantaneous Frequency of a Gaussian Random Process

Let $q(t)$ be a complex Gaussian random process, its spectrum $S(f)$ having zero mean frequency:

$$\int fS(f)\,df = 0$$

Let

$$q(t) = A(t)\, e^{j\Omega(t)}$$

$$\Omega'(t) = \mathrm{Im}\left[\frac{q'(t)}{q(t)}\right]$$

Besides

$$E[q'(t)q^{*}(t)] = r'(0) = \int fS(f)\, df = 0$$

$r(\tau)$: correlation function of the process $q(t)$;
$q(t)$ being Gaussian, $q'(t)$ is also Gaussian, independent of $q(t)$.

Let x_1 and x_2 be two independent circular Gaussian *complex* variables, with respective variance σ_1^2 and σ_2^2. Let us calculate the distribution of x_2/x_1, then $\mathrm{Im}(x_2/x_1)$, before expressing it in the case in which x_2 is the derivative of x_1.

$$p_{x_1,x_2}(x_1, x_2) = \frac{1}{\pi^2\sigma_1^2\sigma_2^2}\, e^{-\left[\frac{|x_1|^2}{\sigma_1^2} + \frac{|x_2|^2}{\sigma_2^2}\right]}$$

Let $u = x_1$ and $v = x_2/x_1$

Then

$$p(u, v) = |u|^2 p_{x_1,x_2}(u, uv)$$

$$p(v) = \int p(u, v)\, du$$

$$p(v) = \frac{1}{\pi^2\sigma_1^2\sigma_2^2}\int |u|^2\, e^{-\left(\frac{1}{\sigma_1^2} + \frac{|v|^2}{\sigma_2^2}\right)|u|^2}\, du$$

$$p(v) = \frac{\sigma_1^2\sigma_2^2}{\pi}\,\frac{1}{(\sigma_2^2 + \sigma_1^2|v|^2)^2}$$

Let

$$v = v_1 + jv_2$$

$$p(v_2) = \int p(v)\, dv_1$$

$$p(v_2) = \frac{\sigma_2^2}{\pi\sigma_1^2} \int \frac{dv_1}{\left(v_1^2 + v_2^2 + \frac{\sigma_2^2}{\sigma_1^2}\right)^2}$$

$$p(v_2) = \frac{\sigma_1}{2\sigma_2} \frac{1}{\left(1 + v_2^2 \frac{\sigma_1^2}{\sigma_2^2}\right)^{3/2}}$$

In our case:

$$\sigma_1^2 = r(0) = \int S(f)\, df$$

$$\sigma_2^2 = -r''(0) = (2\pi)^2 \int f^2 S(f)\, df = (2\pi)^2 \sigma_1^2 (\Delta f)^2$$

$$p(\Omega') = \frac{1}{4\pi\Delta f \left[1 + \left(\frac{\Omega'}{2\pi\Delta f}\right)^2\right]^{3/2}} \qquad \text{(Student law with 2 degrees of freedom)}$$

In the case where the mean frequency of the spectrum is not zero but equal to:

$$f_0 = \int f S(f)\, df$$

Ω' is simply shifted by $2f_0$ (carrier $e^{2i\pi f_0 t}$), and its probability density is given by:

$$p(\Omega') = \frac{1}{4\pi\Delta f \left[1 + \left(\frac{\Omega' - 2\pi f_0}{2\pi\Delta f}\right)^2\right]^{3/2}}$$

About the Author

François Le Chevalier is the scientific director of THALES (formerly Thomson-CSF) Airborne Systems in Elancourt, France.

Mr. Le Chevalier received his degree in telecommunications engineering from the Ecole Nationale Superieure des Telecommunications (ENST), in Paris, in 1974. He began his career at the Office National d'Etudes et de Recherches Aérospatiales (ONERA), where he conducted research and teams on radar target recognition, target and background signatures, radar modeling, and signal processing. He was also employed as an expert in different NATO technical groups, and as an advisor in different industrial competitions.

In 1986, Mr. Le Chevalier worked for Thomson-CSF (now THALES) as a technical director of Le Centre Thomson d'Applications Radar (LCTAR), where he was in charge of upstream studies coordination, initiated adaptive digital beamforming and STAP radar systems demonstrations, and shared apertures and multisensor concept design and validation. In 1998, he joined the Airborne Systems group, and in 2000 was appointed scientific director, in charge of advanced research and developments coordination (airborne radars, electronic warfare, and airborne mission systems).

An author of more than 30 papers and communications, Mr. Le Chevalier holds 16 patents in radar and electronic warfare and gives lectures each year on radar and electronic warfare, adaptive processing, and target analysis and recognition to specialized students and engineers. An Emerite member of the Société des Electriciens et des Electroniciens (SEE) and a senior member of IEEE, he recently chaired the International Scientific Committee of the International Radar Conference in Brest, Paris, May 1999, and acted as a vice chairman of the International Radar Conference in China, Beijing, October 2001.

Index

Recent Titles in the Artech House Radar Library

David K. Barton, Series Editor

Advanced Techniques for Digital Receivers, Phillip E. Pace

Airborne Pulsed Doppler Radar, Second Edition, Guy V. Morris and Linda Harkness, editors

Bayesian Multiple Target Tracking, Lawrence D. Stone, Carl A. Barlow, and Thomas L. Corwin

Computer Simulation of Aerial Target Radar Scattering, Recognition, Detection, and Tracking, Yakov D. Shirman, editor

Design and Analysis of Modern Tracking Systems, Samuel Blackman and Robert Popoli

Digital Techniques for Wideband Receivers, Second Edition, James Tsui

Electronic Intelligence: The Analysis of Radar Signals, Second Edition, Richard G. Wiley

Electronic Warfare in the Information Age, D. Curtis Schleher

EW 101: A First Course in Electronic Warfare, David Adamy

Fundamentals of Electronic Warfare, Sergei A. Vakin, Lev N. Shustov, and Robert H. Dunwell

Handbook of Computer Simulation in Radio Engineering, Communications, and Radar, Sergey A. Leonov and Alexander I. Leonov

High-Resolution Radar, Second Edition, Donald R. Wehner

Introduction to Electronic Defense Systems, Second Edition, Filippo Neri

Introduction to Electronic Warfare, D. Curtis Schleher

Microwave Radar: Imaging and Advanced Concepts, Roger J. Sullivan

Modern Radar System Analysis, David K. Barton

Multitarget-Multisensor Tracking: Applications and Advances Volume III, Yaakov Bar-Shalom and William Dale Blair, editors

Principles of High-Resolution Radar, August W. Rihaczek

Principles of Radar and Sonar Signal Processing, François Le Chevalier

Radar Cross Section, Second Edition, Eugene F. Knott, et al.

Radar Evaluation Handbook, David K. Barton, et al.

Radar Meteorology, Henri Sauvageot

Radar Reflectivity of Land and Sea, Third Edition, Maurice W. Long

Radar Resolution and Complex-Image Analysis, August W. Rihaczek and Stephen J. Hershkowitz

Radar Signal Processing and Adaptive Systems, Ramon Nitzberg

Radar System Performance Modeling, G. Richard Curry

Radar Technology Encyclopedia, David K. Barton and Sergey A. Leonov, editors

Range-Doppler Radar Imaging and Motion Compensation, Jae Sok Son, et al.

Theory and Practice of Radar Target Identification, August W. Rihaczek and Stephen J. Hershkowitz

Time-Frequency Transforms for Radar Imaging and Signal Analysis, Victor C. Chen and Hao Ling

For further information on these and other Artech House titles, including previously considered out-of-print books now available through our In-Print-Forever® (IPF®) program, contact:

Artech House
685 Canton Street
Norwood, MA 02062
Phone: 781-769-9750
Fax: 781-769-6334
e-mail: artech@artechhouse.com

Artech House
46 Gillingham Street
London SW1V 1AH UK
Phone: +44 (0)20 7596-8750
Fax: +44 (0)20 7630-0166
e-mail: artech-uk@artechhouse.com

Find us on the World Wide Web at:
www.artechhouse.com